New Horizons in Nanofillers Based Polymer Composites II

New Horizons in Nanofillers Based Polymer Composites II

Editors

Vineet Kumar
Xiaowu Tang

Basel • Beijing • Wuhan • Barcelona • Belgrade • Novi Sad • Cluj • Manchester

Editors

Vineet Kumar
Department of Mechancial
Engineering
Yeungnam University
Gyeongsan-si
Korea, South

Xiaowu Tang
College of Material and
Chemical Engineering
Zhengzhou University of
Light Industry
Zhengzhou
China

Editorial Office
MDPI
St. Alban-Anlage 66
4052 Basel, Switzerland

This is a reprint of articles from the Special Issue published online in the open access journal *Polymers* (ISSN 2073-4360) (available at: www.mdpi.com/journal/polymers/special_issues/nanofiller_II).

For citation purposes, cite each article independently as indicated on the article page online and as indicated below:

Lastname, A.A.; Lastname, B.B. Article Title. *Journal Name* **Year**, *Volume Number*, Page Range.

ISBN 978-3-7258-0043-8 (Hbk)
ISBN 978-3-7258-0044-5 (PDF)
doi.org/10.3390/books978-3-7258-0044-5

© 2024 by the authors. Articles in this book are Open Access and distributed under the Creative Commons Attribution (CC BY) license. The book as a whole is distributed by MDPI under the terms and conditions of the Creative Commons Attribution-NonCommercial-NoDerivs (CC BY-NC-ND) license.

Contents

About the Editors vii

Preface ix

Vineet Kumar and Xiaowu Tang
New Horizons in Nanofiller-Based Polymer Composites II
Reprinted from: *Polymers* 2023, 15, 4259, doi:10.3390/polym15214259 1

Bülend Ortaç, Saliha Mutlu, Taylan Baskan, Sevil Savaskan Yilmaz, Ahmet Hakan Yilmaz and Burcu Erol
Thermal Conductivity and Phase-Change Properties of Boron Nitride–Lead Oxide Nanoparticle-Doped Polymer Nanocomposites
Reprinted from: *Polymers* 2023, 15, 2326, doi:10.3390/polym15102326 6

Vineet Kumar, Md Najib Alam, Manesh A. Yewale and Sang-Shin Park
Tailoring Triple Filler Systems for Improved Magneto-Mechanical Performance in Silicone Rubber Composites
Reprinted from: *Polymers* 2023, 15, 2287, doi:10.3390/polym15102287 33

Aida Kistaubayeva, Malika Abdulzhanova, Sirina Zhantlessova, Irina Savitskaya, Tatyana Karpenyuk and Alla Goncharova et al.
The Effect of Encapsulating a Prebiotic-Based Biopolymer Delivery System for Enhanced Probiotic Survival
Reprinted from: *Polymers* 2023, 15, 1752, doi:10.3390/polym15071752 51

Alper Kaptan, Fatih Oznurhan and Merve Candan
In Vitro Comparison of Surface Roughness, Flexural, and Microtensile Strength of Various Glass-Ionomer-Based Materials and a New Alkasite Restorative Material
Reprinted from: *Polymers* 2023, 15, 650, doi:10.3390/polym15030650 66

Kun Zhang, Tao Ni, Jin Zhang, Wen Wang, Xi Chen and Mirco Zaccariotto et al.
Experimental and Hybrid FEM/Peridynamic Study on the Fracture of Ultra-High-Performance Concretes Reinforced by Different Volume Fractions of Polyvinyl Alcohol Fibers
Reprinted from: *Polymers* 2023, 15, 501, doi:10.3390/polym15030501 77

Ahmed M. Abdel-Gawad, Adham R. Ramadan, Araceli Flores and Amal M. K. Esawi
Fabrication of Nylon 6-Montmorillonite Clay Nanocomposites with Enhanced Structural and Mechanical Properties by Solution Compounding
Reprinted from: *Polymers* 2022, 14, 4471, doi:10.3390/polym14214471 101

Quimberly Cuenca-Bracamonte, Mehrdad Yazdani-Pedram and Héctor Aguilar-Bolados
Electrical Properties of Polyetherimide-Based Nanocomposites Filled with Reduced Graphene Oxide and Graphene Oxide-Barium Titanate-Based Hybrid Nanoparticles
Reprinted from: *Polymers* 2022, 14, 4266, doi:10.3390/polym14204266 119

Svetoslav Kolev, Borislava Georgieva, Tatyana Koutzarova, Kiril Krezhov, Chavdar Ghelev and Daniela Kovacheva et al.
Magnetic Field Influence on the Microwave Characteristics of Composite Samples Based on Polycrystalline Y-Type Hexaferrite
Reprinted from: *Polymers* 2022, 14, 4114, doi:10.3390/polym14194114 134

Yasser Zare, Kyong Yop Rhee and Soo Jin Park
Advancement of the Power-Law Model and Its Percolation Exponent for the Electrical Conductivity of a Graphene-Containing System as a Component in the Biosensing of Breast Cancer
Reprinted from: *Polymers* **2022**, *14*, 3057, doi:10.3390/polym14153057 . . . . **145**

Vineet Kumar, Md Najib Alam and Sang Shin Park
Soft Composites Filled with Iron Oxide and Graphite Nanoplatelets under Static and Cyclic Strain for Different Industrial Applications
Reprinted from: *Polymers* **2022**, *14*, 2393, doi:10.3390/polym14122393 . . . . **158**

About the Editors

Vineet Kumar

Vineet Kumar has been an Assistant Professor since 2016 at the Department of Mechanical Engineering, Yeungnam University, Gyeongsan-si, Republic of Korea. He completed his Ph.D. in the Department of Material Science at the University of Milan-Bicocca, Milan, Italy, in 2014. He finished his Master of Science in Environment Management from the Forest Research Institute, Dehradun, India, in 2008. He has co-authored more than 75 peer-reviewed articles in reputed *SCI* journals. He has also served as a Topic Editor for *Polymers–MDPI* since 2021. Moreover, he edited as a Guest Editor for more than six Special Issues in *Polymers–MDPI* and *Frontiers in Materials*.

Xiaowu Tang

Xiaowu Tang is presently an Associate Professor since March 2022 at the College of Material and Chemical Engineering, Zhengzhou University of Light Industry, Henan, Zhengzhou, China. He served as an Assistant Professor at Yeungnam University in South Korea from March 2021 to February 2022. He completed his Ph.D. at the Department of Chemical Engineering at Yeungnam University, Republic of Korea, in February 2021. He has co-authored more than 35 peer-reviewed articles in reputed *SCI* journals. Moreover, he edited as a Guest Editor for Special Issues in *Polymers–MDPI*.

Preface

Nanofillers are defined as nanomaterials such as graphene or carbon nanotubes that consist of at least one dimension less than 100 nanometers. These nanofillers are superior to traditional fillers such as carbon black in reinforcing polymer matrices. The combinations of such nanofillers with polymer matrixes are termed "nanocomposites". These nanocomposites are the subject of intensive research due to their outstanding impact on exhibiting robust mechanical, electrical, thermal, or barrier properties. These nanofillers are not only useful for impacting properties in their pristine form but also improve their properties when functionalized. The functionalization process improves the properties by improving interfacial interactions that provide efficient stress transfer within the nanocomposites. With the combined effect of different nanofillers in nanocomposites, they became potentially multi-functional for various applications due to their advanced overall properties. These applications involve their use in automobiles, wearable electronics, flexible and stretchable batteries, energy harvesting, and biomedical applications.

Keeping these objectives in mind, this reprint will provide fundamental information on various aspects of such nanocomposites. The reprint covers all aspects, starting with the type of materials used in fabricating these nanocomposites, fabrication methods, respective mechanical, electrical, or thermal properties, and their multifunctional nature with advanced academic and industrial usefulness. Moreover, the advantages and challenges, along with prospects, are comprehensively reported on these nanocomposites. Finally, the editors would like to thank the reviewers and authors whose efforts contributed to the success of this project.

Vineet Kumar and Xiaowu Tang
Editors

Editorial

New Horizons in Nanofiller-Based Polymer Composites II

Vineet Kumar [1,*] and Xiaowu Tang [2]

1. School of Mechanical Engineering, Yeungnam University, 280 Daehak-Ro, Gyeongsan 38541, Republic of Korea
2. College of Material and Chemical Engineering, Zhengzhou University of Light Industry, Zhengzhou 450001, China; tangxiaowu@naver.com
* Correspondence: vineetfri@gmail.com

1. Introduction

Nanofiller-based polymer composites are a hot-topic research area with significant industrial potential. These polymer composites are reinforced with different classes of filler with at least one dimension and a grain size below 100 nm. These nanoscale fillers are carbon-based, such as carbon nanotubes, graphene, and nano-carbon black [1]. Other nanofiller types include silica and clay minerals. Among these nanofillers, carbon nanotubes and graphene in particular have been extensively explored for their potential to improve mechanical, electrical, and thermal properties. Moreover, clay minerals have been studied for their ability to improve reinforcing properties or barrier properties. Besides nanofillers, polymer matrices are effective for obtaining robust composites [2]. These polymers can be thermoplastic, thermosets, or elastomers. Among them, elastomers such as silicone rubber, natural rubber, and butadiene rubber are frequently used. These elastomers have a versatile role in making stretchable devices, such as wearable electronics. Such composites have the potential to revolutionize industries such as aerospace, automotive, electronics, and healthcare. Key aspects covered by this Special Issue include the following:

(1) **Nanofillers**: Nanofillers are the organic or inorganic additives used in polymer matrices to improve their mechanical, electrical, or thermal properties. These additives confer on the final composite robust properties and make them useful for industrial applications as energy harvesters, strain sensors, etc. [3]. Moreover, the morphological features of nanofiller additives are a matter of interest. For example, carbon nanotubes with one-dimensional morphology and a high aspect ratio are helpful for improving electrical conductivity. Furthermore, two-dimensional graphene with a sheet-like morphology is useful for improving barrier properties. Thus, researchers have different options for nanofillers to fulfill the target property of interest.

(2) **Polymer matrix**: Polymer matrices are composed of macromolecules of different types, such as thermoplastics, thermosets, and elastomers. Among them, elastomers are most promising because of their unique properties, such as their ability to stretch under mechanical deformation and bounce into the original shape once the strain is removed [4]. The properties of a polymer matrix strictly depend on the type of polymer matrix used during composite preparation.

(3) **Polymer composites**: Polymer composites are materials that are based on a combination of nanofillers with a polymer matrix. The properties of these polymer composites are strictly based on the type of polymer matrix and nanofiller additives used during fabrication [5]. Moreover, the fabrication procedure used affects the final properties of these composites. The fabrication method could be melting mixing, solution mixing, or in situ polymerization. Among polymer composites, the composites based on elastomers as a polymer matrix are the most promising. This is due to their usefulness in a wide range of applications, such as energy harvesting and automobile tires.

(4) **Concept of new horizons in polymer composites**: Here, "new horizons" refers to the innovative and novel ideas explored by researchers working in the polymer composite field around the world. These novel ideas include the application of cutting-edge findings to obtain new and improved properties, making composites more robust for engineering applications [6]. These developments may include the emergence of a new class of materials, a new type of polymer matrix, or a new manufacturing technique.

(5) **Industrial applications**: Improving polymer composite properties through the selection of new-generation additive materials in a polymer matrix is a hot topic of research. Their use for the development of particular applications is an interesting subject, e.g., tuning stiffness for high-load applications such as tires or soft applications such as tissue engineering [7]. The properties of these polymer composites can be tuned with respect to a specific application. Thus, the subject of this Special Issue is a matter of interest for the readership of the *Polymers* journal.

(6) **Advantages of nanofillers used in polymer composites**: There are various advantages of using nanofillers in polymer composites, such as (a) enhanced properties of the filler additives at the nanoscale, providing uniform dispersion in the polymer matrices; (b) the nano-effect of these additives provides high reinforcement, higher tensile strength, and higher elongation at break compared to traditional fillers like carbon black; (c) nanofillers are typically lightweight and therefore useful for various smart applications, such as wearable electronics; (d) nanofillers exhibit higher electrical and thermal conductivity due to uniform dispersion and thus are useful for higher engineering applications; (e) the high surface area of these nanofillers provides a higher interfacial area for polymer chains to interact with filler additives and leads to improved properties; and (f) nanofillers can mitigate the shrinkage of polymer composites during the fabrication of polymer composites, thereby leading to improved dimensional stability of the final products [8]. Overall, the careful selection and optimization of the nanofiller additives is necessary to obtain properties and applications of interest.

(7) **Challenges in polymer composites**: There are various advantages of using nanofiller additives in polymer matrices for polymer composites. However, there are a few challenges with respect to properties and final engineering applications. For example, monolayer graphene is a great additive as a nanofiller in polymer composites, but synthesizing it at a large scale is difficult and very expensive. Moreover, in energy-harvesting tests, the mechanical stability of electrodes and substrates based on polymer composites is challenging, and limited mechanical stability influences the long-term durability of the final device [8].

(8) **Environmental impact and sustainability**: With the emergence of new technologies in polymer composites to achieve high performance, the carbon footprint of the process and its sustainability are of the utmost importance. The contribution of non-biodegradable polymers like polyethylene to global warming and pollution needs to be considered [9]. The use of sustainable and environmentally friendly, biologically degradable polymers is the focus of researchers globally. Hence, this Special Issue focuses mostly on environmentally friendly polymer composites.

Keeping all these aspects in mind, the present Special Issue presents a collection of articles (Communication, Research Articles, Review Articles) on polymer composites reinforced with nanofillers and their usefulness for different applications. More specifically, the key areas addressed include but are not limited to (a) all types of organic or inorganic nanofillers in the pristine or functionalized state, (b) all types of polymer matrices, (c) all types of properties, multi-scale modeling with theoretical aspects in polymer composites, and (d) finally self-healing, aging, and durability of composites. This Special Issue contains 10 articles from researchers across the globe covering the diverse topic of polymer composites, as summarized below.

2. Overview of Published Articles

Ortaç et al. [1] present a novel route of obtaining high thermal conductivity and phase change properties of a polymer nanocomposite. These nanocomposites were filled with boron nitride and doped with lead oxide nanoparticles to achieve enhanced properties. At 13 wt%, the thermal conductivity (λ) of the nanocomposites was 18.874 W/(mK). Moreover, the crystallization fraction (FC) with different co-polymers was 0.032, 0.034, and 0.063. The key takeaway from this study is that the composites present in this study can be used as energy storage materials due to their versatile nature. Kumar et al. [2] present an interesting method of reinforcing silicone rubber with a triple-filler hybrid system (carbon nanotubes, clay minerals, and iron particles) to obtain a robust magneto–mechanical performance. The compressive modulus was 1.73 MPa (control), 3.9 MPa (MWCNT, 3 phr), 2.2 MPa (clay mineral, 2 phr), 3.2 MPa (iron particle, 80 phr), and 4.1 MPa (hybrid system, 80 phr). Moreover, the results show that the triple-filler system emerged as the best candidate due to its high mechanical stiffness, which is useful for high load applications; optimum fracture strain; and tensile strength. Finally, the triple-filler system exhibits efficient magnetic sensitivity and significant output voltage as an energy-harvesting device.

Kistaubayeva et al. [3] investigated the influence of encapsulating a prebiotic with a biological origin polymer system on probiotic survival. The average size of the hybrid symbiotic beads was 3401 μm (wet) and 921 μm (dry). Moreover, the bacterial titer was 10^9 CFU/g. These results show the promising prospect of encapsulating prebiotics for a delivery system. Kaptan et al. [4] evaluated the surface roughness and flexural and micro-tensile strength of composites based on glass-ionomer-based composites in vitro. Among the different samples, the highest roughness achieved was 0.33 ± 0.1 while the lowest roughness obtained was 0.17 ± 0.04. Similarly, the highest flexural strength obtained was 86.32 ± 15.37, while the lowest one was 41.75 ± 10.05. Moreover, bonding among the materials was noticed between self-cured Cention N and other traditional composites. Zhang et al. [5] conducted a predynamic study on concentrate composites reinforced with polyvinyl alcohol. At a center deflection of 40 mm, the peak force increased from 3700 (PVA content 0.5 wt%) to 4700 (PVA content 2 wt%). These improvements are beneficial for the use of these composites for high-load-bearing applications. Abdel-Gawad et al. [6] fabricated composites based on Nylon 6 and clay minerals and reported their mechanical and structural properties. The results show that the solution mixing method used was more promising for the dispersion of clay minerals than traditional melt mixing. Moreover, the crystallinity of the control sample, which was originally 36%, increased to 58% after the addition of clay minerals in the Nylon-6 matrix. These prospects make these composites useful for high-performance applications.

Cuenca-Bracamonte et al. [7] conducted a study on composites based on a polyetherimide matrix reinforced with reduced graphene oxide or a hybrid of graphene oxide and barium titanate. The results show that the electrical conductivity, using 20 wt% filler, was ~10^{-9} S/cm for the hybrid filler, ~10^{-7} S/cm for rGO, and ~5×10^{-6} S/cm for rGO and the hybrid filler combined. Kolev et al. [8] investigated composites based on polycrystalline Y-type hexaferrate and the influence of the magnetic field on the properties of such composites. The highest microwave reflection, 35.4 dB, was achieved at 5.6 GHz without a magnetic field. However, under a magnetic field, it was 21.4 dB at 8.2 GHz. Zare et al. [9] present a power law model and percolation threshold in electrical conductivity for biosensing applications. Their study supports the use of graphene-filled nanocomposites, and the optimized model shows their use as biosensors. Finally, Kumar et al. [10] present an interesting study that deals with the fabrication of composites based on silicone rubber under cyclic and static strain for energy harvesting and magnetic sensitivity. The composites showed promise in their use for energy generation, with an output voltage around 10 volts and a durability of more than 0.5 million cycles. Similarly, the stretchability of the energy-harvesting device was 89% (control), with higher values found for GNP (109%), iron oxide (105%), and the hybrid (133%).

3. Summary and Future Outlook of Nanofillers in Polymer Composites

In the last three decades since the discovery of carbon nanotubes in 1991, the demand for nanofillers has increased significantly [10]. Additionally, when comparing them with traditional fillers like carbon black, using carbon nanotubes as nanofillers results in better properties at 3-5 phr compared to carbon black at a 60 phr loading. Thus, carbon nanotubes emerge as a promising additive for polymer composites. The addition of carbon nanotubes as an additive in a polymer matrix improves all properties except the barrier properties. Moreover, graphene emerged as a nanofiller due to its tremendous potential to improve the properties of polymer composites [11]. Moreover, the use of graphene as an additive also improves the barrier properties, which makes it advantageous over carbon nanotubes [8,10]. Thus, the increasing demand for such nanofillers has a tremendous impact on polymer matrices, especially as a reinforcing filler. Moreover, the influence of nanofillers is not only limited to their reinforcing effect; it also makes them suitable for various industrial applications as strain sensors and energy harvesters [8,12]. Keeping these aspects in mind, this Special Issue analyzes the potential impacts of using nanofillers for polymer composites and supports a promising future for materials scientists globally. Overall, the contributions of researchers to this Special Issue present a good quality of research work covering all possible topics within the scope of this field. Key aspects to consider in future research include the following:

(1) **Novel and robust nanofillers as additives**: As supported by the literature and the articles covered in this Special Issue, nanofillers show unique properties and multi-functionality. This aspect allows researchers to tailor the composites with respect to properties and applications of interest. Thus, nanofillers hold promise for scientists working in the polymer composite field.

(2) **Advanced processing and manufacturing**: The continuous efforts of scientists in improving manufacturing techniques have led to better filler dispersion. Additionally, other aspects related to the fabrication of polymer composite will continue to evolve. This will help in optimizing properties and indicates the promising multifunctionality of such composites.

(3) **Medical compatibility**: The use of nanofiller-reinforced polymer composites also indicates a promising future in the nanomedicine field. For example, polymer composites with nanofiller additives with improved biocompatibility and functionality have promising applications in drug delivery and medical implants.

(4) **Eco-friendly and green polymer composites**: With the advances in polymer composite science, the emergence of bio-based nanofillers as additives for polymer composites holds promise. These so-called green polymers are not only environmentally friendly materials but also possess high performance and multifunctionality.

(5) **Promising technologies and integrated performance**: Polymer composites reinforced by nanofillers could find use in new-generation applications such as self-powered electronic devices, flexible and stretchable electronic devices, advanced energy storage, and lightweight, cost-effective additive materials.

Overall, the use of nanofillers as additives represents a subject of interest in the polymer composite field. With continuous efforts from scientists, this field is expected to grow further, providing new cutting-edge technologies for multifunctional applications. Finally, the articles presented in this Special Issue provide insight into new advances and routes for further research and development in this area.

Acknowledgments: The authors thanks to all the contributors and reviewers for their valuable contributions and support from section editors of this special issue.

Conflicts of Interest: The authors declare no conflict of interest.

List of Contributions:

1. Ortaç; B; Mutlu, S.; Baskan, T.; Savaskan Yilmaz, S.; Yilmaz, A.H.; Erol, B. Thermal Conductivity and Phase-Change Properties of Boron Nitride–Lead Oxide Nanoparticle-Doped Polymer Nanocomposites. *Polymers* **2023**, *15*, 2326.
2. Kumar, V.; Alam, M.N.; Yewale, M.A.; Park, S.S. Tailoring Triple Filler Systems for Improved Magneto-Mechanical Performance in Silicone Rubber Composites. *Polymers* **2023**, *15*, 2287.
3. Kistaubayeva, A.; Abdulzhanova, M.; Zhantlessova, S.; Savitskaya, I.; Karpenyuk, T.; Goncharova, A.; Sinyavskiy, Y. The Effect of Encapsulating a Prebiotic-Based Biopolymer Delivery System for Enhanced Probiotic Survival. *Polymers* **2023**, *15*, 1752.
4. Kaptan, A.; Oznurhan, F.; Candan, M. In Vitro Comparison of Surface Roughness, Flexural, and Microtensile Strength of Various Glass-Ionomer-Based Materials and a New Alkasite Restorative Material. *Polymers* **2023**, *15*, 650.
5. Zhang, K.; Ni, T.; Zhang, J.; Wang, W.; Chen, X.; Zaccariotto, M.; Yin, W.; Zhu, S.; Galvanetto, U. Experimental and Hybrid FEM/Peridynamic Study on the Fracture of Ultra-High-Performance Concretes Reinforced by Different Volume Fractions of Polyvinyl Alcohol Fibers. *Polymers* **2023**, *15*, 501.
6. Abdel-Gawad, A.M.; Ramadan, A.R.; Flores, A.; Esawi, A.M. Fabrication of Nylon 6-Montmorillonite Clay Nanocomposites with Enhanced Structural and Mechanical Properties by Solution Compounding. *Polymers* **2022**, *14*, 4471.
7. Cuenca-Bracamonte, Q.; Yazdani-Pedram, M.; Aguilar-Bolados, H. Electrical Properties of Polyetherimide-Based Nanocomposites Filled with Reduced Graphene Oxide and Graphene Oxide-Barium Titanate-Based Hybrid Nanoparticles. *Polymers* **2022**, *14*, 4266.
8. Kolev, S.; Georgieva, B.; Koutzarova, T.; Krezhov, K.; Ghelev, C.; Kovacheva, D.; Vertruyen, B.; Closset, R.; Tran, L.M.; Babij, M.; et al. Magnetic field influence on the microwave characteristics of composite samples based on polycrystalline Y-type hexaferrite. *Polymers* **2022**, *14*, 4114.
9. Zare, Y.; Rhee, K.Y.; Park, S.J. Advancement of the Power-Law Model and Its Percolation Exponent for the Electrical Conductivity of a Graphene-Containing System as a Component in the Biosensing of Breast Cancer. *Polymers* **2022**, *14*, 3057.
10. Kumar, V.; Alam, M.N.; Park, S.S. Soft composites filled with iron oxide and graphite nanoplatelets under static and cyclic strain for different industrial applications. *Polymers* **2022**, *14*, 2393.

References

1. Srivastava, S.K.; Mishra, Y.K. Nanocarbon reinforced rubber nanocomposites: Detailed insights about mechanical, dynamical mechanical properties, payne, and mullin effects. *Nanomaterials* **2018**, *8*, 945. [CrossRef] [PubMed]
2. Li, Y.; Huang, X.; Zeng, L.; Li, R.; Tian, H.; Fu, X.; Wang, Y.; Zhong, W.-H. A review of the electrical and mechanical properties of carbon nanofiller-reinforced polymer composites. *J. Mater. Sci.* **2019**, *54*, 1036–1076.
3. Surmenev, R.A.; Orlova, T.; Chernozem, R.V.; Ivanova, A.A.; Bartasyte, A.; Mathur, S.; Surmeneva, M.A. Hybrid lead-free polymer-based nanocomposites with improved piezoelectric response for biomedical energy-harvesting applications: A review. *Nano Energy* **2019**, *62*, 475–506. [CrossRef]
4. Donnet, J.B. Nano and microcomposites of polymers elastomers and their reinforcement. *Compos. Sci. Technol.* **2003**, *63*, 1085–1088. [CrossRef]
5. Miedzianowska, J.; Masłowski, M.; Rybiński, P.; Strzelec, K. Modified nanoclays/straw fillers as functional additives of natural rubber biocomposites. *Polymers* **2021**, *13*, 799. [CrossRef]
6. Das, P.P.; Chaudhary, V.; Ahmad, F.; Manral, A. Effect of nanotoxicity and enhancement in performance of polymer composites using nanofillers: A state-of-the-art review. *Polym. Compos.* **2021**, *42*, 2152–2170. [CrossRef]
7. Mollajavadi, M.Y.; Tarigheh, F.F.; Eslami-Farsani, R. Self-healing polymers containing nanomaterials for biomedical engineering applications: A review. *Polym. Compos.* **2023**, *44*, 6869–6889.
8. Kumar, V.; Lee, D.J.; Park, S.S. Multi-functionality prospects in functionalized and pristine graphene nanosheets reinforced silicone rubber composites: A focused review. *FlatChem* **2023**, *41*, 100535.
9. Peng, C.; Wang, J.; Liu, X.; Wang, L. Differences in the plastispheres of biodegradable and non-biodegradable plastics: A mini review. *Front. Microbiol.* **2022**, *13*, 849147. [CrossRef] [PubMed]
10. Iijima, S. Helical microtubules of graphitic carbon. *Nature* **1991**, *354*, 56–58. [CrossRef]
11. Geim, A.K.; Novoselov, K.S. The rise of graphene. *Nat. Mater.* **2007**, *6*, 183–191. [CrossRef] [PubMed]
12. Noh, J.S. Conductive elastomers for stretchable electronics, sensors and energy harvesters. *Polymers* **2016**, *8*, 123. [CrossRef]

Disclaimer/Publisher's Note: The statements, opinions and data contained in all publications are solely those of the individual author(s) and contributor(s) and not of MDPI and/or the editor(s). MDPI and/or the editor(s) disclaim responsibility for any injury to people or property resulting from any ideas, methods, instructions or products referred to in the content.

Thermal Conductivity and Phase-Change Properties of Boron Nitride–Lead Oxide Nanoparticle-Doped Polymer Nanocomposites

Bülend Ortaç [1], Saliha Mutlu [1,2], Taylan Baskan [3], Sevil Savaskan Yilmaz [1,2], Ahmet Hakan Yilmaz [3,*] and Burcu Erol [4]

1. UNAM-National Nanotechnology Research Center and Institute of Materials Science and Nanotechnology, Bilkent University, Ankara 06800, Turkey
2. Department of Chemistry, Faculty of Sciences, Karadeniz Technical University, Trabzon 61080, Turkey
3. Department of Physics, Faculty of Sciences, Karadeniz Technical University, Trabzon 61080, Turkey
4. Department of Physics, Faculty of Arts and Sciences, Recep Tayyip Erdoğan University, Rize 53100, Turkey
* Correspondence: hakany@ktu.edu.tr; Tel.: +90-462-377-2552; Fax: +90-462-325-3197

Abstract: Thermally conductive phase-change materials (PCMs) were produced using the crosslinked Poly (Styrene-block-Ethylene Glycol Di Methyl Methacrylate) (PS-PEG DM) copolymer by employing boron nitride (BN)/lead oxide (PbO) nanoparticles. Differential Scanning Calorimetry (DSC) and Thermogravimetric Analysis (TGA) methods were used to research the phase transition temperatures, the phase-change enthalpies (melting enthalpy (ΔH_m), and crystallization enthalpies (ΔH_c)). The thermal conductivities (λ) of the PS-PEG/BN/PbO PCM nanocomposites were investigated. The λ value of PS-PEG/BN/PbO PCM nanocomposite containing BN 13 wt%, PbO 60.90 wt%, and PS-PEG 26.10 wt% was determined to be 18.874 W/(mK). The crystallization fraction (F_c) values of PS-PEG (1000), PS-PEG (1500), and PS-PEG (10,000) copolymers were 0.032, 0.034, and 0.063, respectively. XRD results of the PCM nanocomposites showed that the sharp diffraction peaks at 17.00 and 25.28 °C of the PS-PEG copolymer belonged to the PEG part. Since the PS-PEG/PbO and the PS-PEG/PbO/BN nanocomposites show remarkable thermal conductivity performance, they can be used as conductive polymer nanocomposites for effective heat dissipation in heat exchangers, power electronics, electric motors, generators, communication, and lighting equipment. At the same time, according to our results, PCM nanocomposites can be considered as heat storage materials in energy storage systems.

Keywords: thermal conductivity; phase-change materials; boron nitride–lead oxide polymer nanocomposite; polystyrene–polyethyleneglycol block copolymer; nanocomposite

1. Introduction

Water and phase-change materials have been extensively studied in the literature as potential thermal energy storage media in construction applications. Water-based and PCM-based glass systems have been found to have much greater temperature-damping qualities than standard air-based glass systems [1]. By using the right cavity thickness, the storage system can be tailored for a certain climate zone [2]. Experimental studies found temperature damping to be promising in water-based systems [3]. Thermal energy storage (TES) is critical for the conservation of fossil fuels. New technologies, such as solar energy storage systems, are being introduced and studied in order to lower the energy demand of buildings [4]. In addition to batteries, Akr et al. investigated mechanical energy storage, thermal energy storage, magnetic energy storage, fuel cells, and energy storage technologies. A preliminary study and cost analysis, as well as appropriate building storage methods, will boost the efficiency of storage technology [5]. TES is a cutting-edge energy technology that is gaining traction in applications such as air and water heating, refrigeration, and air conditioning. TES appears to be the most appropriate mechanism for correcting imbalances between energy supply and demand [6]. Microencapsulated PCMs typically have a wall construction, whereas macroencapsulated PCMs can be embedded in floors and ceilings. Different researchers employ different approaches to studying the thermo-physical properties of new phase-change materials [7].

Phase-change Materials (PCMs) have a high capacity to store thermal energy. However, they have low thermal conductivity and poor heat transfer properties [8,9]. Therefore, heat transfer improvement techniques such as fins [10–12], metal foams [13–15], nano-additives [16–18] and encapsulation [19] are used to improve the heat transfer capabilities of PCMs. Khedher et al. investigated the effect of heat transfer on the thermal behavior of a closed environment filled with a neopentyl glycol/CuO solid–solid PCM nanocomposite and demonstrated that increasing the heat transfer rate using a fixed amount of material is an important task to improve fine performance [20]. In general, the addition of CNT nanoparticles, which have greater conductivity than Al_2O_3, to PCMs increases the effective thermal conductivity and surface area for heat conduction [21]. Meng et al. studied PCMs based on sodium sulfate decahydrate ($Na_2SO_4 \cdot 10H_2O$, SSD) as a thermal energy storage material. Alginate/SSD composite PCMs have been prepared by mixing SSD with different concentrations of alginate polymer [22]. Microcapsules have the ability to increase the thermal and mechanical performance of PCMs used in thermal energy storage, as they increase the heat transfer area and prevent leakage of melted materials [23]. Mohaddes et al. used a melamine formaldehyde (MF) resin as the shell material to encapsulate n-eicosane and showed that the latent heats of melting and crystallization of MF-based microcapsules were 166.6 J/g and 162.4 J/g, respectively [24]. They found that fabrics doped with such microcapsules exhibited a lower thermal lag efficiency and a higher thermoregulation capacity. PCMs are recognized as promising candidates for thermal energy storage that can improve energy efficiency in building systems. Li et al. designed and developed a new salt hydrate-based PCM composite with high energy storage capacity, relatively higher thermal conductivity, and excellent thermal cycling stability. The composite's energy storage capacity and thermal conductivity are enhanced by the addition of various graphitic materials along with the Borax nucleator [18]. The use of PCMs provides higher heat storage capacity and more isothermal behavior during the charging and discharging state compared with sensible heat storage [25–27]. PCMs are chosen because of their use in various energy storage areas such as solar panels, waste heat recovery, and other heat energy storage systems [28]. Because the low thermal conductivity of PEG is undesirable in energy storage processes, many different studies have been carried out to overcome this disadvantageous situation [28–30]. These properties are fascinating for thermal interface materials [31,32]. Because PCMs are functional materials that can store and release large amounts of latent heat energy within a slight temperature change [33,34], they have been frequently used in solar energy storage [35], smart textiles [36–39], thermal protection of electronic devices [40], waste heat recovery [41] and smart housing [42–44].

This study used DSC and TGA to examine the thermal changes in the phase transitions of PS-PEG copolymers and PS-PEG/BN/PbO PCM nanocomposites. The ΔH_m, ΔH_c, T_m, T_c, and the decomposition temperatures of the PCM nanocomposites were investigated. ΔH_m and ΔH_c enthalpies of the PCM nanocomposites were investigated between −20–250 °C. The addition of PbO nanoparticles and BN nanoparticles to the copolymers increased the degradation temperature and residual amount of the polymers. For example, the PS-PEG (1000) polymer, which remained at 30.3 wt% at the 380 °C decomposition temperature, increased the decomposition temperature of the PCM nanocomposite to 402.9 °C, and that of the remaining composite amount to 51.820 wt%, as a result of 90% PbO nanoparticle addition. As a result of interactions between PbO nanoparticles and BN nanoparticles in PCM nanocomposites, F_c values were calculated to see how the crystallization rate changed [45–47]. The value of β in the equation is the mass fraction of the PS-PEG block copolymer [48]. Sun et al. [49] found that, as the ratio of additive materials in the

PEG/CMPs composite increased, the F_c values were greater than the F_c value of PEG and were in the range of 102–105%.

Thermally conductive composite materials are produced using polymer materials with good machinability, low cost, and light weight [50–52]. Alizadeh et al. synthesized graft semi-interpenetrating polymer networks out of polyethylene glycol PEG 8000-based polyurethane and acrylic copolymers, and these graft-IPN samples can be used for thermal energy storage due to their high thermal properties [53]. Commercial grade polyethylene glycol (PEG) with a molecular weight of 6000 was tested by Sharma et al., and reliability tests of the PEG 6000 combined with techno-economic analysis have shown that this PCM can be used as a thermal energy storage system [54]. By creating nanodispersion polyethylene glycol (PEG)/PMMA/GnPs composites, Zhang et al. investigated the thermal and electrical properties of FSPCMs as well as their effects on morphology, structure, and form-stable performance [55]. The sol–gel coating method by Mo et al. obtained ternary lithium, sodium, and potassium carbonate/silica microcomposites as phase-change materials. It was concluded that microcomposites have an important place in high-temperature thermal energy storage [56]. Increased thermal conductivity of the PCMs was achieved by adding expanded graphite carbon nanotubes, graphene nanomaterials, activated carbon, carbon fiber, and metallic/oxide nanoparticles. BN is a universally accepted ceramic filler, especially for thermally conductive composites, due to its thermal conductivity and electrical insulator properties [57–67]. Qi and coworkers [61] increased the λ value of the PEG/graphene oxide (GO)/graphene nanoplatelet (GNP) PCM composite to 1.72 W/(mK) when they filled PEG using GO 2 wt% and GNP 4 wt% filler. Jia et al. added polyethylene glycol (PEG) to BN@CS scaffolds and showed that this increased the thermal conductivity value up to 2.77 W/(mK) [68]. While the thermal conductivity of pure PEG was measured at the level of 0.285 W/(mK), the thermal conductivity value for PEG@MXene was increased up to 2.052 W/(mK), with a 7.2-fold increase determined by Lu et al. [69]. The thermal conductivity of pure PCM and EHS/BNF composite PCMs containing 4 wt% were compared by Han et al., and the conductivity value of the EHS/BNF composite was determined to be 10.37 times higher than that of pure PCMs [26]. The thermal conductivities of pure PEG, PVA, PPVA, and the composite of GA/PEG as noted by Shen et al. [62] were 0.493, 0.152, 0.112 and 0.687 W/(mK). Polymers are flexible, light, durable, cheap, and resistant to abrasion and heat energy, and their usage has been increasing in every field from clothing to buildings and vehicles. In terms of developing the needed properties in structures, these characteristics highlight the important role of polymers in multidisciplinary scientific research [70–72]. The ability of a material to transfer heat energy is defined as thermal conductivity. We performed thermal conductivity calculations according to Equation (2) [73]. The thermal conductivity and the thermal properties of Portland Cement-HB (PAE-b-PCL)-PU plaster and HB (PAE-b-PCL)-PU/PbO-BN nanocomposites have been investigated by Cinan et al. [74]. When HB (PAE-b-PCL)-PU plaster, PbO, and AsO were added to Portland Cement, we found that they increased the properties of the cement, based on the thermal conductivity values. The thermal conductivity values of these PCMs were between 3.22 W/(mK) and 3.90 W/(mK) [72].

We have shown that PS-PEG copolymers doped with BN nanoparticles and PbO nanoparticles (i.e., PS-PEG PCM nanocomposites) are promising in terms of improving λ values and energy use efficiency. This article presents the preparation and thermal/physical characterizations of nano-enhanced PCMs with BN, a PbO nano blend, or single BN and PbO using crosslinked PS-PEG copolymers that we have previously synthesized and characterized [75–77]. The particle sizes of block copolymers and PCM nanocomposites were investigated using SEM and TEM analyses. At the same time, the XRD technique was used to determine the crystallographic structure of PCM nanocomposites.

Because of PCMs' low thermal conductivity, their practical usage in latent heat storage units is limited. PS-PEG copolymers, PbO, and BN nanoparticles were utilized in this study, not only to boost thermal conductivity but also to develop PCMs with optimal compositions that can reduce latent heat. This study provides an outline of how phase-transition materials can be used in melting and solidification. The melting temperatures of the examined PCM nanocomposites ranged from 55.5 °C to 200 °C. As a result, the PCM nanocomposites can be used in high-temperature-operated absorption applications, cooling, waste energy production, and heat recovery operations. The results demonstrated strong intermolecular interactions between the PS-PEG copolymer, the BN, and the PbO nanoparticles and demonstrated that nanoparticle dispersion inside the PCM had no effect. The chemical structure of the nanoparticles was altered, but their thermal and chemical stability was improved. PCM nanocomposites were discovered to be more stable and to perform better thermally than PS-PEG copolymers. PS-PEGs have low thermal conductivity, which limits heat storage and release rates and limits their applicability. Greater thermal conductivity in PCMs reduces melting and solidification times and speeds heat transfer throughout these processes. The NCPSPB3 PCM nanocomposite has H_m, T_m, H_c, and T_c values of 67.6 J g^{-1}, 81.8 °C, 5.12 J g^{-1}, and 188.4 °C, respectively. The ΔH_m, T_m, ΔH_c and T_c values of PEG/CNF [1], PEG/90CNF + 10rGONP [77], PEG1000 (45 wt%)/HNT-Ag^{-1} [1], PEG1000 (45 wt%)/HNT-Ag^{-3} [31] nanocomposites are 84.3 J g^{-1}, 25.8 °C, 79.3 J g^{-1}, 23.2 °C; 69.5 J g^{-1}, 24.3 °C, 62.2 J g^{-1}, 22.3 °C; 72.5 J g^{-1}, 35.2 °C, 28.1 °C; 71.3 J g^{-1}, 33.6 °C, and 25.7 °C, respectively. The enthalpy values of the NCPSPB3 PCM nanocomposite obtained in the study, PEG/CNF, PEG/90CNF + 10rGONP, the PEG1000 (45% wt%)/HNT-Ag^{-1} nanocomposites investigated by Zeighampour et al. [78], and the PEG1000 (45% wt%)/HNT-Ag^{-3} composite investigated by Song et al. [31] are near these values. PEG/CNF, PEG/90CNF + 10rGONP, PEG1000 (45 wt%)/HNT-Ag^{-1}, and PEG1000 (45 wt%)/HNT-Ag^{-3} have thermal conductivity values of 0.68 Wm^{-1}K^{-1}, 0.85 Wm^{-1}K^{-1}, 0.73 Wm^{-1}K^{-1}, and 0.90 Wm^{-1}K^{-1}, respectively. The thermal conductivity value of the NCPSPB3 PCM nanocomposite is 27.30 times, 21.84 times, 25.43 times, and 20.63 times greater than the values of PEG/CNF, PEG/90CNF + 10rGONP, PEG1000 (45 wt%)/HNT-Ag^{-1}, and PEG1000 (45 wt%)/HNT@Ag^{-3} composites, respectively. The H_m, T_m, H_c, and T_c values of the NCPSPb8 and NCPSPbBN17 PCM nanocomposites are 43.8 J g^{-1}, 83.8 °C, 4.94 J g^{-1}, 185.5 °C, and 34.6 J g^{-1}, 83.8 °C, and 5.36 J g^{-1}, 190.2 °C, respectively. The values of PEG/50CNF + 50rGONP nanocomposite are 55.1 J g^{-1}, 17.3 °C, 48.8 J g^{-1}, 16.8 °C, which are comparable to those of the PEG/50CNF/50rGONP nanocomposite. The NCPSPb8 PCM nanocomposite, NCPSPbBN17 PCM nanocomposite, and PEG/50CNF + 50rGONP nanocomposite have thermal conductivity values of 17.14 Wm^{-1}K^{-1}, 15.71 Wm^{-1}K^{-1}, and 2.39 Wm^{-1}K^{-1}, respectively. PEG/50CNF + 50rGONP has a thermal conductivity value that is 7.17 times and 6.57 times lower than that of the NCPSPb8 and NCPSPbBN21 PCM nanocomposites, respectively. Song et al. [31] developed the PEG/HNT-Ag^{-3} nanocomposite PCM as a unique kind of stable nanocomposite PCM with a suitable phase-change temperature (33.6 °C), relatively significant latent heat (71.3 J g^{-1}), outstanding thermal reliability, and increased thermal conductivity and conversion. Taking this into account, scientists demonstrated that it has a high potential for thermal energy storage and can be utilized as a building material to reduce indoor temperature changes, improve thermal comfort, and conserve electrical energy. According to Zeigampour et al. [78], SSPCNs with and without rGONP loadings have advantageous phase-transition temperatures, with latent heat values ranging from 55.1 to 84.3. They created SSPCNs that have shown excellent high-tech applications in accurate temperature control and quick temperature regulation [78]. The latent temperatures of various PCM nanocomposites ranged from 34.6 to 67.6 in this investigation. As a result, as described in the literature [31,78], these PCM composites can be used as building materials to reduce indoor temperature variations, increase indoor thermal comfort, conserve electrical energy, and provide precise temperature control and fast temperature regulation under certain conditions.

2. Materials and Methods

2.1. Materials

PbO Merck & Co. Inch. is produced and manufactured in Kenilworth, NJ, USA. BN particles of 1 µm size are an Aldrich product. PEG DM macrocrosslinkers were obtained from PEG polymers with molecular weights of 1000 gmol^{-1}, 1500 gmol^{-1}, and 10,000 gmol^{-1} by using methacrylic acid chloride [75–77].

2.2. Polymers

2.2.1. Synthesis of the PEG DM Macrocrosslinkers and the PS-PEG Block Copolymers

The PEG DM macrocrosslinkers and the PS-PEG copolymers were synthesized according to [75–77].

2.2.2. Preparation of the BN- and PbO-Doped PS-PEG PCM Nanocomposites

Table 1 shows the content of the PCM nanocomposites examined.

Table 1. The content of the PCM nanocomposites.

PCM ID	PS-PEG	PS-PEG (wt%)	BN (wt%)	PbO (wt%)	Volume (mm^3)
NCPS1	1000	100	0	0	575.3
NCPSPb2	1000	50	0	50	579.1
NCPSPb3	1000	30	0	70	523.9
NCPSPb4	1000	10	0	90	357.5
NCPSPb5	1000	46.2	0	53.8	753.8
NCPS6	1500	100	0	0	631.8
NCPSPb7	1500	50	0	50	734.5
NCPSPb8	1500	30	0	70	631.8
NCPSPb9	1500	10	0	90	463.6
NCPSPb10	1500	46.2	0	53.8	858.3
NCPS11	10,000	100	0	0	782.3
NCPSPb12	10,000	50	0	50	399.4
NCPSPb13	10,000	30	0	70	519.8
NCPSPb14	10,000	10	0	90	511.2
NCPSPb15	10,000	46.2	0	53.8	657.6
NCPSBN16	1000	50	50	0	696.5
NCPSPbBN17	1000	15	15	70	599.6
NCPSPbBN18	1000	5	5	90	551.1
NCPSPbBN19	1000	26.1	13	60.9	513.5
NCPSBN20	1500	50	50	0	591.3
NCPSPbBN21	1500	15	15	70	515.9
NCPSPbBN22	1500	5	5	90	638.3
NCPSPbBN23	1500	26.1	13	60.9	607.6
NCPSBN24	10,000	50	50	0	741.6
NCPSPbBN25	10,000	15	15	70	413.5
NCPSPbBN26	10,000	5	5	90	528.8
NCPSPbBN27	10,000	26.1	13	60.9	842.4

Known weights of PS-PEG block copolymers, PbO nanoparticles, and BN nanoparticles were mixed in an agar mortar and homogenized before being compressing into tablets. The tablets of the PCM nanocomposites were formed by hydraulic pressure at 10 MPa stress for 20 min at 22 °C. The tablet's thickness was measured by using a BTS of 12,051 µm. The thickness of the tablets with a diameter of 12 mm ranges between 0.8–5 mm.

Figure 1 shows the chemical structure of the polymers and the interaction between the polymers and nanoparticles.

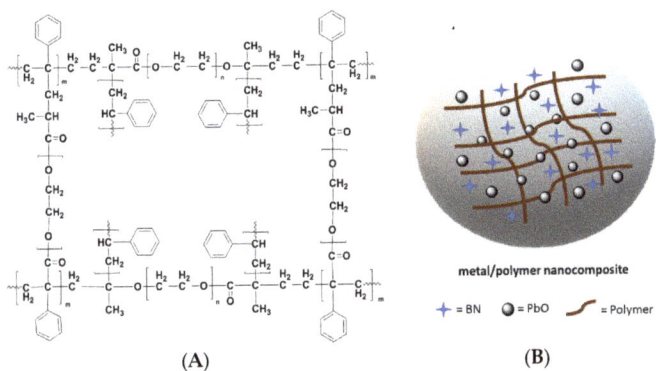

Figure 1. Schematic representation of the molecular formula (**A**) [74] of the PS-PEG block copolymer and (**B**) the PCM nanocomposite tablets.

2.3. Characterizations

The characterization of the macrocrosslinker synthesized according to the literature [67–69] was investigated by FT-IR, NMR, and GPC methods. PS-PEG block copolymers were investigated with FT-IR, SEM, and TGA instruments. The characteristic FT-IR peaks and properties of PEG DMs and PS-PEG block copolymers are similar to the results in the literature [67–69].

2.3.1. Thermal Properties

TGA Method

The thermal decomposition process of the PEG-DM macrocrosslinkers, the PS-PEG block copolymers, and the PCM nanocomposites was implemented via the Seiko II Exstar 6000 TG/DTA (Seiko Instruments Inc., Chiba, Japan) analysis instrument. TGA thermograms of the macrocrosslinkers, block copolymers and PCM nanocomposites were obtained in a nitrogen gas atmosphere (200 mL/min) between 30–500 °C. The heating rate was taken as approximately 20 °C/min.

DSC Method

The T_m, T_c, ΔH_m, and ΔH_c values of the PS-PEG copolymers and the PCM nanocomposites were obtained by using DSC (Perkin-Elmer Jade model, Perkin-Elmer Inc., Waltham, MA, USA). DSC measurements were made under nitrogen gas. The samples were examined at a heating rate of 10 degrees per minute from −20 °C to 300 °C and a cooling rate of 10 °C per minute from 300 °C to −20 °C. The weight of the PCM nanocomposites was approximately 3.0 mg. F_c values were calculated by using Equation (1) using DSC data [45–47].

$$F_c = \frac{\Delta H_{PCMS}}{\Delta H_{pure} \beta} \quad (1)$$

Here, ΔH_{pcms} and ΔH_{pure} are the latent heat of the nanocomposite and the PS homopolymer, respectively. The value of ΔH_{pure} is 22.5 J/g. β is the mass fraction of the PS-PEG block copolymer in the nanocomposite.

2.4. Thermal Conductivity Method

By measuring the temperature difference between the two ends of the PCM nanocomposites, we calculated the thermal conductivity from Equation (2) [73]. We used a resistor that could go up to 50 W for 12 V to energize one surface of the sample. We determined the inlet temperature between 35–90 °C by applying 22.2 W of power.

$$\lambda = \frac{q(x_2 - x_1)}{A(T_2 - T_1)} \quad (2)$$

Here, λ is the thermal conductivity, and its unit is given as W/(mK). The parameter q is the power of the resistor, given in Watts. x_1 and x_2 are the distance between the beginning and ends of the sample exposed to heat and the end of the heat flow, respectively. Equation (2), $q = -\lambda \cdot A \cdot \nabla T$, is based on Fourier's law. $T_1 > T_2$, since T_1 is the initial temperature value applied to the x_1 point of the sample. T_2 is the temperature measured on the other end of the sample after the heat has passed through the sample. Therefore, $T_2 - T_1 < 0$.

2.5. Morphology

2.5.1. SEM Analysis

The surface properties of the polymers and the PCM nanocomposites were elucidated with the SEM method. The SEM photographs were pictured by the JEOL JXA-840 brand model SEM instrument (Tokyo, Japan). The PS-PEG copolymers and the PCM nanocomposites were frozen with liquid nitrogen and then broken with an Edwards S 150 B model spray-coater. Broken specimens were plated with gold (300 Angstroms). SEM photos were taken at 10 kV, high vacuum, ESEM at 30 kV, and 3.0 nm resolution. The standard detectors used were ETD, low vacuum SED (LVD), gas SED for ESEM mode (GSED), and IR camera. Electron images from the cathode ray tube were recorded on a Polaroid film.

2.5.2. TEM Analysis

The structural analyses of PbO and BN nanoparticles and the PCM nanocomposites were investigated by using the FEI-Tecnai G2F30 model TEM tool. The samples of which TEM pictures were taken were examined by fixing on carbon-coated TEM grids. The nanoparticles and PCM nanocomposites were imaged at 300 kV. The histogram of PbO nanoparticle sizes was obtained by counting more than 450 particles from the TEM image and using Image J Processing and Analysis software.

2.5.3. XRD Method

The XRD data of the PCM nanocomposites were obtained with an X-Ray Diffractometer with trademark DMAX 2400 (Rigaku, Japan). The measurements were made under Copper K_α radiation with properties of 1.541 Å, 40 kV, and 100 mA. The measurements were taken at a scan rate of 8/min over 2 h and between 5–90 °C. The 2θ value and Miller indices of the PS-PEG block copolymers and the nanocomposite PCMs were investigated.

3. Results and Discussion

3.1. TGA Measurements of the PS-PEG-PbO and -BN Nanocomposite PCMs

Table 2 shows the degradation temperatures and the remaining mass (wt%) of the PCM nanocomposites. Figure 2 and Figure S1 show thermograms of PS-PEG (1000) PCMs with PbO additives. Figure 2 presents TGA thermograms of NCPSPb3, NCPSPb8, NCPSPb13, NCPSPbBN17, NCPSPbBN21, and NCPSPbBN25. Thermograms of other PCM nanocomposites are presented in Figures S1–S6. The thermograms demonstrate thermal degradations of all PCMs which were investigated between 40.3–402.9 °C. As a result, the thermal stability of the PCM nanocomposite with the addition of PbO nanoparticles is higher than the value of the polymers. This situation is due to increased physical interactions between PbO nanoparticles and PS-PEG polymer, such as van der Waals force and hydrophobic–hydrophobic interactions. The initial degradation temperature of NCPSPb3 PCM nanocomposite is higher than the value of NCPS1 and NCPSPb2. When the PbO content of the NCPSPb3 is increased and the polymer content is decreased, the moisture evaporates and the temperature decreases to 49.0 °C. NCPSPb3 has more PbO nanoparticles than the polymer, so its degradation temperature increases. The remaining masses of NCPSPb3 at its first and second degradation temperature are 96,900 wt% and 83,100 wt%, respectively. As a result, when PbO nanoparticles are doped into the PS-PEG (1000) copolymer (NCPS1), this increases the thermal stability of the polymer. As seen in Figure 2, Figures S1C, S2B, S3C, S4B, S5B and S6B, when NCPSPb4 contains PbO nanoparticles

70 wt% and 15 wt% BN, the amount of remaining mass is higher. TGA graphs of NCPSPb4 and NCPSPb5 are shown in Figure S1D and Figure S1E, respectively. The results of the NCPSPb5 are very different from the results of the NCPSPb2. Although the initial decay temperature is high in NCPSPb5, the remaining mass amounts are lower than in NCPSPb2. As a result, 53.8 wt% PbO nanoparticles in the NCPSPb5 decreased the thermal stability of the PCM nanocomposite.

Table 2. The degradation temperatures and the remaining mass (wt%) of the PCM nanocomposites.

PCM ID	First Stage of Degradation		Second Stage of Degradation		Third Stage of Degradation		Fourth Stage of Degradation	
	t (°C)	Remaining Mass, %wt	t (°C)	Remaining Mass, %wt	t (°C)	Remaining Mass, %wt	t (°C)	Remaining Mass, %wt
NCPS1	46.7	76.7	254.2	62.5	343.5	49.4	381.2	36.4
NCPSPb2	40.3	99.2	249.6	88.4	335.4	71.4	382.8	45.1
NCPSPb3	49.0	99.9	244.4	83.1	289.0	71.8	375.3	42.3
NCPSPb4	43.3	98.7	85.9	88.73	241.8	83.7	402.9	51.8
NCPSPb5	80.8	71.5	257.9	66.3	383.7	21.2	-	-
NCPS6	53.4	93.5	322.4	82.5	417.9	0.5	-	-
NCPSPb7	50.7	92.7	257.6	64.8	280.7	58.4	386.9	22.7
NCPSPb8	39.9	97.5	243.3	83.4	374.0	38.0	428.7	33.4
NCPSPb9	47.2	98.1	165.9	85.9	349.0	61.0	370.1	53.9
NCPSPb10	47.2	90.51	85.7	81.9	253.4	76.7	380.0	30.4
NCPS11	31.2	96.1	66.5	76.0	293.8	69.3	401.8	3.0
NCPSPb12	25.3	94.3	75.2	49.7	259.9	40.4	403.3	−54.0
NCPSPb13	41.6	97.3	239.1	89.5	326.4	71.2	371.9	48.8
NCPSPb14	44.0	97.6	231.9	87.2	287.7	77.2	372.0	56.2
NCPSPb15	45.1	98.4	245.1	91.4	372.2	64.6	424.9	61.1
NCPSBN16	46.5	98.8	254.8	96.4	366.9	63.4	424.8	60.0
NCPSPbBN17	54.7	97.04	233.7	93.2	331.7	77.1	381.5	69.2
NCPSPbBN18	37.1	97.5	228.3	91.6	370.0	65.2	420.1	62.7
NCPSPbBN19	47.2	87.9	230.8	84.6	331.1	12.6	393.5	3.9
NCPSBN20	37.0	99.0	281.9	95.3	376.6	63.4	418.7	61.2
NCPSPbBN21	39.2	97.8	237.4	88.7	344.0	68.5	371.9	61.5
NCPSPbBN22	53.8	99.0	152.2	97.0	357.9	72.0	421.0	69.4
NCPSPbBN23	41.2	97.1	242.8	86.9	365.1	58.6	427.0	53.2
NCPSBN24	51.8	99.2	272.8	96.6	385.3	75.9	417.3	74.8
NCPSPbBN25	64.7	96.5	235.5	92.6	361.9	68.7	410.8	66.7
NCPSPbBN26	65.8	99.0	236.7	95.1	377.2	68.8	412.4	68.0
NCPSPbBN27	28.3	99.2	238.5	95.1	371.7	54.7	421.8	50.8

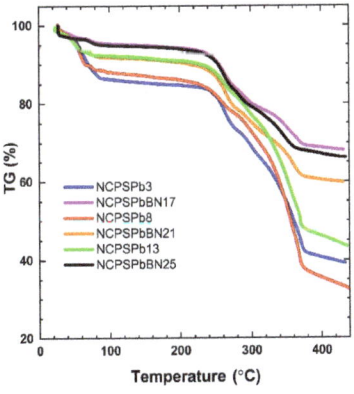

Figure 2. TGA thermograms of the NCPSPb3, NCPSPb8, NCPSPb13, NCPSPbBN17, NCPSPbBN21, and NCPSPbBN25 PCM nanocomposites.

The thermogram of the NCPS6 PCM nanocomposite is shown in Figure S2A. For NCPS6, the thermal stability, initial degradation temperature, and the amount of mass remaining after decomposition decreases as the molecular weight of the crosslinker increases, while the final degradation temperature increases. NCPSPb7 NCPSPb8, NCPSPb9 and NCPSPb10 graphs are presented in Figure S2B–E. When the PbO ratio is 50 wt% in the NCPSPb7, the initial degradation temperature is 50.7 °C, which is higher than the value of the NCPSPb2. As a result, thermal stability increased when the molecular weight of the crosslinker PEG was increased from 1000 to 1500, when the amount of PbO was the same. The initial decomposition temperatures and the amount of evaporated water of NCPSPb8 and NCPSPb9 containing 70 wt% and 90 wt% of PbO nanoparticles were decreased compared with the values of the NCPS6. As a result, the addition of a large amount of PbO for the PS-PEG-1500 polymer reduced thermal stability. The results for the NCPSPb10 are in agreement with the other nanocomposites.

Thermograms of NCPS11, NCPSPb12, NCPSPb13, NCPSPb14, and NCPSPb15 are shown in Figure S3A–E. Considering the degradation temperatures and the residual mass obtained for NCPSPb11-15 PCM nanocomposites, thermal stability decreased as the molecular weight of the macrocrosslinker increased from 1000 and 1500 to 10,000. However, an increased amount of PbO in PEG-10,000 indicates that it increased thermal stability.

The thermograms of NCPSBN16, NCPSPbBN17, NCPSPbBN18, and NCPSPbBN19 are presented in Figure S4A–D. When BN was added to the NCPS1 polymer in NCPSPb16, the thermal stability was observed to be higher than the thermal stability of NCPS1. As observed in NCPSPb17, thermal stability appears to be better than in NCPSPb16 when PbO is added. Increasing the PbO ratio to 90 wt% in the NCPSPb18 reduced the thermal stability of the composite to below the value of NCPSPb17.

Figure S1 and Figure 2 and show the thermal plot of PS-PEG (1500) doped with BN and PbO nanoparticles. The thermograms of NCPSBN20, NCPSBN21, NCPSBN22, and NCPSBN23 are shown in Figure S5A–D. When the TGA graphs of NCPSPb20, NCPSPb21 and NCPSPb22 nanocomposites are examined, it is clearly seen that increasing the amount of PbO nanoparticles in the PEG 1500 polymer increases its thermal stability. Considering the composition of the NCPSPb23 nanocomposite, its thermal stability is higher than that of NCPSPb21, because the amount of polymer is higher in its structure. However, its thermal stability is decreased a little due to the amount of PbO being lower than that in NCPSPb23.

The PCM nanocomposite thermograms (containing PS-PEG (10,000) block copolymer, and BN and PbO nanoparticles) are presented in Figure 2 and Figure S6. The graph for NCPSBN24 is shown in Figure S6A. The thermograms of NCPSPbBN25, NCPSPbBN26, and NCPSPbBN27 are shown in Figure S6B–D. It was observed that the addition of BN and PbO at different rates to the NCPSP11 nanocomposite (for NCSPb25 and NCPSPb26) increased the thermal stability. When the BN nanoparticle and PbO nanoparticle is below 15 wt% and 70 wt% (for NCPSPb27), respectively, the first decomposition temperature of the PCM nanocomposite is decreased, but the amount of the remaining mass does not change significantly. The remaining mass at the final decomposition temperature ranges from 68–74% in other samples, while it is around 50% for NCPSPb27.

The TGA results were examined in detail to illustrate the degradation of PbO- and BN-doped PS-PEG PCMs. As seen in all figures, there was little degradation when the PCMs were heated to about 430 °C, indicating that the PCMs are thermally stable. Three degradation temperatures were observed: 40 °C, 250 °C and 380 °C. The initial degradation of PS-PEG (1000) PCMs started at about 40 °C, and the final degradation was observed at 380 °C. In addition, the nanostructured BN particle additive yielded three degradation temperatures at 40 °C, 230 °C, and 330 °C. The TGA curves of all the PbO–BN-doped nanocomposite PCMs exposed thermal degradation actions parallel to that of the PbO-doped PS-PEG (1000) PCMs. A few degradation points were between 50–380 °C. The initial loss began at 50 °C and the last degradation at 380 °C originated from the degradation of the PS-PEG (1500) PCMs. In addition, the nanostructured BN nanoparticle additive showed three degradation temperatures around 40–240 °C and

420 °C. The first stage of degradation started at 40 °C and the final decomposition temperature was 430 °C, which caused the degradation of the PS-PEG copolymer in PCMs. Furthermore, the addition of BN nanoparticles showed several degradation temperatures in the range of 40 °C to 420 °C. As a result, it was seen that losses between 40 °C and 80 °C in all composites were caused by the evaporation of water. In addition, PCMs with higher PbO and BN nanoparticle ratios represent thermal stability. This is due to the large number of nanoparticles in PCMs that prevent the degradation of polymer chains. The present results show that the PS-PEG polymer becomes more thermally stable when the PCM nanocomposites are incorporated with PbO and BN nanoparticles. Because of these uses, PCM nanocomposites can exhibit excellent thermal persistence for a variety of energy application systems.

3.2. DSC Results of the PS-PEG/BN/PbO PCM Nanocomposites

Phase-change temperatures of PS-PEG/PbO, PS-PE/BN, and PS-PEG/PbO/BN PCM nanocomposites were measured by the DSC technique. Melting and solidification DSC curves of PS-PEG PS-PEG/PbO, PS-PE/BN, and PS-PEG/PbO/BN PCM nanocomposites are shown in Figures 3 and 4. The T_m, ΔH_m, T_c, ΔH_c values of the PS-PEG PCM nanocomposites were determined from the endothermic and exothermic curves during phase change. The melting/solidification temperature (T_m/T_c), endothermic/exothermic enthalpy ($\Delta H_m/\Delta H_c$) and F_c values are given in Table 3. T_m-T_c, ΔH_m-ΔH_c and F_c values are the most effective ways to interpret the thermal behavior of PCM nanocomposites. Endothermic peaks were observed between 55.5–205.8 °C due to melting (as seen in Table 3 and Figures 3 and 4).

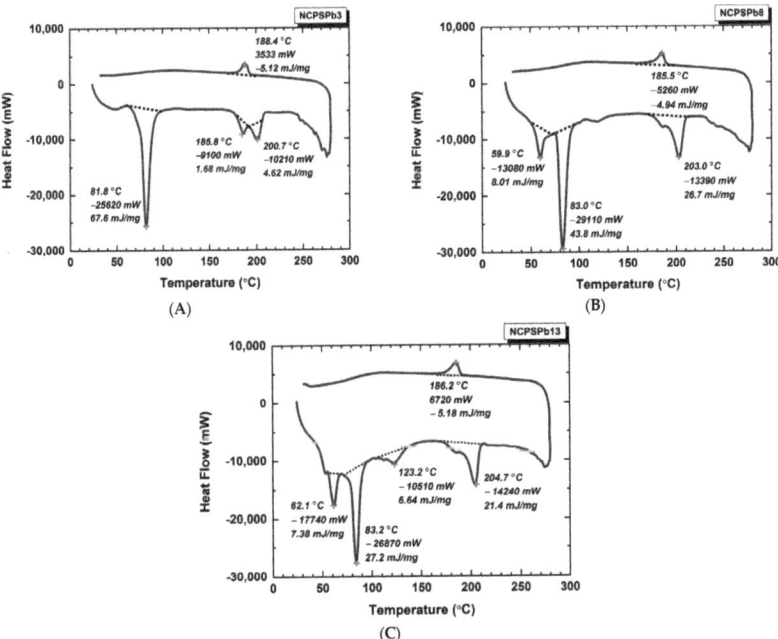

Figure 3. Thermal behavior of PS-PEG PCMs: (A) NCPSPb3, (B) NCPSPb8, (C) NCPSPb13.

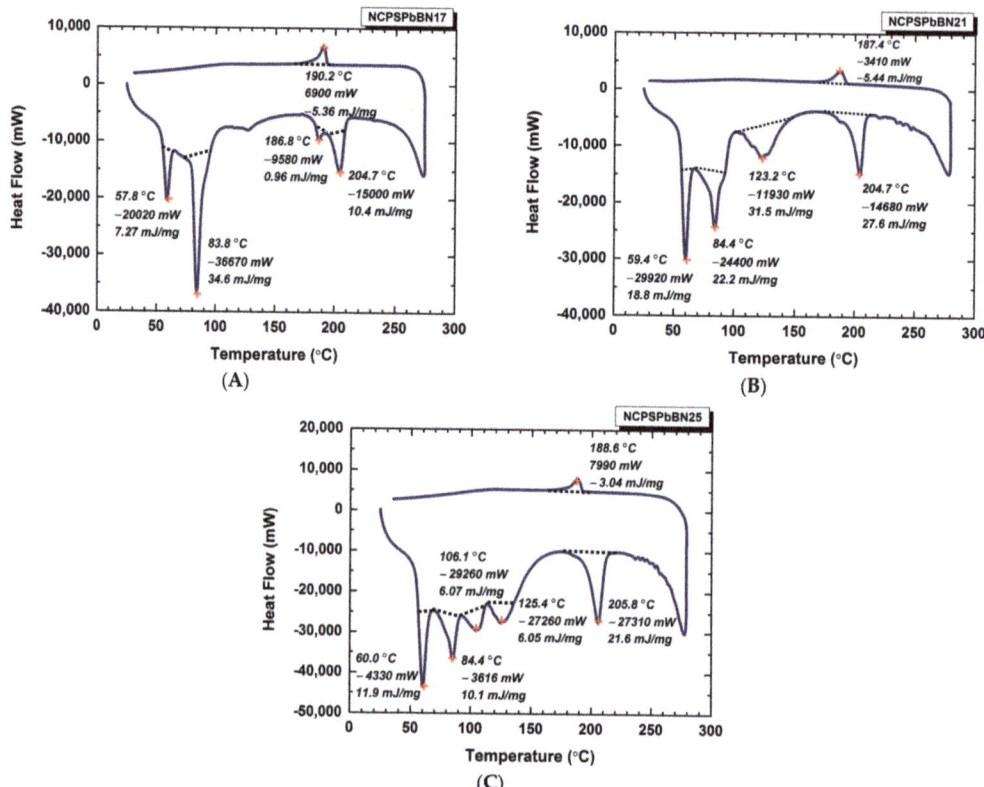

Figure 4. Thermal behavior of PS-PEG PCM nanocomposites: (**A**) NCPSPbBN17, (**B**) NCPSPbBN21, (**C**) NCPSPbBN25.

Figure 3 shows the thermal behavior of NCPSPb3, NCPSPb8, and NCPSPb13 PCM nanocomposites, while those of other PCM nanocomposites are given in Supplementary S2. Figure S7 presents the thermal behavior of NCPS1, NCPS6 and NCPS11. Compared with the NCPS1 (PS-PEG 1000) polymer, the T_c of the NCPSPb3 PCM nanocomposite slightly increased. A significant increase was observed in ΔH_m values (Figure 3A and Figure S2A). DSC curves of NCPSPb8 and NCPS6 are seen in Figure 3B and Figure S7B. As can be seen from these figures, the addition of 70% by weight of PbO to the NCPS6 block copolymer caused a partial increase in T_m values of NCPSPb8. It also greatly increased the ΔH_m values. T_m and ΔH_m values of NCPSPb13 and NCPS11 are shown in Figure 3C and Figure S7C. The average ΔH_m value corresponding to melting temperatures of the NCPS11 in the range of 58.8–190.1 °C was 10.3 J g^{-1}. The ΔH_c value at 31.5 °C was found to be -7.98 J g^{-1}. T_m-ΔH_m values of NCPS13 were 62.1 °C–7.38 J g^{-1}, 83.2 °C–27.2 J g^{-1}, 123.2 °C–6.64 J g^{-1}, and 204.7 °C–21.4 J g^{-1}. T_c-ΔH_c values of NCPS13 were 186.2 °C, -5.18 J g^{-1}.

T_m values of NCPSBN16 (in Figure S8A) are shown in Table 3. When the NCPS1 results are compared with those of NCPSBN16, their T_m values were found to be not much different. As can be seen from the data in Table 3, the addition of BN nanoparticles to the PS-PEG polymer increases the ΔH_m values. T_m-ΔH_m values of NCPSPbBN17 can be seen in Figure 4A. It is evident that the incorporation of the filler into the matrix increases the T_m of the PCM nanocomposites. T_m values of the NCPSBN20 (in Figure S8B) showed minor fluctuation changes compared with the NCPS6 block copolymer. DSC thermograms of the NCPSPbBN21, NCPSBN24, and NCPSPbBN25 PCMs are shown in Figure 4B, Figure S8C, and Figure 4C, respectively. The average ΔH_m value for the four T_m

values corresponding to the 59.4–204.7 °C range of the NCPSPbBN21 is 25.0 J g^{-1}, and the ΔH_c value corresponding to the T_c (187.4 °C) temperature is -5.44 J g^{-1}. For NCPSBN24, the mean ΔH_m value corresponding to the temperatures of 55.5–184.8 °C was found to be 15.2 J g^{-1}. The ΔH_c value at the T_m of 112.1 °C is 0.62 J g^{-1}. For NCPSPbBN25, the mean ΔH_m value corresponding to five melting temperatures was found to be 11.1 J g^{-1} and the ΔH_c value was found to be 3.04 J g^{-1} at T_c = 188.6 °C. Adding a certain amount of BN nanoparticles to the PCM nanocomposite causes the composite to shift to lower melting and crystallization transition temperatures. T_m and T_c values of the NCPS1, NCPS6, and NCPS11 copolymers were found to increase with increasing molecular weight of the PEGs. It was observed that T_c and T_m values increased with the addition of PbO nanoparticles in NCPSPb3, NCPSPb8, and NCPSPb13. When BN particles are added to NCPS1, NCPS6, and NCPS11 block copolymers, T_c values decrease while T_m values are almost the same. The melting and cooling curves of the selected sample composites were formed in almost the same regions as the DSC curves in Figure 3, Figure 4, Figures S7 and S8. In this case, it can be concluded that PCM nanocomposites have a similar phase change.

Table 3. Thermal Properties of the PS-PEG PCM, PS-PEG/PbO PCM, and PS-PEG/BN/PbO PCM nanocomposites.

PCM ID	T_m (°C)	ΔH_m (J g^{-1})	T_c (°C)	ΔH_c (J g^{-1})	F_c
NCPS1	73.2	3.7	7.9	1.24	0.032
	185.2	7.31	76.8	1.31	0.034
NCPSPb3	81.8	67.6			
	185.8	1.68	188.4	5.12	0.728
	200.7	4.62			
NCPS6	80.6	4.16			
	133.1	0.69	78	2.27	0.063
	159.5	0.31			
	185.4	0.40			
NCPSPb8	59.9	8.01			
	83.0	43.8	185.5	4.94	0.724
	203.0	26.7			
NCPS11	58.8	12.6			
	137.2	13.3	31.5	7.98	0.305
	190.1	4.96			
NCPSPb13	62.1	7.38			
	83.2	27.2	186.2	5.18	0.698
	123.2	6.64			
	204.7	21.4			
NCPSBN16	73.8	2.75	4.7	1.67	0.089
	137.3	0.58	89.4	1.14	0.087
	186.1	4.69			
NCPSPbBN17	57.8	7.27			
	83.8	34.6	190.2	5.36	1.985
	186.8	0.96			
	204.7	10.4			
NCPSBN20	80.3	1.81	11.5	1.04	0.094
	185.3	1.15	93.2	0.67	0.092
NCPSPbBN21	59.4	18.8			
	84.4	22.2	187.4	5.44	1.777
	123.2	31.5			
	204.7	27.6			
NCPSBN24	55.5	6.73	112.1	0.62	0.053
	184.8	8.49			
NCPSPbBN25	60.0	11.9			
	84.1	10.01			
	106.1	6.07	188.6	3.04	0.860
	125.4	6.05			
	205.8	21.6			

Table 3. Cont.

PCM ID	T_m (°C)	ΔH_m (J g^{-1})	T_c (°C)	ΔH_c (J g^{-1})	F_c
PEO-CMC [79]	58.4–62.5	52.8–140.2	35.3–41.3	5.2–138	-
PEO-CEL [79]	62.5–63.4	40.6–134.7	32.5–39.5	40.2–127.3	-
AMPD/TAM [80]	114.7–122.6	5.1–181.5	19.3–187.1	20–203.8	-
NPG/TAM/PE/AMPD [80]	171.6	14.1	169.6	17.4	-
NPG/PE [81]	160.3–169.8	18.8–26.2	-	-	-
PE-TAM [82]	184.6	14.2	188.6	14.3	

ΔH_m is important for PCMs containing PEG [41]. ΔH_m values decreased for BN/PS-PEG nanocomposites. This decrease led to a decrease in ΔH_m as a result of steric effects that change the structure of the polymer chains and the increase in the mobility and free volume of the polymer matrix at temperatures above T_m. These effects are due to the decrease in T_m of the PS-PEG/BN nanocomposite. In addition, the interactions between the nanoparticles and the polymer matrix reduced the free volume of the polymer chain [49,72]. When PbO and BN nanoparticles are homogeneously dispersed in the PS-PEG matrix, close interactions such as surface tension forces, π–π interactions and capillary forces between nanoparticles and PS-PEG will limit the mobility of the PS-PEG polymer. This causes a decrease in the phase-change temperature. The highest phase-change enthalpy of the PCM nanocomposite containing PbO 70 wt% and PS-PEG (1000) 30 wt% is 67.6 J g^{-1}. In addition, the thermal conductivity of composites can be significantly increased by the addition of PbO and BN nanoparticles, which leads to a fast thermal response.

As a result, the T_m, ΔH_m, T_c and ΔH_c values of the PCM nanocomposites that we examined fall within the range of the PEO-CMC, PEO-CEL [79], AMPD/TAM, NPG/TAM/PE/AMPD [80], NPG/PE [81], and PE-TAM [82] composites.

In addition, F_c values were calculated for different compounds using the crystallization enthalpy values obtained from DSC (Table 3). It was observed that the F_c values calculated from the DSC result increased. The F_c value of the PS-PEG copolymer increases as the ratio of BN nanoparticles and PbO nanoparticles increases. The F_c value of NCPSPb8 increases to 0.724 with the addition of PbO nanoparticles into NCPS6, whose F_c value is 0.063. When BN nanoparticles are added to NCPS6, the F_c value of 0.063 increases to 0.094 and 0.092 (NCPSBN20), and when PbO is added to this composite, the ratio increases even more, with a value of 1.777 obtained for NCPSPbBN21. We observed the same effect in all our other samples as follows: F_c values are 0.032 and 0.034 for pure NCPS1. These values increased with the contribution of PbO and reached 0.728. Only with the BN additive does the F_c value rise to 0.089 and 0.087, and when pure PS-PEG is added to BN nanoparticles and PbO nanoparticles, the F_c value reaches 1.985. The F_c value of the NCPS11 copolymer is 0.305. With the addition of PbO nanoparticles to the NCPS11 copolymer, the F_c value increases to 0.698. At the same time, if the BN nanoparticle and PbO nanoparticle are used together, the F_c value reaches 0.860.

The measured latent heat capacity of the 70 wt% PbO-nanoparticle-doped NCPSPb3 composite is 89.5% larger than the value of NCPS1. The latent heat capacity of NCPSPb8 doped with 70 wt% PbO nanoparticles is 92% higher than that of its polymer. This possibility is attributed to the scarcity of physical interactions between nanoparticles and PS-PEG, such as van der Waals force and hydrophobic–hydrophobic interactions, which can restrict the mobility of PS-PEG molecular chains during the crystallization process. As a result, the phase-change enthalpy of NCPSPB3, NCPSPb8, and NCPSPb13 containing PS-PEG and PbO nanoparticles increases. As the PbO and BN nanoparticle content increases in PCM nanocomposites, the thermal conductivity increases. Also, the latent heat gradually increases. This implies that increasing thermal conductivity using PbO and BN nanoparticles will be accompanied by increased latent heat in the nanocomposites. In this study, PbO and BN nanoparticles led to an increase of 61.9% in thermal conductivity and 93.8% in latent heat of PS-PEG block copolymers. Therefore, it would be beneficial to add PbO and

BN nanoparticles at a low loading rate to obtain the appropriate latent heat and to increase the thermal conductivity of the composites.

3.3. Thermal Conductivity

λ values were investigated for nanocomposites prepared from different amounts of PS-PEG copolymer, BN nanoparticles, and PbO nanoparticles. To determine the thermal conductivity of PCM nanocomposites, 5 wt%, 13 wt%, 15 wt%, and 50 wt% of BN nanoparticles and 10 wt%, 53.8 wt%, 70 wt%, 90 wt% of PbO nanoparticles were used.

The results of the λ values according to the additive ratios are given in Figure 5A–C. The λ values of PS-PEG copolymers and the PCM nanocomposites were calculated according to Equation (2). When the λ values of PS-PEG copolymers and PCM nanocomposites are examined from Figure 5, it is seen that the λ values of PS-PEG/PbO PCM nanocomposites are higher than that of the PS-PEG block copolymer. The greater value of PCM nanocomposites compared with their copolymers is due to the increase in free volume with the addition of PbO nanoparticles. The λ value of the NCPS1 block copolymer (PS-PEG (1000) was found to be 5.77 W/(mK)). The λ values of NCPSPb2, NCPSPb3, NCPSPb4, and NCPSPb5 containing the PS-PEG (1000) copolymer were 234%, 222%, 210%, and 332% higher, respectively, than the λ value of the copolymer (Figure 5A). The λ values increased with the contribution of 50–90 weight PbO nanoparticles to the NCPS1 polymer. It was observed that the increase in λ value of the NCPSPb5 nanocomposite was higher than that of the polymer. The λ value of NCPS6 (PS-PEG (1500) block copolymer) was found to be 5.70 W/(mK). The λ values of NCPSPb7, NCPSPb8, NCPSPb9, and NCPSPb10 were 196%, 201%, and 286% higher than the λ value of its copolymer. Among the PbO nanoparticle-doped PS-PEG (1500) PCM nanocomposites, NCP-SPb10 (46.2 wt% PS-PEG (1500) and 53.8 wt% PbO nanoparticles) had the highest λ value. The λ value of NCPS11 (PS-PEG (10,000) block copolymer) was 5.65 W/(mK). The λ values of NCPSPb12, NCPSPb13, NCP-SPb14, and NCPSPb15 (50 wt%, 70 wt%, 90 wt%, and 53.8 wt% of PbO nanoparticles) were 212%, 211%, 210%, and 305% higher, respectively, than the λ value of NCPS11 (Figure 5C). The λ value of the NCPSPb15 nanocomposite (22.91 W/mK) was much higher than that of the NCPS11 block copolymer. The λ value of NCPSBN16 was calculated as 11.54 W/(mK). When 50 wt% BN nanoparticles were added to the thermal conductivity of the NCPS1 copolymer, the λ value increased from 5.77 W/(mK) to 11.54 W/(mK). The λ graphs of NCPSPbBN17, NCPSPbBN18, and NCPSPbBN19 are presented in Figure 5A, and the λ values of these PCM nanocomposites are n 15.71 W/(mK), 16.9 W/(mK) and 18.87 W/(mK), respectively. NCPSBN16, NCPSBN20, and NCPSBN24 contain the addition of 50 wt% BN to PS-PEG (1000), PS-PEG (1500), and PS-PEG (10,000) block copolymers. When the λ values of these composites were compared with the λ values of the copolymers, BN nanoparticle addition increased the thermal conductivity of the block copolymers by 100%, 61%, and 75%, respectively. It is the NCPSPbBN23 nanocomposite which has the highest thermal conductivity (17.44 W/(mK)) among NCPSPbBN21, NCPSPbBN22, and NCP-SPbBN23. As a result of the incorporation of PbO and BN nanoparticles into NCPSPbBN21, NCPSPbBN22, and NCPSPbBN23 PCM nanocomposites, the λ values of the composites increased by 163%, 193%, and 206% from the λ values of their copolymers (Figure 5B). The λ values of the NCPSPbBN25, NCPSPbBN26, and NCPSPbBN27 PCM nanocomposites were 15.21 W/(mK), 16.74 W/(mK), 17.87 W/(mK). NCPSPbBN27 produced the highest λ value among NCPSPbBN25, and NCPSPbBN26. The λ values of NCPSPbBN25, NCPSPbBN26, and NCPSPbBN27 were increased by 169%, 196%, and 216%, respectively, from the thermal conductivity values of their copolymers when PbO and BN nanoparticles are added (Figure 5). As a result, the BN additive increased the thermal conductivity values of the nanocomposites. Composites with added PS-PEG and PbO produce the best value every time (with PS-PEG 46.2 wt% and PbO 53.8 wt%). The PCM nanocomposites prepared from synthesized PS-PEG (1000, 1500, 10,000) copolymers can be used for thermal conductivity. Using BN and PbO nanoparticles thermal conductive fillers, Zheng et al. produced a new type of energy storage material with high thermal conductivity by adding different masses of hydroxylated multi-walled carbon nanotubes (MWCNTs) to the stable form of the PEG1500·CaCl$_2$ phase-change material [82]. We observed that the phase change and thermal conductivity values of the PS-PEG copolymers that we synthesized here were as high as in those studies.

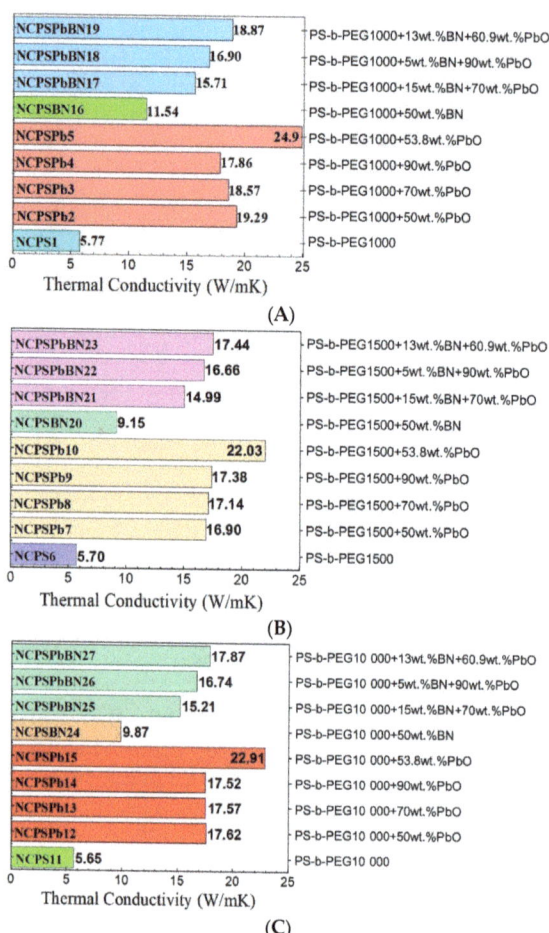

Figure 5. λ values of the (**A**) PS-PEG (1000) copolymers and PS-PEG (1000)/PbO and PS-PEG (1000)/PbO/BN PCM nanocomposites; (**B**) PS-PEG (1500) copolymers, PS-PEG (1500)/PbO and PS-PEG (1500)/BN PCM nanocomposites; and (**C**) PS-PEG (10,000) copolymers and PS-PEG (10,000)/PbO and PS-PEG (10,000)/PbO/BN PCM nanocomposites.

We also used BN nanoparticles and PbO nanoparticles to solve the low heat dissipation problem of PS-PEG polymer materials. We show that PS-PEG block copolymers are important in the thermal conductivity of PCM nanocomposites. Pure Pb metal has a high λ value and was found as 35 W/(mK) in the temperature range of 20 °C–85 °C [83]. As a result, since oxides are useful materials for thermal barriers and form rich structures, this allows them to increase the thermal conductivity of the high compositions and the high efficiency in forming composites. We obtained high conductivity values by using PbO nanoparticles in our study. The λ value for PbO nanoparticles was 17.5 W/(mK) [84]. Based on the work of Zhou et al., the λ value for BN was 2 W/(mK) [72].

Since the λ value of polymers can be increased by using various materials [84–87], it was concluded that the λ value of nanocomposites formed by adding PS-PEG copolymers and BN and PbO nanoparticles increased, Lebedev's study presented that the thermal conductivity values of LLDPE and PLA-based composites increased 1.9 and 3.5 times with 40% filler [72,88]. The thermal conductivity of materials is also important for heat performance, which is a way to conserve energy to increase the efficiency of the systems. Phase-change energy storage technologies are developing. Polymers with phase-change properties are far from the desired performance due to their thermal conductivity. It is common to use additional fillers to improve the performance of polymers. Therefore, the composites we prepared have reached the desired level.

The λ value of the PS-PEG/BN PCM nanocomposite was found to be increased by 63% compared with the λ value of the PS-PEG copolymer. In addition, the λ value of the PS-PEG (1000)/PbO PCM containing PbO 53.8 wt% and PS-PEG 46.20 wt% is 24.90 W/(mK). The λ value of the PCM nanocomposite doped with PbO nanoparticles was found to be 332% higher than that of the PS-PEG (1000) block copolymer. The thermal conductivity of PS-PEG/PbO PCM nanocomposites containing PbO nanoparticles 50 wt%, 70 wt%, and 90 wt% were higher than the thermal conductivity values of their copolymers. It was found that the thermal conductivity values of PS-PEG (1000, 1500, 10,000)/PbO PCM nanocomposites (containing PbO nanoparticles 53.8 wt%) were increased by 332%, 291%, and 305% from their block copolymer, respectively. The results also showed that BN nanoparticles and PbO nanoparticles significantly increased the thermal conductivity of the PCM nanocomposites.

3.4. Morphology Results

3.4.1. SEM Images of the PCM Nanocomposites

The polymer chains were linked together. The surface of the polymer is granular (Figure 6A) and porous (Figure S9) (magnified images from 1000 to 10,000) [46]. PS blocks linked with the macrocrosslinker PEG are continuous, forming the polymer phase and a branched structure. A structure was observed in which PS and PEG formed a homogeneous continuous matrix. The surface of NCPSPb3 has granular, rough, porous, voids and clusters (in Figure 6B and Figure S10). The size of the PbO nanoparticles on the NCPSPb3 surface is 352.11–456 nm. The morphological images of the PS-PEG/BN/PbO PCM nanocomposites are shown in Figures S3 and S9–S18. Structural (SEM, TEM) studies were conducted to examine the morphology, distribution of the nanoparticles, and changes in chemical structures of the PCM nanocomposites after thermal conductivity measurements. As a result of the analysis, it was observed that there were no structural aggregates in the nanocomposites and that the nanoparticles were homogeneously dispersed. The morphologies of PS-PEG/PbO, PS-PEG/BN, and the PS-PEG/BN/PbO PCM nanocomposites are presented using SEM images. In Figures 7–10, the PS-PEG matrix showed smooth particles, cracks, voids, and porous surfaces, which took place in the presence of BN and PbO nanoparticles. In contrast, PS-PEG/BN/PbO PCM nanocomposites showed a rough and crumpled fracture structure, which was the result of local polymer deformation due to cracking from the addition of the BN nanoparticles and PbO nanoparticles. In Figures 7 and 10, the BN nanoparticles and PbO nanoparticles were well dispersed in the PS-PEG matrix, and the compatibility between fillers and matrix was observed to be fine. When the content of BN nanoparticles and PbO nanoparticles was increased, the fillers formed a well-interconnected network in the PS-PEG matrix. In addition, at higher magnifications, the presence of cavities at the interactive interface of the BN/PbO nanoparticles and the polymer matrix was confirmed. In addition, these properties are effective factors that improve heat transfer in thermally conductive polymer composites.

(**A**)

(**B**)

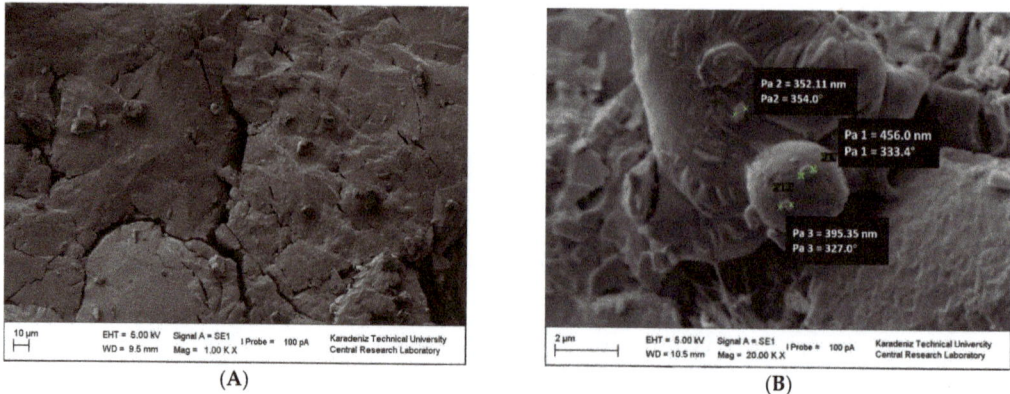

Figure 6. SEM images of (**A**) NCPS6 and at (**B**) NCPSPb3.

(**A**)

(**B**)

(**C**)

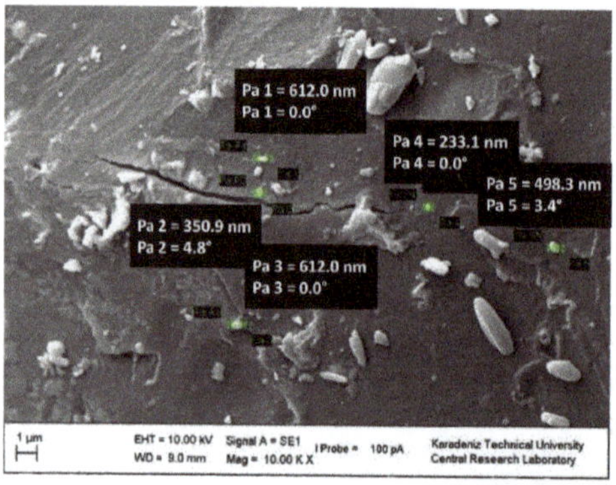

Figure 7. SEM images of NCPSPb8 (**A**) and NCPSPb13 (**B**). EDS mapping and analysis indicating the element distribution on the surface of the NCPSPb8 nanocomposite (**C**).

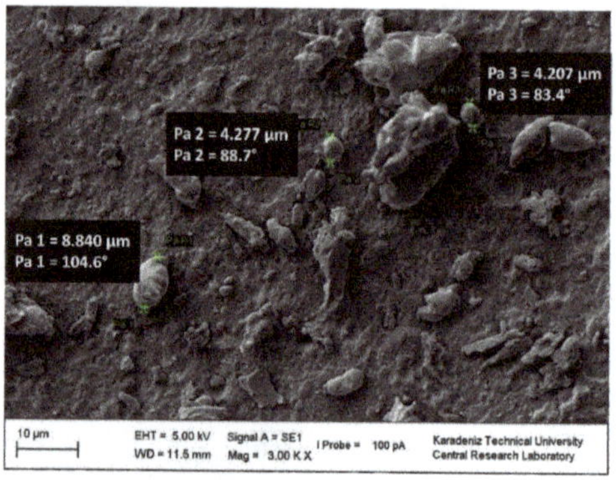

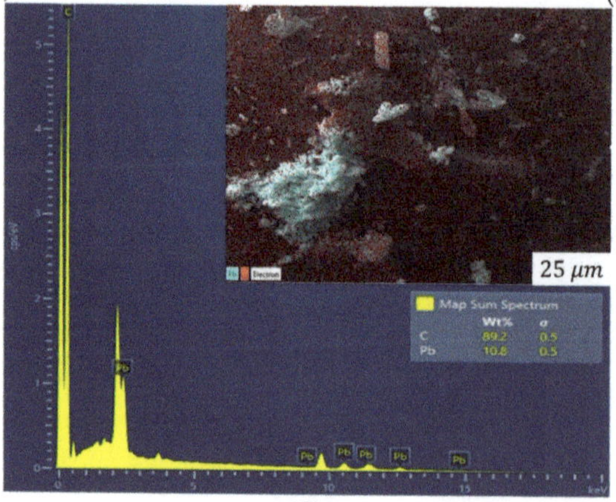

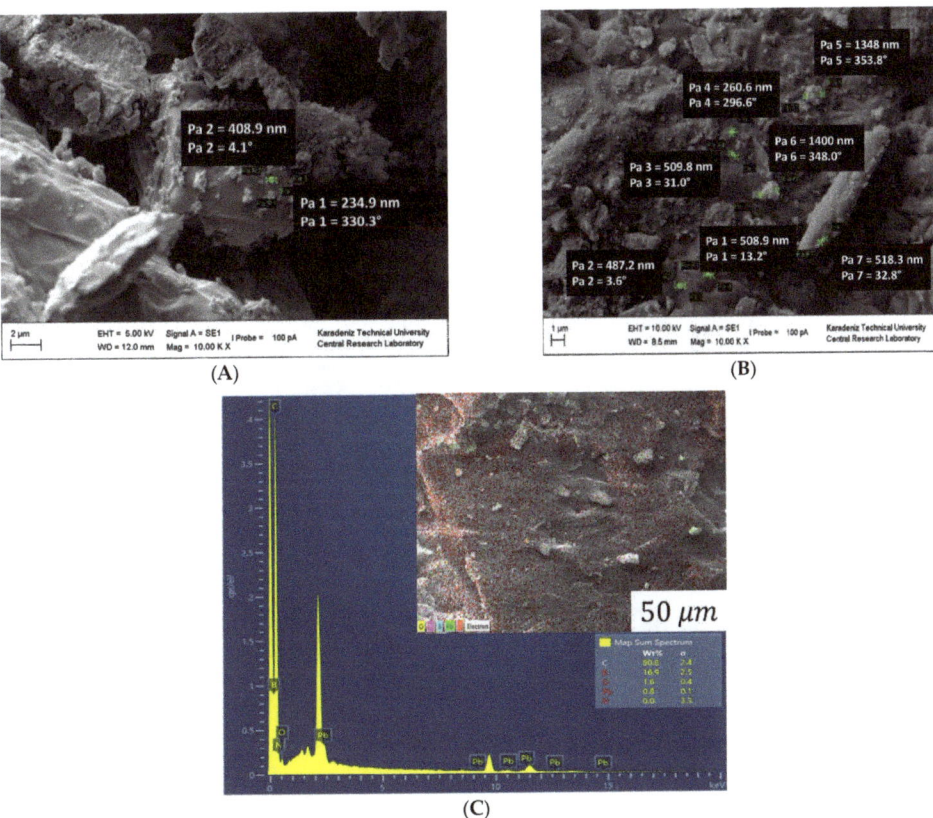

Figure 8. SEM images of NCPSPbBN17 (**A**) and NCPSPbBN21 (**B**). EDS mapping and analysis indicating the element distribution on the surface of the PS-PEG-BN-PbO nanocomposite and SEM images of NCPSPbBN21 at (**C**).

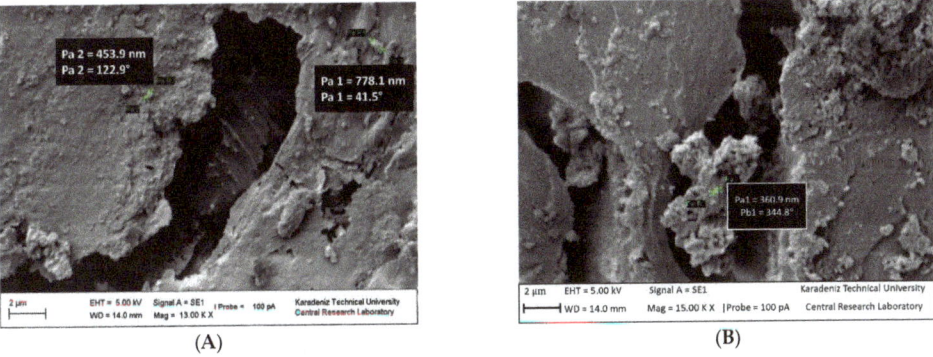

Figure 9. SEM images (**A**,**B**) of the NCPSPbBN25 nanocomposite.

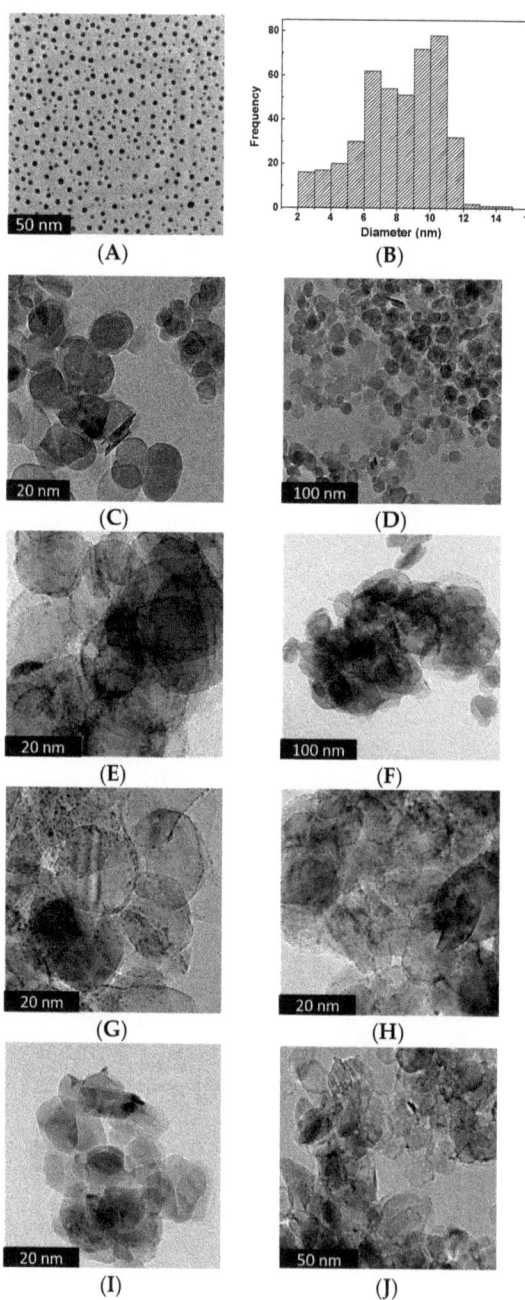

Figure 10. (**A**) TEM images of PbO nanoparticles and (**B**) the calculated size distribution of PbO nanoparticles from the TEM images. (**C–J**) TEM images of BN nanoparticles (**C,D**), NCPSPbBN17 (**E,F**), NCPSPbBN21 (**G,H**), and NCPSPbBN25 (**I,J**).

SEM images of the NCPSPb8 nanocomposite are presented in Figure 7A and Figure S11 (magnified images from 1000 to 14,000). As the molecular weight of the crosslinker PEG in the PS-PEG (1500) copolymer increased, the roughness on the NCPSPb8 surface decreased, and a flatter surface was formed. When SEM photographs of 70 wt% PbO-doped nanocomposites are compared, the effect of the macrocrosslinker PEG molecular weight increase is seen. The size of the PbO particle on the NCPSPb8 surface is 233–720.7 nm (Figure 7A and Figure S11A–C). SEM images of NCPSPb13 have determined that pores (Figure 7B and Figure S12, magnified images from 1000 to 5000), granulation, branching, and agglomeration (in Figure 7B and Figure S12) are observed from the surface films of the composite, where the percentage of PbO is higher than the percentage of polymer. The size of the PbO particles on the surface was 9.266 nm in Figure 7B and 2.288–8.840 nm in Figure S12C. EDS images of the NCPSPb8 nanocomposite is presented in Figure 7C. EDS mapping and analysis of element distribution on the surface verified the presence of carbon (C) and lead (Pb) elements throughout the surface of the PS-PEG-PbO nanocomposite. The EDS spectra further showed the corresponding peaks of C and Pb in NCPSPb8 (Figure 7C).

SEM images of NCPSPbBN17 are shown in Figure 8A (magnified images from 5000 to 12,000). As seen in Figure 8A and Figure S13, the surfaces of the NCPSPbBN17 nanocomposite exhibit a morphological structure containing granular particles and metal compound particles (Figure 8A and Figure S13), pores, branching (Figure 8A and Figure S13A,B), aggregates, and cavities (in Figure 8A and Figure S13B,C). Nano-sized BN and PbO particles appear on the surface of the NCPSPbBN17 PCM nanocomposite. The magnitude of the BN and PbO particles is 235–409 nm. As seen from the SEM images shown in Figure 8B and Figure S14 (magnified images from 3000 to 10,000), the NCPSPbBN21 surface also has pores, particles, layers, branches, and voids, similar to the surface of the NCPS6 copolymer. BN and PbO nanoparticles appear on the surface of NCPSPbBN21; these have a particle size of 260–1400 nm (Figure 8B). EDS analysis of the NCPSPbBN21 composite shows Pb, B, C, O, and N elements on the surface of the composite (Figure 8C).

SEM images of the NCPSPbBN25 nanocomposite are shown in Figure 9A,B and Figure S15.

As seen in Figure 9A,B and Figure S15 (magnified images from 2500 to 15,000), when the amount of PbO added to PS-PEG (10,000) copolymer is increased to 70 wt%, PbO particles on the surface appear more intense. The particles, pores, and clusters on the surface of the composite can be seen from the SEM photographs in Figure 9A,B and Figure S15. The sizes of the BN and PbO nanoparticles on NCSPBBN25 are 361–778 nm.

3.4.2. TEM Results

Figure 10A,B shows the TEM images of PbO and BN nanoparticles. The PbO nanoparticles were in the range of 2–15 nm and the size distribution was very narrow. The TEM images showed the regular spherical shape and confirmed both their size and the homogenous size distribution. In addition, BN nanoparticles with a mean diameter as small as 100 nm could be obtained, as shown by TEM analysis (Figure 10C,D). The TEM photographs of the NCPSPbBN PCM nanomaterial are shown in Figure 10E–J. The small dark spots in the TEM images indicate the presence of BN and PbO nanoparticles that were bound to the copolymers. The TEM images of NCPSPbBN17 in Figure 10E,F are similar to the TEM images of NCPSPbBN21 and NCPSPbBN25, with BN nanoparticles and PbO nanoparticles dispersed in the interlocking spherical and rod-shaped copolymer structures, which appear as bright objects that are light in color.

3.4.3. XRD Patterns of the BN Nanoparticle, PbO Nanoparticle, and the PS-PEG/BN/PbO PCM Nanocomposites

X-ray studies were taken to examine the nanoparticles and changes in chemical structures of the PCM nanocomposites after thermal conductivity measurements. It was observed that there were no chemical changes.

Figure 11 shows the XRD patterns of PbO, NCPS1, NCPSPb4, NCPS6, NCPSPb9, NCPS11, and NCPSPb14. The PbO nanoparticles were further analyzed by powder XRD. The diffractogram shown in Figure 11A is consistent with the nanostructure of PbO. The two strong peaks with 2θ values of 29.20° and 30.42° correspond to the (211) and (002) planes, respectively. The distance values between the planes corresponding to these values were determined as 3.056 and 2.936 Å, respectively. The other peaks of the nanostructure of PbO are 32.73°, 37.97°, 45.24°, and 53.24°, corresponding to Miller indices of (220), (003), (222), and (213), respectively. The distance values between the planes corresponding to 2θ = 32.73°, 37.97°, 45.24°, and 53.24° values are 2.733 Å, 2.367 Å, 2.002 Å, and 1.719 Å, respectively. The XRD patterns obtained for PbO are compatible with the literature [88,89].

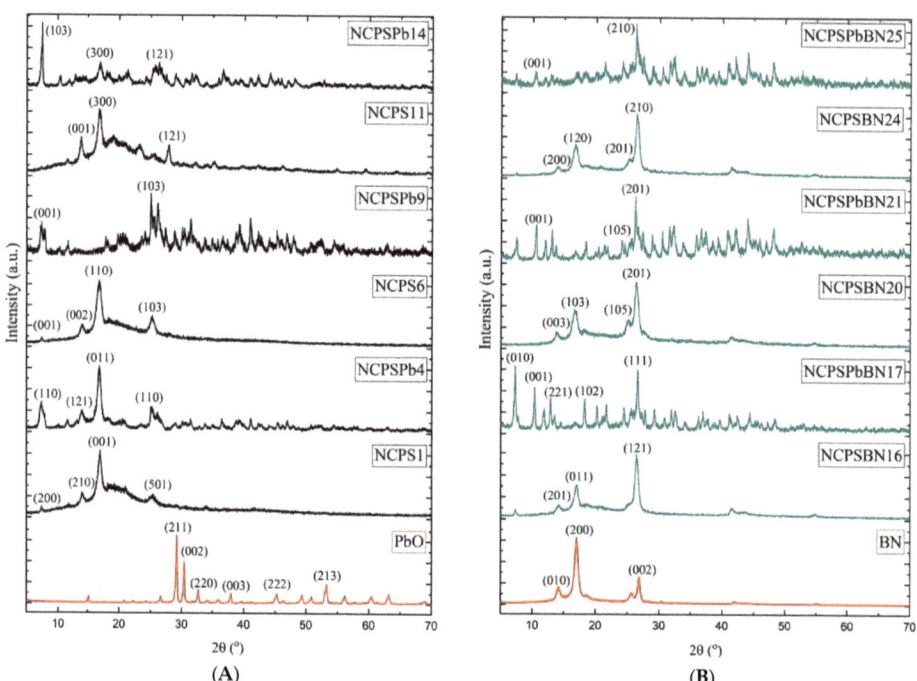

Figure 11. (A) XRD patterns of PbO, NCPS1, NCPSPb4, NCPS6, NCPSPb9, NCPS11, and NCPSPb14 PCM nanocomposites. (B) XRD patterns of BN, NCPSBN16, NCPSPbBN17, NCPSBN20, NCPSPbBN21, NCPSBN24, and NCPSPbBN25 PCM nanocomposites.

The XRD patterns coded NCPS1, NCPS6, and NCPS11 in Figure 11A belong to the PS-PEG (1000, 1500, 10,000) block copolymer. Figure 11 shows X-ray diffraction patterns of the PS-PEG block copolymer, revealing spectra of a broad amorphous peak that appeared at 2θ = 20–24°. The sharp diffraction peak of the PEG part of the PS-PEG (1000) copolymer appeared at 17.00° (d = 5.211 Å) and 25.28° (d = 3.520 Å), which indicates a polymer with crystallinity. Miller indices of the sharp diffraction peaks of the PEG corresponding to 17.00 Å and 25.28° are (001), (501), respectively. X-ray diffraction patterns corresponding to 14.04° (d-spacing = 6.309 Å) belong to the PS part of the PS-PEG block copolymer, and Miller indices are (210) [90]. PEG, a semicrystalline polymer, and the PS network, an amorphous crosslinked polymer, produce a semicrystalline mixture when the molecular weight of PEG is changed in our experiments. For the molecular weight used in this study (PEG-1000, -1500, and -10,000), the two-component system is expected to be composed of amorphous blended regions, with some crystalline regions made up entirely of PEG [91]. The XRD patterns of BN, NCPSBN16, NCPSPbBN17, NCPSBN20, NCPSPbBN21, NCPSBN24, and NCPSPbBN25 nanocomposites are presented in Figure 11B. XRD diffraction peaks of BN nanostructures of (010), (200), and (002) h-BN are observed at 2θ = 14.24°, 17.02°, 26.88°. XRD structural analysis of the BN nanoparticles shows the highest peak at 17.02°, corresponding to Miller indices of (200) (Figure 11B). We determined that the 2θ value of the 100% peak value of BN is 17.02°, the d-spacing distance corresponding to this value is

5.205 Å, and the hkl value is (200). In some of the other reflections at $2\theta = 30.40°$, $41.86°$, and $55.18°$, the d-spacing distances detected in the XRD pattern corresponding to these values are 2.938 Å, 2.156 Å, and 1.663 Å, respectively, and this result correlates with the literature [92]. Looking at the graph in Figure 11B, it is seen that BN nanoparticles have a more crystalline structure compared with the PS-PEG polymer, and the BN nanoparticle has peak values close to the PS-PEG polymer. The XRD diffraction peaks of the NCPSBN16 nanocomposite PCM containing 50 wt% BN and 50 wt% PS-PEG block copolymer are $2\theta = 7.29°$, $14.20°$, $17.08°$, $26.44°$, $54.66°$, $75.86°$, and $82.02°$.

The XRD diffraction peaks belonging to BN nanoparticles are at $26.44°$ (d-spacing = 3.314 Å, and (121) Miller indices) and $55.18°$ d = 1.663 Å (see Figure 11B). The XRD patterns at $14.20°$ (d-spacing = 6.231 Å) and $17.08°$ (d-spacing = 5.187 Å) belong to the PEG macrocrosslinker part of the NCPSBN16 nanocomposite PCM. NCPSPbBN17 contains 15 wt% BN, 15 wt% PS-PEG, and 70 wt% PbO. In the XRD image of this sample, it is seen that the NCPSPbBN17 has a (100) peak value of $7.24°$, and the corresponding interplanetary distance value is 12.201 Å while the hkl value is (010). As a result, when the SEM, TEM, and XRD results of the nanocomposites are analyzed, the size analysis and XRD patterns of the nanoparticles in the PCM nanocomposites gave results consistent with the literature. At the same time, F_c values varying in the range of 0.032–1.985% calculated from the DSC results of the PEG macrocrosslinker correspond to $2\theta = 17.00°$ and $25.28°$ peaks on XRD charts. The XRD results show that the composites retain the crystallization structure of the PS-PEG polymer and nanoparticles; furthermore, there are only physical rather than chemical reactions between the PSPEG polymer and PBO and BN nanoparticles.

4. Conclusions

In the study, PCM nanocomposites containing PbO nanoparticles, BN nanoparticles, and PS-PEG copolymers were prepared. The characteristic analysis of materials was performed using DSC, TGA, SEM, and XRD methods. By examining the DSC curves, it is seen that the phase-change temperatures and enthalpy values of each material change with the change in the percentage contribution of the PS-PEG, BN nanoparticles, and PbO nanoparticles. These differences can be explained by the fact that phase-change materials provide new heat conduction paths, thereby changing the phase transition rate of the samples. In this study, PCM nanocomposites with latent heats ranging from 34.6 J g^{-1}–67.6 J g^{-1} can be used to reduce indoor temperature fluctuations, improve indoor thermal comfort, and save electrical energy. At the same time, they can be used as building materials for precise temperature control and rapid temperature regulation conditions. The roles of particle/polymer and particle/particle interfaces on the thermal conductivity of PS-PEG DM/BN and PbO nanocomposites are discussed in detail, as well as the relationship between the thermal conductivity and the micro- and nano-structure of the composites. Recently, studies to improve the thermal conductivity of polymers have been directed toward the selective addition of nanofillers with high thermal conductivity properties. The thermal conductivity of the PS-PEG polymers is increased by PbO nanoparticle doping. It can be seen from the results that the thermal conductivity increases with increasing values of 0–60.90–70–90% of the PbO ratio. Higher thermal conductivity values were obtained for nanocomposites by adding nanoparticles such as BN and PbO to the PS-PEG polymer. Thermal conductivity values obtained from our materials show that they can be widely used in various engineering applications based on energy storage/release. PS-PG/PbO and PS-PG/PbO/BN PCM nanocomposites have emerged as versatile functional materials. PCM nanocomposites can be used for electrical and magnetic materials, EMI and Gamma radiation shielding, and reflecting and absorbing materials. The X-ray diffraction pattern corresponding to $2\theta = 14.04°$ belongs to the PS part of the PS-PEG block copolymer and the Miller indices are (210). Two strong peaks of PbO nanoparticles were $2\theta = 29.20°$ and $2\theta = 30.42°$, and the 2θ value of the 100% peak value of BN is $17.02°$.

Supplementary Materials: The following supporting information can be downloaded at: https://www.mdpi.com/article/10.3390/polym15102326/s1. Supplementary S1: TGA Measurements of the PS-PEG-PbO and -BN Nanocomposite PCMs; Supplementary S2: DSC Results of the PS-PEG/BN/PbO PCM Nanocomposites; Supplementary S3: The morphological images of the PS-PEG/BN/PbO PCM Nanocomposites.

Author Contributions: Conceptualization, B.O., S.S.Y. and A.H.Y.; methodology, B.O., S.S.Y. and A.H.Y.; validation, B.O., S.S.Y. and A.H.Y.; formal analysis, B.O., S.S.Y. and A.H.Y.; investigation, S.M., B.O. and S.S.Y.; resources, A.H.Y.; data curation, S.M. and T.B.; writing—original draft preparation, B.O., S.S.Y., A.H.Y., S.M. and T.B.; writing—review and editing, B.O., S.S.Y. and A.H.Y.; visualization, S.M., T.B. and B.E.; supervision, S.S.Y. and A.H.Y.; project administration, S.S.Y., funding acquisition, A.H.Y. All authors have read and agreed to the published version of the manuscript.

Funding: This research received no external funding.

Institutional Review Board Statement: Not applicable.

Informed Consent Statement: Not applicable.

Data Availability Statement: The data presented in this study are available on request from the corresponding author.

Acknowledgments: The authors thank to Karadeniz Technical University and UNAM-National Nanotechnology Research Center and Institute of Materials Science and Nanotechnology, Bilkent University for their support. We would like to thank Ayşegul Yilmaz for correcting the grammar of the article. The authors would like to thank Idris Altunel, the Chairman of the Board of Directors of Private Ordu Umut Hospital, Sahincili, Ordu-Turkey and Altunel Hazelnut Integrated Inc. Küçükköy District on Dereli Rod 1 Km, No: 60 Giresun-Turkey; Handstrasse 212, Bergish Gladbach-Germany, for his contribution to the scientific study (Tel.: +90-452-234-9030, 444-90-52; +90-533-155-9552; Fax: +90-452-234-3497; emails: info@umuthastanesi.com.tr and info@altunelnut.com).

Conflicts of Interest: The authors declare no competing interest.

References

1. Duraković, B.; Mešetović, S. Thermal performances of glazed energy storage systems with various storage materials: An experimental study. *Sustain. Cities Soc.* **2019**, *45*, 422–430. [CrossRef]
2. Duraković, B.; Torlak, M. Experimental and numerical study of a PCM window model as a thermal energy storage unit. *Int. J. Low-Carbon Technol.* **2017**, *12*, 272–280. [CrossRef]
3. Durakovic, B.; Torlak, M. Simulation and experimental validation of phase change material and water used as heat storage medium in window applications. *J. Mater. Environ. Sci.* **2017**, *8*, 1837–1846.
4. Farhat, N.; Inal, Z. Solar thermal energy storage solutions for building application: State of the art. *Herit. Sustain. Dev.* **2019**, *1*, 1–13. [CrossRef]
5. Çakır, A.; Kurmuş, E.F. Energy storage technologies for building applications. *Herit. Sustain. Dev.* **2019**, *1*, 41–47. [CrossRef]
6. Ermiş, K.; Findik, F. Thermal energy storage. *Sustain. Eng. Innov.* **2020**, *2*, 66–88. [CrossRef]
7. Duraković, B. *PCM-Based Building Envelope Systems Innovative Energy Solutions for Passive Design*; Springer Nature Switzerland AG: Basel, Switzerland, 2020; pp. 63–87. [CrossRef]
8. Teggar, M.; Arıcı, M.; Mezaache, E.H.; Mert, M.S.; Ajarostaghi, S.S.M.; Niyas, H.; Tuncbilek, E.; Ismail KA, R.; Younsi, Z.; Benhouia, A. A comprehensive review of micro/nano-enhanced phase change materials. *J. Therm. Anal. Calorim.* **2022**, *147*, 3989–4016. [CrossRef]
9. Williams, J.D.; Peterson, G. A review of thermal property enhancements of low temperature nano-enhanced phase change materials. *Nanomaterials* **2021**, *11*, 2578. [CrossRef]
10. Najim, F.T.; Mohammed, H.I.; Taqi Al-Najjar, H.M.; Thangavelu, L.; Mahmoud, M.Z.; Mahdi, J.M.; Tiji, M.E.; Yaïci, W.; Talebizadehsardari, P. Improved melting of latent heat storage using fin arrays with non-uniform dimensions and distinct patterns. *Nanomaterials* **2022**, *12*, 403. [CrossRef]
11. Chen, K.; Mohammed, H.I.; Mahdi, J.M.; Rahbari, A.; Cairns, A.; Talebizadehsardari, P. Effects of non-uniform fin arrangement and size on the thermal response of a vertical latent heat triple-tube heat exchanger. *J. Energy Storage* **2022**, *45*, 103723–103747. [CrossRef]
12. Hosseinzadeh, K.; Erfani Moghaddam, M.A.; Asadi, A.; Mogharrebi, A.R.; Jafari, B.; Hasani, M.R.; Ganji, D.D. Effect of two different fins (longitudinal-tree like) and hybrid nano-particles (MoS_2-TiO_2) on solidification process in triplex latent heat thermal energy storage system. *Alex. Eng. J.* **2021**, *60*, 1967–1979. [CrossRef]

3. Talebizadehsardari, P.; Mohammed, H.I.; Mahdi, J.M.; Gillott, M.; Walker, G.S.; Grant, D.; Giddings, D. Effect of airflow channel arrangement on the discharge of a composite metal foam-phase change material heat exchanger. *Int. J. Energy Res.* **2021**, *45*, 2593–2609. [CrossRef]
4. Ali, H.M.; Janjua, M.M.; Sajjad, U.; Yan, W.M. A critical review on heat transfer augmentation of phase change materials embedded with porous materials/foams. *Int. J. Heat Mass Transf.* **2019**, *135*, 649–673. [CrossRef]
5. Moghaddam, M.E.; Abandani, M.H.S.; Hosseinzadeh, K.; Shafii, M.B.; Ganji, D. Metal foam and fin implementation into a triple concentric tube heat exchanger over melting evolution. *Theor. Appl. Mech. Lett.* **2022**, *12*, 100332–100341. [CrossRef]
6. Bondareva, N.S.; Buonomo, B.; Manca, O.; Sheremet, M.A. Heat transfer performance of the finned nano-enhanced phase change material system under the inclination influence. *Int. J. Heat Mass Transf.* **2019**, *135*, 1063–1072. [CrossRef]
7. Hosseinzadeh, K.; Montazer, E.; Shafii, M.B.; Ganji, A. Solidification enhancement in triplex thermal energy storage system via triplets fins configuration and hybrid nanoparticles. *J. Energy Storage* **2021**, *34*, 102177–102183. [CrossRef]
8. Hosseinzadeh, K.; Moghaddam, M.E.; Asadi, A.; Mogharrebi, A.; Ganji, D. Effect of internal fins along with hybrid nano-particles on solid process in star shape triplex latent heat thermal energy storage system by numerical simulation. *Renew. Energy* **2020**, *154*, 497–507. [CrossRef]
9. Ho, C.; Siao, C.-R.; Yang, T.-F.; Chen, B.-L.; Rashidi, S.; Yan, W.-M. An investigation on the thermal energy storage in an enclosure packed with micro-encapsulated phase change material. *Case Stud. Therm. Eng.* **2021**, *25*, 100987–100996. [CrossRef]
20. Khedher, N.B.; Ghalambaz, M.; Alghawli, A.S.; Hajjar, A.; Sheremet, M.; Mehryan SA, M. Study of tree-shaped optimized fins in a heat sink filled by solid-solid nanocomposite phase change material. *Int. Commun. Heat Mass Transf.* **2022**, *136*, 106195. [CrossRef]
21. Khedher, N.B.; Bantan, R.A.; Kolsi, L.; Omri, M. Performance investigation of a vertically configured LHTES via the combination of nano-enhanced PCM and fins: Experimental and numerical approaches. *Int. Commun. Heat Mass Transf.* **2022**, *137*, 106246. [CrossRef]
22. Meng, L.; Ivanov, A.S.; Kim, S.; Zhao, X.; Kumar, N.; Young-Gonzales, A.; Saito, T.; Bras, W.; Gluesenkamp, K.; Bocharova, V. Alginate–Sodium Sulfate Decahydrate Phase Change Composite with Extended Stability. *ACS Appl. Polym. Mater.* **2022**, *4*, 6563–6571. [CrossRef]
23. Peng, G.; Dou, G.; Hu, Y.; Sun, Y.; Chen, Z. Review Article Phase Change Material (PCM) Microcapsules for Thermal Energy Storage. *Adv. Polym. Technol.* **2020**, *2020*, 9490873. [CrossRef]
24. Mohaddes, F.; Islam, S.; Shanks, R.; Fergusson, M.; Wang, L.; Padhye, R. Modification and evaluation of thermal properties of melamine-formaldehyde/n-eicosane microcapsules for thermo-regulation applications. *Appl. Therm. Eng.* **2014**, *71*, 11–15. [CrossRef]
25. Yuzhan, L.; Navin, K.; Jason, H.; Damilola, O.A.; Kai, L.; Turnaoglu, T.; Monojoy, G.; Rios, O.; Tim JLaClair Samuel, G.; Kyle, R.G. Stable salt hydrate-based thermal energy storage materials. *Compos. Part B* **2022**, *233*, 109621. [CrossRef]
26. Han, W.; Ge, C.; Zhang, R.; Ma, Z.; Wang, L.; Zhang, X. Boron nitride foam as a polymer alternative in packaging phase change materials: Synthesis, thermal properties and shape stability. *Appl. Energy* **2019**, *238*, 942–951. [CrossRef]
27. Sundararajan, S.; Samui, A.B.; Kulkarni, P.S. Synthesis and characterization of poly(ethylene glycol) acrylate (PEGA) copolymers for application as polymeric phase change materials (PCMs). *React. Funct. Polym.* **2018**, *130*, 43–50. [CrossRef]
28. Anuar Sharif, M.K.; Al-Abidi, A.A.; Mat, S.; Sopian, K.; Ruslan, M.H.; Sulaiman, M.Y.; Rosli, M.A.M. Review of the application of phase change material for heating and domestic hot water systems. *Renew. Sustain. Energy Rev.* **2015**, *42*, 557–568. [CrossRef]
29. Deng, Y.; Li, J.; Qian, T.; Guan, W.; Li, Y.; Yin, X. Thermal conductivity enhancement of polyethylene glycol/expanded vermiculite shape-stabilized composite phase change materials with silver nanowire for thermal energy storage. *Chem. Eng. J.* **2016**, *295*, 427–435. [CrossRef]
30. Wang, W.; Yang, X.; Fang, Y.; Ding, J.; Yan, J. Enhanced thermal conductivity and thermal performance of form-stable composite phase change materials by using β-Aluminum nitride. *Appl. Energy* **2009**, *86*, 1196–1200. [CrossRef]
31. Song, S.; Qiu, F.; Zhu, W.; Guo, Y.; Zhang, Y.; Ju, Y.; Feng, R.; Liu, Y.; Chen, Z.; Zhou, J.; et al. Polyethylene glycol/halloysite@Ag nanocomposite PCM for thermal energy storage: Simultaneously high latent heat and enhanced thermal conductivity. *Sol. Energy Mater. Sol. Cells* **2019**, *193*, 237–245. [CrossRef]
32. Chen, J.; Huang, X.; Sun, B.; Jiang, P. Highly Thermally Conductive Yet Electrically Insulating Polymer/Boron Nitride Nanosheets Nanocomposite Films for Improved Thermal Management Capability. *ACS Nano* **2019**, *13*, 337–345. [CrossRef] [PubMed]
33. Jiang, Y.; Shi, X.; Feng, Y.; Li, S.; Zhou, X.; Xie, X. Enhanced thermal conductivity and ideal dielectric properties of epoxy composites containing polymer modified hexagonal boron nitride. *Compos. Part A Appl. Sci. Manuf.* **2018**, *107*, 657–664. [CrossRef]
34. Wu, M.Q.; Wu, S.; Cai, Y.F.; Wang, R.Z.; Li, T.X. Form-stable phase change composites: Preparation, performance, and applications for thermal energy conversion storage and management. *Energy Storage Mater.* **2021**, *42*, 380–417. [CrossRef]
35. Meng, Q.; Hu, J. A poly(ethylene glycol)-based smart phase change material. *Sol. Energy Mater. Sol. Cells* **2008**, *92*, 1260–1268. [CrossRef]
36. Liu, B.; Zhang, X.; Ji, J. Review on solar collector systems integrated with phase-change material thermal storage technology and their residential applications. *Int. J. Energy Res.* **2021**, *45*, 8347–8369. [CrossRef]
37. Dash, L.; Mahanwar, P.A. A Review on Organic Phase Change Materials and Their Applications. *Int. J. Eng. Appl. Sci. Technol.* **2021**, *5*, 268–284. [CrossRef]
38. Chen, C.; Wang, L.; Huang, Y. Crosslinking of the electrospun polyethylene glycol/cellulose acetate composite fibers as shape-stabilized phase change materials. *Mater. Lett.* **2009**, *63*, 569–571. [CrossRef]

39. Chen, C.; Wang, L.; Huang, Y. Ultrafine electrospun fibers based on stearyl stearate/polyethylene terephthalate composite as form stable phase change materials. *Chem. Eng. J.* **2009**, *150*, 269–274. [CrossRef]
40. Chen, C.; Wang, L.; Huang, Y. A novel shape-stabilized PCM: Electrospun ultrafine fibers based on lauric acid/polyethylene terephthalate composite. *Mater. Lett.* **2008**, *62*, 3515–3517. [CrossRef]
41. Al-Migdady, A.K.; Jawarneh, A.M.; Ababneh, A.K.; Dalgamoni, H.N. Numerical Investigation of the Cooling Performance of PCM-based Heat Sinks Integrated with Metal Foam Insertion. *Jordan J. Mech. Ind. Eng.* **2021**, *15*, 191–197.
42. Nomura, T.; Okinaka, N.; Akiyama, T. Waste heat transportation system, using phase change material (PCM) from steelworks to chemical plant. *Resour. Conserv. Recycl.* **2010**, *54*, 1000–1006. [CrossRef]
43. Gracia, D.A.; Rincón, L.; Castell, A.; Jiménez, M.; Boer, D.; Medrano, M.; Cabeza, L.F. Life cycle assessment of the inclusion of phase change materials (PCM) in experimental buildings. *Energy Build.* **2010**, *42*, 1517–1523. [CrossRef]
44. Tyagi, V.V.; Buddhi, D. PCM thermal storage in buildings: A state of art. *Renew. Sustain. Energy Rev.* **2007**, *11*, 1146–1166. [CrossRef]
45. Jin, J.; Lin, F.; Liu, R.; Xiao, T.; Zheng, J.; Qian, G.; Liu, H.; Wen, P. Preparation and thermal properties of mineral-supported polyethylene glycol as form-stable composite phase change materials (CPCMs) used in asphalt pavements. *Sci. Rep.* **2017**, *7*, 16998. [CrossRef] [PubMed]
46. Tang, B.; Qiu, M.; Zhang, S. Thermal conductivity enhancement of PEG/SiO$_2$ composite PCM by in situ Cu doping. *Sol. Energy Mater. Sol. Cells* **2012**, *105*, 242–248. [CrossRef]
47. Meng, X.; Zhang, H.; Sun, L.; Xu, F.; Jiao, Q.; Zhao, Z.; Zhang, J.; Zhou, H.; Sawada, Y.; Liu, Y. Preparation and thermal properties of fatty acids/CNTs composite as shape-stabilized phase change materials. *J. Therm. Anal. Calorim.* **2013**, *111*, 377–384. [CrossRef]
48. Feng, L.; Song, P.; Yan, S.; Wang, H.; Wang, J. The shape-stabilized phase change materials composed of polyethylene glycol and graphitic carbon nitride matrices. *Thermochim. Acta* **2015**, *612*, 19–24. [CrossRef]
49. Sun, Q.; Yuan, Y.; Zhang, H.; Cao, X.; Sun, L. Thermal properties of polyethylene glycol/carbon microsphere composite as a novel phase change material. *J. Therm. Anal. Calorim.* **2017**, *130*, 1741–1749. [CrossRef]
50. Pielichowska, K.; Pielichowski, K. Phase change materials for thermal energy storage. *Prog. Mater. Sci.* **2014**, *65*, 67–123. [CrossRef]
51. Yuan, Y.; Zhang, H.; Zhang, N.; Sun, Q.; Cao, X. Effect of water content on the phase transition temperature, latent heat and water uptake of PEG polymers acting as endothermal-hydroscopic materials. *J. Therm. Anal. Calorim.* **2016**, *126*, 699–708. [CrossRef]
52. Wensel, J.; Wright, B.; Thomas, D.; Douglas, W.; Mannhalter, B.; Cross, W.; Hong, H.; Kellar, J. Enhanced thermal conductivity by aggregation in heat transfer nanofluids containing metal oxide nanoparticles and carbon nanotubes. *Appl. Phys. Lett.* **2008**, *92*, 023110. [CrossRef]
53. Alizadeh, N.; Broughton, R.M.; Auad, M.L. Graft Semi-Interpenetrating Polymer Network Phase Change Materials for Thermal Energy Storage. *ACS Appl. Polym. Mater.* **2021**, *3*, 1785–1794. [CrossRef]
54. Sharma, R.K.; Ganesan, P.; Tyagi, V.V.; Mahlia, T.M.I. Accelerated thermal cycle and chemical stability testing of polyethylene glycol (PEG) 6000 for solar thermal energy storage. *Sol. Energy Mater. Sol. Cells* **2016**, *147*, 235–239. [CrossRef]
55. Zhang, L.; Zhu, J.; Zhou, W.; Wang, J.; Wang, Y. Thermal and electrical conductivity enhancement of graphite nanoplatelets on form-stable polyethylene glycol/polymethyl methacrylate composite phase change materials. *Energy* **2012**, *39*, 294–302. [CrossRef]
56. Mo, S.; Mo, B.; Wu, F.; Jia, L.; Chen, Y. Preparation and thermal performance of ternary carbonates/silica microcomposites as phase change materials. *J. Sol-Gel Sci. Technol.* **2021**, *99*, 220–229. [CrossRef]
57. Zhang, C.; Shi, Z.; Li, A.; Zhang, Y.F. RGO-Coated Polyurethane Foam/Segmented Polyurethane Composites as Solid–Solid Phase Change Thermal Interface Material. *Polymers* **2020**, *12*, 3004. [CrossRef] [PubMed]
58. Chen, B.; Han, M.; Zhang, B.; Ouyang, G.; Shafei, B.; Wang, X.; Hu, S. Efficient Solar-to-Thermal Energy Conversion and Storage with High-Thermal-Conductivity and Form-Stabilized Phase Change Composite Based on Wood-Derived Scaffolds. *Energies* **2019**, *12*, 1283. [CrossRef]
59. Cui, Y.; Liu, C.; Hu, S.; Yu, X. The experimental exploration of carbon nanofiber and carbon nanotube additives on thermal behavior of phase change materials. *Sol. Energy Mater. Sol. Cells* **2011**, *95*, 1208–1212. [CrossRef]
60. Wu, B.; Chen, R.; Fu, R.; Agathopoulos, S.; Su, X.; Liu, H. Low thermal expansion coefficient and high thermal conductivity epoxy/Al2O3/T-ZnOw composites with dual-scale interpenetrating network structure. *Compos. Part A Appl. Sci. Manuf.* **2020**, *137*, 105993. [CrossRef]
61. Qi, G.Q.; Liang, C.L.; Bao, R.Y.; Liu, Z.Y.; Yang, W.; Xie, B.H.; Yang, M.B. Polyethylene glycol based shape-stabilized phase change material for thermal energy storage with ultra-low content of graphene oxide. *Sol. Energy Mater. Sol. Cells* **2014**, *123*, 171–177. [CrossRef]
62. Shen, J.; Zhang, P.; Song, L.; Li, J.; Ji, B.; Li, J.; Chen, L. Polyethylene glycol supported by phosphorylated polyvinyl alcohol/graphene aerogel as a high thermal stability phase change material. *Compos. Part B Eng.* **2019**, *179*, 107545. [CrossRef]
63. Barani, Z.; Mohammadzadeh, A.; Geremew, A.; Huang, C.Y.T.; Coleman, D.; Mangolini, L.; Kargar, F.; Baladin, A.A. Thermal Properties of the Binary-Filler Hybrid Composites with Graphene and Copper Nanoparticles. *Adv. Funct. Mater.* **2019**, *30*, 1904008. [CrossRef]
64. Tu, J.; Li, H.; Cai, Z.; Zhang, J.; Hu, X.; Huang, J.; Xiong, C.; Jiang, M.; Huang, L. Phase change-induced tunable dielectric permittivity of poly(vinylidene fluoride)/polyethylene glycol/graphene oxide composites. *Compos. Part B Eng.* **2019**, *173*, 106920. [CrossRef]

55. Hu, H. Recent advances of polymeric phase change composites for flexible electronics and thermal energy storage system. *Compos. Part B Eng.* **2020**, *195*, 108094. [CrossRef]
56. Yang, J.; Tang, L.S.; Bao, R.Y.; Bai, L.; Liu, Z.Y.; Yang, W.; Xie, B.H.; Yang, M.B. Largely enhanced thermal conductivity of poly(ethylene glycol)/boron nitride composite phase change materials for solar-thermal-electric energy conversion and storage with very low content of graphene nanoplatelets. *Chem. Eng. J.* **2017**, *315*, 481–490. [CrossRef]
57. Guerra, V.; Wan, C.; McNally, T. Thermal conductivity of 2D nano-structured boron nitride (BN) and its composites with polymers. *Prog. Mater. Sci.* **2019**, *100*, 170–186. [CrossRef]
58. Jia, X.; Li, Q.; Ao, C.; Hu, R.; Xia, T.; Xue, Z.; Wang, Q.; Deng, X.; Zhang, X.; Lu, C. High thermal conductive shape-stabilized phase change materials of polyethylene glycol/boron nitride@chitosan composites for thermal energy storage. *Compos. Part A Appl. Sci. Manuf.* **2020**, *129*, 105710. [CrossRef]
59. Lu, X.; Huang, H.; Zhang, X.; Lin, P.; Huang, J.; Sheng, X.; Zhang, L.; Qu, J. Novel light-driven and electro-driven polyethylene glycol/two-dimensional MXene form-stable phase change material with enhanced thermal conductivity and electrical conductivity for thermal energy storage. *Compos. Part B Eng.* **2019**, *177*, 107372. [CrossRef]
70. Kenisarin, M.; Mahkamov, K. Solar energy storage using phase change materials. *Renew. Sustain. Energy Rev.* **2007**, *11*, 1913–1965. [CrossRef]
71. Huang, C.; Qian, X.; Yang, R. Thermal conductivity of polymers and polymer nanocomposites. *Mater. Sci. Eng. R Rep.* **2018**, *132*, 1–22. [CrossRef]
72. Zhou, W.; Qi, S.; An, Q.; Zhao, H.; Liu, N. Thermal conductivity of boron nitride reinforced polyethylene composites. *Mater. Res. Bull.* **2007**, *42*, 1863–1873. [CrossRef]
73. Presley, M.A.; Christensen, P.R. Thermal conductivity measurements of particulate materials 1. A review. *J. Geophys. Res. Planets* **1997**, *102*, 6535–6550. [CrossRef]
74. Cinan, Z.M.; Baskan, T.; Erol, B.; Mutlu, S.; Misirlioglu, Y.; Savaskan Yilmaz, S.; Yilmaz, A.H. Gamma irradiation, thermal conductivity, and phase change tests of the cement-hyperbranched poly amino-ester-block-poly cabrolactone-polyurathane plaster-lead oxide and arsenic oxide composite for development of radiation shielding material. *Int. J. Energy Res.* **2021**, *45*, 20729–20762. [CrossRef]
75. Cinan, Z.M.; Erol, B.; Baskan, T.; Mutlu, S.; Savaskan Yilmaz, S.; Yilmaz, A.H. Gamma Irradiation and the Radiation Shielding Characteristics: For the Lead Oxide Doped the Crosslinked Polystyrene-b-Polyethyleneglycol Block Copolymers and the Polystyrene-b-Poly ethylene glycol-Boron Nitride Nanocomposites. *Polymers* **2021**, *13*, 3246. [CrossRef] [PubMed]
76. Savaskan Yilmaz, S. Synthesis and Investigation of Ion Exchange Properties of New Ion Exchangers. Ph.D. Thesis, Karadeniz Technical University, Trabzon, Turkey, 1994.
77. Savaşkan, S.; Besirli, N.; Hazer, B. Synthesis of some new cation-exchanger resins. *J. Appl. Polym. Sci.* **1996**, *59*, 1515–1524. [CrossRef]
78. Zeighampour, F.; Khoddami, A.; Hadadzadeh, H.; Ghane, M. Thermal conductivity enhancement of shape-stabilized phase change nanocomposites via synergistic effects of electrospun carbon nanofiber and reduced graphite oxide nanoparticles. *J. Energy Storage* **2022**, *51*, 104521–104534. [CrossRef]
79. Pielichowska, K.; Pielichowski, K. Biodegradable PEO/cellulose-based solidsolid phase change materials. *Polym. Adv. Technol.* **2011**, *22*, 1633–1641. [CrossRef]
80. Gao, W.; Lin, W.; Liu, T.; Xia, C. An experimental study on the heat storage performances of polyalcohols NPG, TAM, PE, and AMPD and their mixtures as solid-solid phase-change materials for solar energy applications. *Int. J. Green Energy* **2007**, *4*, 301–311. [CrossRef]
81. Wang, X.; Lu, E.; Lin, W.; Liu, T.; Shi, Z.; Tang, R.; Wang, C. Heat storage performance of the binary systems neopentyl glycol/pentaerythritol and neopentyl glycol/trihydroxy methylaminomethane as solid–solid phase change materials. *Energy Convers. Manag.* **2000**, *41*, 129–134. [CrossRef]
82. Fallahi, A.; Guldentops, G.; Tao, M.; Granados-Focil, S.; Van Dessel, S. Review on solid-solid phase change materials for thermal energy storage: Molecular structure and thermal properties. *Appl. Therm. Eng.* **2017**, *127*, 1427–1441. [CrossRef]
83. Tritt, T.M. *Physics of Solids and Liquids, Thermal Conductivity Theory, Properties, and Applications*; Kluwer Academic/Plenum Publishers: New York, NY, USA, 2004.
84. Pan, W.; Phillpot, S.R.; Wan, C.; Chernatynskiy, A.; Qu, Z. Low thermal conductivity oxides. *Camb. Univ. Press* **2012**, *37*, 917–922. [CrossRef]
85. Pierson, H.O. *Handbook of Carbon, Graphite, Diamond and Fullerences: Properties, Processing and Applications*; Noyes Publications: Park Ridge, NJ, USA, 1993.
86. Wypych, G. *Handbook of Fillers: Physical Properties of Filled Materials*; ChemTec Publishing: Toronto, ON, Canada, 2000; Chapter 2.
87. Fischer, J.E. Carbon nanotubes: Structure and properties. In *Carbon Nanomaterials*; Taylor and Francis Group: New York, NY, USA, 2006; pp. 51–58.
88. Lebedev, S.M. A comparative study on thermal conductivity and permittivity of composites based on linear low-density polyethylene and poly(lactic acid) filled with hexagonal boron nitride. *Polym. Compos.* **2021**, *43*, 111–117. [CrossRef]
89. Osman, A.F.; El Balaa, H.; El Samad, O.; Awad, R.; Badawi, M.S. Assessment of X-ray shielding properties of polystyrene incorporated with different nano-sizes of PbO. *Radiat. Environ. Biophys.* **2023**, 1–17. [CrossRef] [PubMed]

90. Zeng, S.; Liang, Y.; Lu, H.; Wang, L.; Dinh, X.Y.; Yu, X.; Ho, H.P.; Hu, X.; Yong, K.T. Synthesis of symmetrical hexagonal-shape PbO nanosheets using gold nanoparticles. *Mater. Lett.* **2012**, *67*, 74–77. [CrossRef]
91. Hussein, A.M.; Dannoun, E.M.A.; Aziz, S.B.; Brza, M.A.; Abdulwahid, R.T.; Hussen, S.A.; Rostam, S.; Mustafa, D.M.T.; Muhammad, D.S. Steps Toward the Band Gap Identification in Polystyrene Based Solid Polymer Nanocomposites Integrated with Tin Titanate Nanoparticles. *Polymers* **2020**, *12*, 2320. [CrossRef] [PubMed]
92. Li, Y.; Ma, Q.; Huang, C.; Liu, G. Crystallization of Poly (ethylene glycol) in Poly (methyl methacrylate) Networks. *Express Polym. Lett.* **2013**, *7*, 416–430. [CrossRef]

Disclaimer/Publisher's Note: The statements, opinions and data contained in all publications are solely those of the individual author(s) and contributor(s) and not of MDPI and/or the editor(s). MDPI and/or the editor(s) disclaim responsibility for any injury to people or property resulting from any ideas, methods, instructions or products referred to in the content.

Article

Tailoring Triple Filler Systems for Improved Magneto-Mechanical Performance in Silicone Rubber Composites

Vineet Kumar, Md Najib Alam, Manesh A. Yewale and Sang-Shin Park *

School of Mechanical Engineering, Yeungnam University, 280, Daehak-ro, Gyeongsan 38541, Republic of Korea; vineetfri@gmail.com (V.K.); mdnajib.alam3@gmail.com (M.N.A.); maneshphd@gmail.com (M.A.Y.)
* Correspondence: pss@ynu.ac.kr

Abstract: The demand for multi-functional elastomers is increasing, as they offer a range of desirable properties such as reinforcement, mechanical stretchability, magnetic sensitivity, strain sensing, and energy harvesting capabilities. The excellent durability of these composites is the key factor behind their promising multi-functionality. In this study, various composites based on multi-wall carbon nanotubes (MWCNT), clay minerals (MT-Clay), electrolyte iron particles (EIP), and their hybrids were used to fabricate these devices using silicone rubber as the elastomeric matrix. The mechanical performance of these composites was evaluated, with their compressive moduli, which was found to be 1.73 MPa for the control sample, 3.9 MPa for MWCNT composites at 3 per hundred parts of rubber (phr), 2.2 MPa for MT-Clay composites (8 phr), 3.2 MPa for EIP composites (80 phr), and 4.1 MPa for hybrid composites (80 phr). After evaluating the mechanical performance, the composites were assessed for industrial use based on their improved properties. The deviation from their experimental performance was studied using various theoretical models such as the Guth–Gold Smallwood model and the Halpin–Tsai model. Finally, a piezo-electric energy harvesting device was fabricated using the aforementioned composites, and their output voltages were measured. The MWCNT composites showed the highest output voltage of approximately 2 milli-volt (mV), indicating their potential for this application. Lastly, magnetic sensitivity and stress relaxation tests were performed on the hybrid and EIP composites, with the hybrid composite demonstrating better magnetic sensitivity and stress relaxation. Overall, this study provides guidance on achieving promising mechanical properties in such materials and their suitability for various applications, such as energy harvesting and magnetic sensitivity.

Keywords: multi-wall carbon nanotube; silicone rubber; stretchability; energy harvesting; magnetic sensitivity

1. Introduction

In magneto-rheological elastomers (MREs), an important constituent is elastomers. There are different types of elastomers used, such as natural rubber (NR) [1], styrene-butadiene rubber (SBR) [2], nitrile butadiene rubber (NBR) [3], and silicone rubber (SR) [4]. Among them, SR is frequently used as an elastomer matrix in MREs. Various studies have shown that SR is a fascinating matrix that is well-suited for use in MREs due to its soft nature, low viscosity, ease of curing, and ease of processing [4,5]. There are various possible types of silicone rubbers depending on the type of vulcanization used, with single components, such as room-temperature silicone rubber [6], and two-component silicone rubber or high-temperature vulcanized silicone rubber [7]. Among them, RTV-SR is more promising due to its ease of processing and soft nature, and is thus explored in this work.

The properties of MREs are affected by the types of additives used [8]. Magnetic fillers and reinforcing fillers are commonly used as additives in MREs [9,10]. Magnetic fillers can be classified based on their particle size, shape, or surface area [11]. Among the various types of magnetic fillers used in MREs, carbonyl iron particles (CIP) with different morphologies and sizes are most commonly used [12,13]. Studies have shown that CIP can act as a favorable magnetic filler due to its favorable oval morphology and small particle size [13,14]. Other types of magnetic particles include iron oxides ranging from micron-sized to nano-sized [15]. In addition to magnetic fillers, reinforcing fillers from various classes are also used. The most promising reinforcing fillers reported in the literature over the last two to three decades are nanocarbon black (NCB) [16], carbon nanotubes (CNTs) [17], graphene (GR) [18], and clay minerals [19]. Studies have shown that CNTs are a fascinating reinforcing filler that lead to a drastic increase in mechanical and electrical properties at loadings lower than 5 phr, especially in elastomer matrixes [20]. Several studies have reported the use of CNTs as reinforcing additives in MREs [21,22]. In a few studies, GR was used as a reinforcing agent in MREs [22,23]. However, the reinforcement provided by using clay minerals in MREs is not yet fully understood and is thus explored in the present work.

The mechanical stiffness of the composites used in MREs depends upon the formation of the microstructure under a magnetic field [24]. The non-magnetic fillers are dispersed randomly while the magnetic filler is oriented in the direction of the magnetic field [25,26]. The orientation of the magnetic fillers depends upon the magnitude of the magnetic field, the time of exposure to the magnetic field, and the type of magnetic filler used in such composites [27]. In addition to these parameters, the mechanical properties also depend upon the type of non-magnetic filler, its morphology, its shape, size, and the aspect ratio of the non-magnetic filler [28]. In some cases, a hybrid filler containing both magnetic and non-magnetic fields was found to be promising [29], and is thus explored in the present work.

Numerous studies have been conducted on the use of hybrid fillers in MREs [30]. These hybrid fillers can be either both magnetic or a combination of one magnetic and one reinforcing filler [31]. However, the use of triple hybrid fillers, which consist of two reinforcing and one magnetic filler, is not fully understood in MREs, and, therefore, this study aims to explore their properties. Additionally, the stress–strain curves of composites containing these hybrid fillers require further investigation, which is also explored in this study. The present work assesses the synergistic effect of these triple hybrid fillers. It should be noted that MWCNT is a promising reinforcing filler; however, its use in high amounts significantly reduces the stretchability of composites. Therefore, the addition of MT-Clay is proposed to improve this mechanical property without significantly affecting the modulus. Furthermore, EIP was added to make the composites magnetically active. Hence, the use of these three fillers is justified and presented in this work. This study also investigates the magneto-mechanical behavior of individual fillers and their hybrid filler systems.

2. Materials and Methods

2.1. Materials

The RTV-Silicone rubber used in this work was obtained from Shin-Etsu Chemical Corporation Ltd., Tokyo, Japan. It was purchased under the commercial name "KE-441-KT" and has a transparent appearance. The vulcanizing material used was also obtained from Shin-Etsu Chemical Corporation Ltd., Tokyo, Japan, and its commercial name is "CAT-RM." The MWCNT used, which has the commercial name CM-100, was purchased from Hanwha Nanotech Corporation Ltd., Seoul, Republic of Korea. The clay minerals used (Montmorillonite K10) have a surface area of 220–270 m^2/g and were purchased from Sigma Aldrich, St. Louis, MO, USA. The electrolyte iron particles (EIP) used, which have the commercial name "Fe#400," were purchased from Aometal Corporation Limited, Gomin-si, Republic of Korea. The EIP particles were irregular in shape and had a greyish color with micron-sized particles in the range of 10–12 µm. The elemental composition of the EIP was 98.8% iron with traces of nitrogen, oxygen, and carbon. The mold-releasing agent was purchased from Nabakem, Pyeongtaek-si, Republic of Korea.

2.2. Characterizations of Fillers and Composites

The morphology of the nanofillers used in this study was investigated using a SEM microscope (S-4800, Hitachi, Japan). Prior to imaging, the samples were sputtered with platinum for 2 min to make their surface conductive. The dispersion of fillers in the composite samples was evaluated using an optical micrograph (Sometech Inc., Seoul, Republic of Korea). To study filler dispersion in the rubber matrix using SEM, the cylindrical samples used for measuring the compressive mechanical properties were sectioned into approximately 0.2 mm thick slices. These slices were then mounted on an SEM stub and their surfaces were coated with platinum to make them conductive. Finally, SEM measurements were taken. The mechanical properties under compressive and tensile strain were evaluated using a universal testing machine (UTS, Lloyd instruments, West Sussex, UK). The mechanical properties under compressive strain were determined using cylindrical samples at a strain rate of 4 mm per minute from 0 to 35% strain. Similarly, mechanical properties under tensile strain were determined using a UTS machine at a strain rate of 200 mm per minute using a dumbbell-shaped specimen. The thickness of the dumbbell-shaped specimen was 2 mm and the gauge length was 25 mm. These mechanical tests were performed according to DIN 53 504 standards. Piezoelectric tests were performed using a UTS machine under cyclic loads (Lloyd Instruments, West Sussex, UK). The output voltage generated through the specimen was recorded using a digital multi-meter (Agilent 34401A, Santa Rosa, CA, USA). The energy harvesting sample was composed of MWCNT, MT-Clay, EIP, and their hybrid. The magneto-mechanical properties were tested at 30% strain using UTS under compressive strain. The procedure for magnetic sensitivity and stress relaxation under a magnetic field involved am investigation using cylindrical samples (10 mm thickness and 20 mm diameter) under 10% compressive strain. The strain rate was 1 mm/min for 5 s of magnetic switching to complete one cycle. Both magnetic sensitivity and stress relaxation were studied under on–off switching of the magnetic field at 100 mT.

2.3. Preparation of Rubber Nanocomposites

The fabrication process of the MREs was initiated by following the optimized procedure from a previous study [32]. First, a predetermined amount of liquid silicone rubber was taken in a beaker, and then a known amount of different grades of nanofillers (Table 1) were added to the liquid rubber. The mixture was then stirred for 10 min. After the nanofiller–rubber mixing phase, 2 phr of the vulcanizing agent was added, and the final rubber composite was poured into molds. The molds were manually pressed and left for 24 h for vulcanization at room temperature (25 °C). Finally, the samples (Scheme 1) were removed from the molds and tested for various properties to assess their suitability for industrial MRE applications.

Table 1. Fabrication of the different rubber composites.

Samples	RTV-SR (phr)	MWCNT (phr)	MT-Clay (phr)	EIP (phr)	Vulcanizing Solution (phr)
Control	100	-	-	-	2
RTV-SR/MWCNT	100	1, 2, 3	-	-	2
RTV-SR/MT-Clay	100		2, 4, 6, 8		2
RTV-SR/EIP	100			40, 60, 80, 100	2
RTV-SR/Hybrid *	100	1	4	35, 55, 75, 95	2

* The formulation of the filler loadings in the hybrid sample was based on their near-to-percolation value, at which the properties are improved significantly and a dominating effect of the filler can be observed.

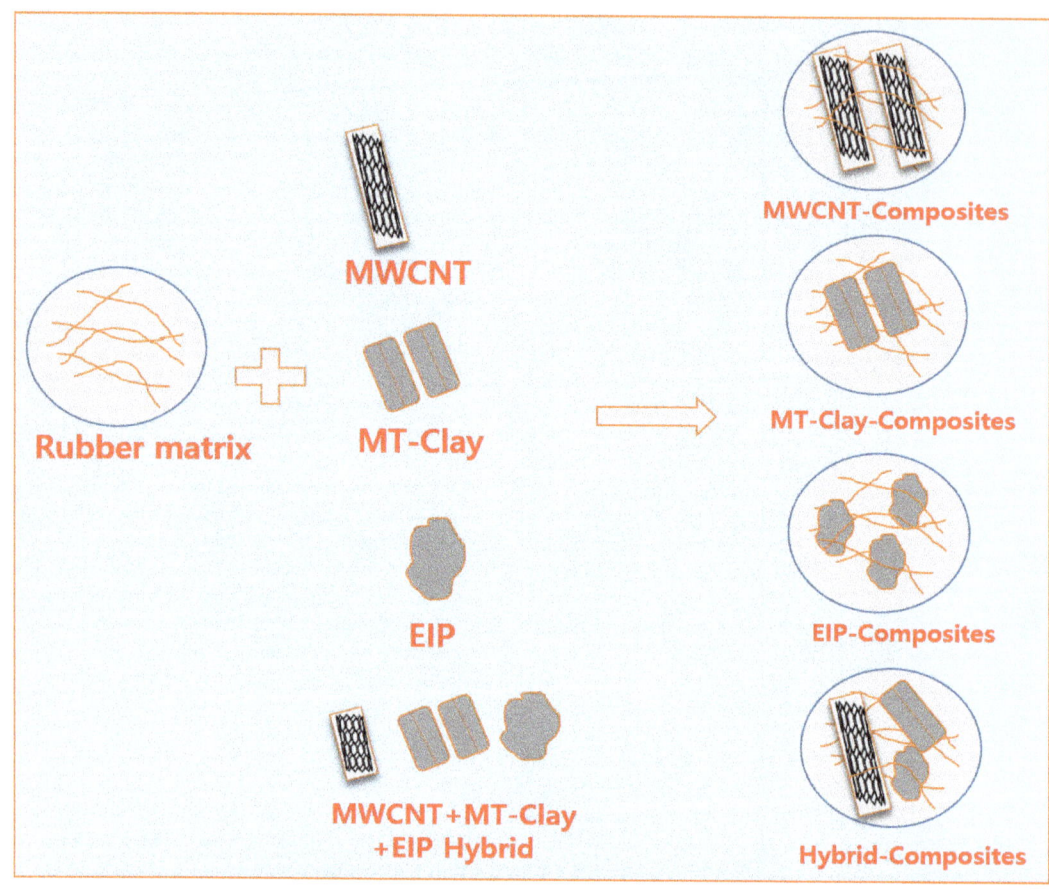

Scheme 1. Fabrication of different types of MREs.

3. Results and Discussion

3.1. Morphology of the Filler

It is well known that the morphology of nanofillers greatly affects the properties of composites [33]. Fillers with small particle sizes and favorable shapes have better and more uniform dispersion, leading to a greater impact on the composites [34]. Figure 1 illustrates the morphology of the different nanofillers used as fillers in this study. The morphologies range from one-dimensional (1D) MWCNT to 2D MT-Clay and 3D EIP. MWCNT has a tube-shaped morphology, which allows for easy dispersion and formation of continuous filler–filler contacts with a much lower MWCNT content in a rubber matrix. Furthermore, its high surface area and small particle size provide a large interfacial area, allowing more polymer chains to adsorb to its surface [35]. MT-Clay has a sheet-like morphology, making it easy to disperse in the rubber matrix. It is considered a nanofiller since its particle size is in the nanometer range. Both MWCNT and MT-Clay are ideal fillers and significantly improve composite properties in small amounts in the rubber matrix. Lastly, EIP has an irregular morphology and large particle size, likely in the micrometer range. Due to its large particle size, it has relatively poor reinforcing abilities in lower amounts and is thus used in higher amounts to achieve optimal reinforcement in a composite.

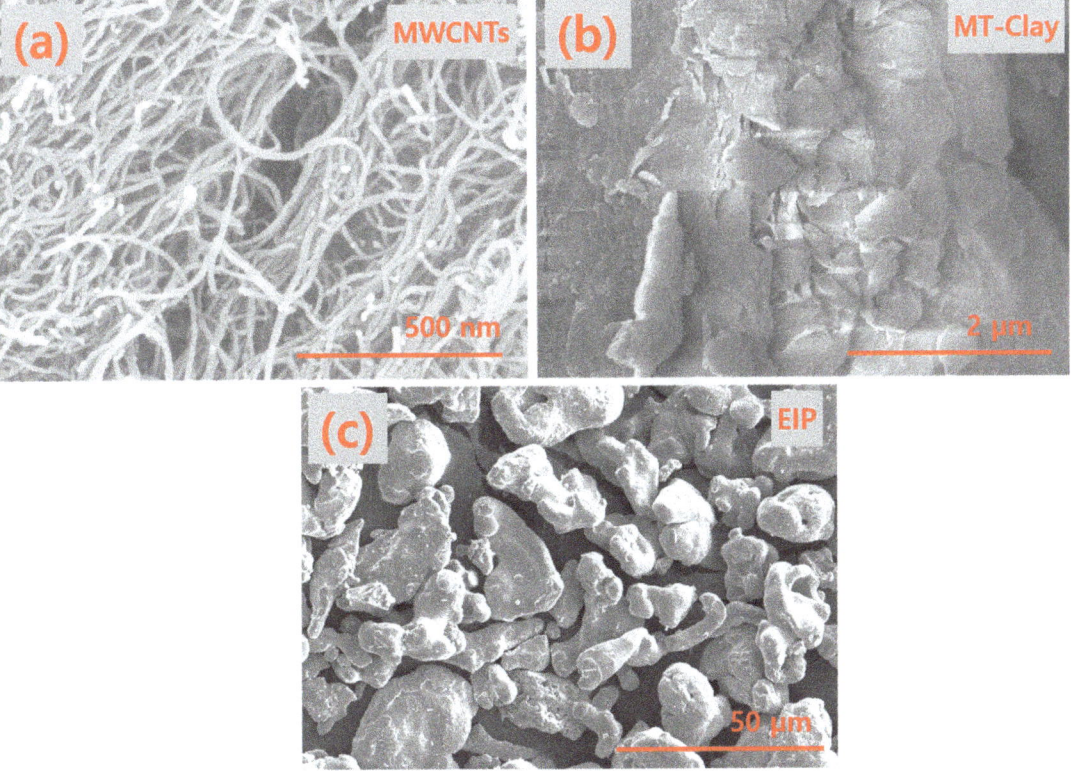

Figure 1. SEM images of (**a**) MWCNT particles; (**b**) MT-Clay particles; (**c**) EIP particles.

3.2. Filler Dispersion in SR Matrix Using Optical and SEM Micrographs

The dispersion of filler in composites is known to affect their properties [36]. A uniform filler dispersion leads to improved overall properties, while composites with poorly dispersed filler have poorer properties [37]. Therefore, this study investigates filler dispersion using optical microscopy and reports its correlation with mechanical properties. The presented optical micrographs in Figure 2 show good filler dispersion for all fillers except for MT-Clay and the hybrid filler. Figure 2a displays the micrographs of the control sample without any filler [38], indicating the absence of filler. Figure 2b shows the uniform dispersion of the MWCNT filler in the rubber matrix, and filler-rich zones with no aggregation are observed, justifying the promotion of MWCNT-based composites as having better properties. Similarly, Figure 2d shows the optical micrographs of the EIP-filled rubber matrix, showing the uniform dispersion of EIP particles and their correlation with improved mechanical properties such as modulus. As reported earlier, Figure 2c,d shows the optical micrographs of the MT-Clay and hybrid composites, respectively. The images also show improved filler dispersion, as in other filled composites, but few filler aggregates or filler-rich zones are reported [39]. The optical micrographs alone do not provide convincing evidence for studying filler dispersion, particularly due to the lack of high-resolution information about the fillers. As a result, filler dispersion was further analyzed using SEM microscopy at both lower and higher resolutions. Figure 2f–h displays SEM images of the control sample at different resolutions, where the absence of filler particles with a smooth surface can be seen. Figure 2i–k shows SEM images of MWCNT-filled composites, where the low-resolution images indicate an increase in the roughness of the rubber matrix. Moreover, at a higher resolution, the CNTs can be seen protruding out from the rubber matrix. Figure 2l–n displays SEM micrographs for MT-Clay-filled composites, where both lower and higher-resolution images show the presence of filler aggregates, supporting the conclusion that these composites have poorer mechanical properties. Next, Figure 2o–q exhibits the dispersion of EIP particles in rubber composites, where micron-sized EIP particles were uniformly dispersed. Furthermore, the high-resolution image shows good adhesion between the EIP and rubber particles. Finally, the study of hybrid fillers is presented in Figure 2r–t, where the different filler particles are uniformly dispersed in the composite, resulting in better properties in the hybrid composites.

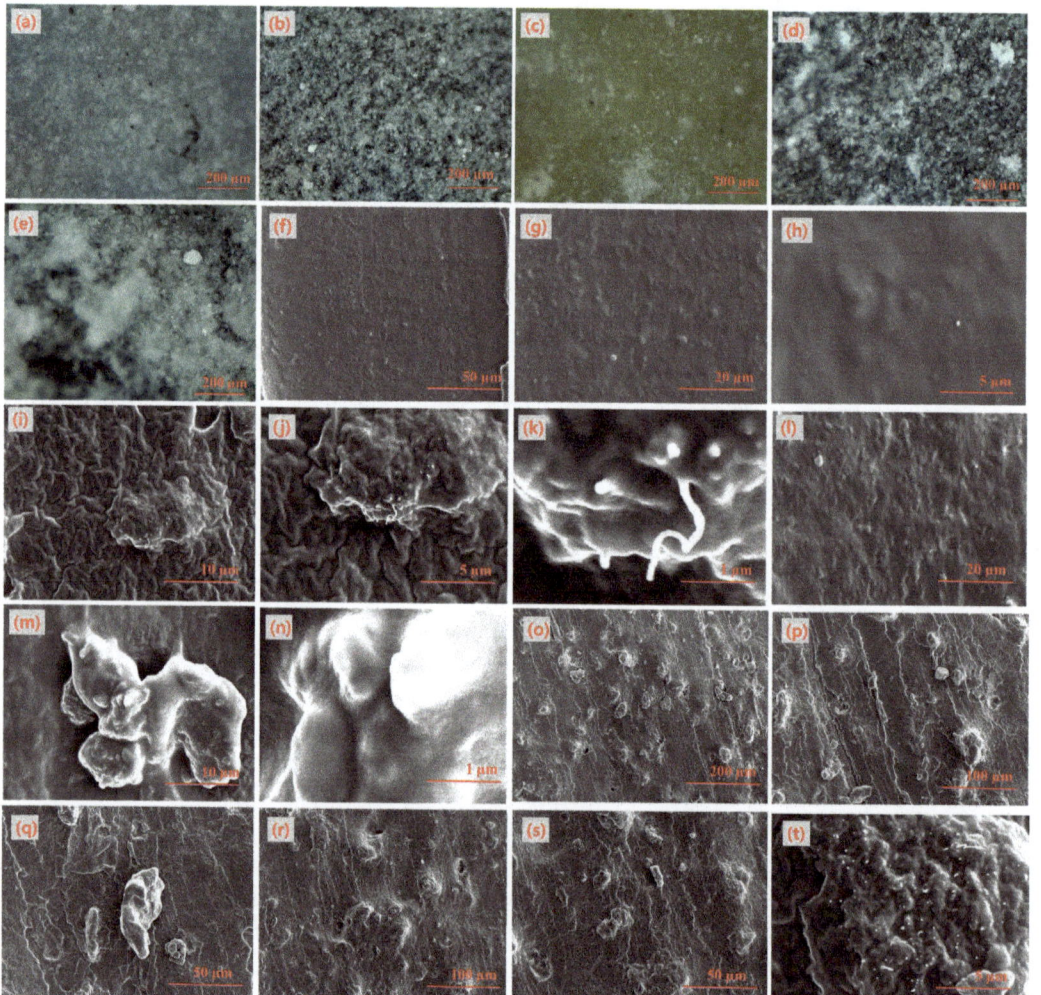

Figure 2. Optical micrographs of different specimens: (**a**) control sample; (**b**) 3 phr MWCNT; (**c**) 6 phr MT-Clay, (**d**) 60 phr EIP, (**e**) 60 phr Hybrid. SEM micrographs at different resolutions: (**f–h**) control sample; (**i–k**) MWCNT-filled composites; (**l–n**) MT-Clay-filled composites; (**o–q**) EIP-filled composites; (**r–t**) Hybrid-filled composites.

3.3. Mechanical Properties of Rubber Nanocomposites under Compressive Strain

The mechanical properties of composites depend on various parameters, such as the type of filler, the type of polymer matrix, and the type of applied strain during testing [40,41].

Certain mechanical properties, such as stretchability and stiffness, play an important role in prospective applications, such as flexible electronics [42]. The stress–strain curves under compressive strain from 0–35% are shown in Figure 3a–d. The maximum compressive strain of 35% was chosen due to the fracture of the cylindrical sample after 35% compressive strain. The stress–strain behavior of different composites indicates that the stress increases linearly up to 15% and then increases exponentially. This behavior is attributed to the increase in packing fractions of the filler and rubber particles under higher compressive strain [43]. Additionally, the stress increases in all composites with an increase in filler content, which is attributed to improved filler networking, filler–filler, and rubber–filler interactions [44,45].

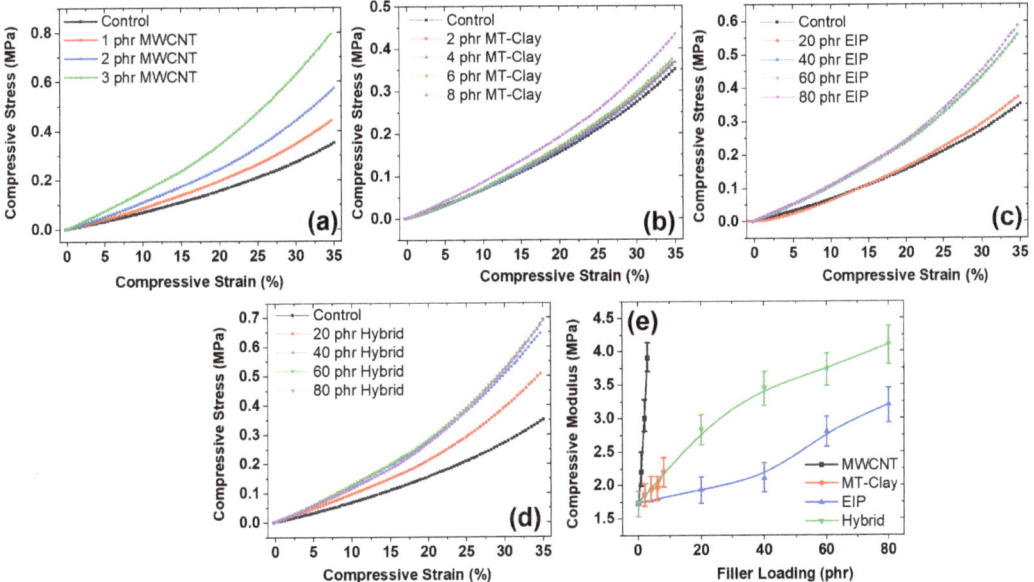

Figure 3. Mechanical properties under compressive strain: (**a**) stress–strain of MWCNT; (**b**) stress–strain of MT-Clay, (**c**) stress–strain of EIP; (**d**) stress–strain of hybrid filler; (**e**) compressive modulus of different composites under different loading conditions.

In Figure 3e, the impact of filler loading and filler type on compressive modulus is illustrated. Firstly, the effect of different filler types on mechanical properties was examined. It was observed that MWCNT, with its small particle size, high surface area, and high aspect ratio, demonstrated a promising reinforcing effect on the silicone rubber matrix. These MWCNT features, such as (a) the large aspect ratio, which helps to improve filler–filler interconnection at a lower filler loading [46]; (b) small particle size and large surface area, which provide a greater interfacial area for more rubber polymer chains to get adsorbed onto the filler surface [47]; and (c) higher interfacial area, which allows improved stress transfer at the polymer–filler interface [48]. It was also observed that MT-Clay provides a medium level of reinforcement which is higher than EIP and much lower than MWCNT. The poor reinforcement of EIP particles is due to their micron-sized particles and small surface area, which translates to a lower interfacial area. All of these EIP qualities make it an inferior source of reinforcement. Additionally, a higher amount of EIP filler is required to obtain optimum reinforcement, which is an order of magnitude higher (40 phr) than that of MWCNT (1 phr).

3.4. Mechanical Properties under Tensile Strain

The effect of filler concentration and tensile strain on the mechanical properties was investigated and is presented in Figure 4a–d. The stress–strain curves reveal that the tensile stress increases with increasing strain until it reaches its maximum at the point of failure. This behavior can be attributed to the re-orientation of filler–filler and filler–polymer microstructures in the direction opposing the tensile strain, leading to an increase in stiffness and, consequently, higher tensile stress.

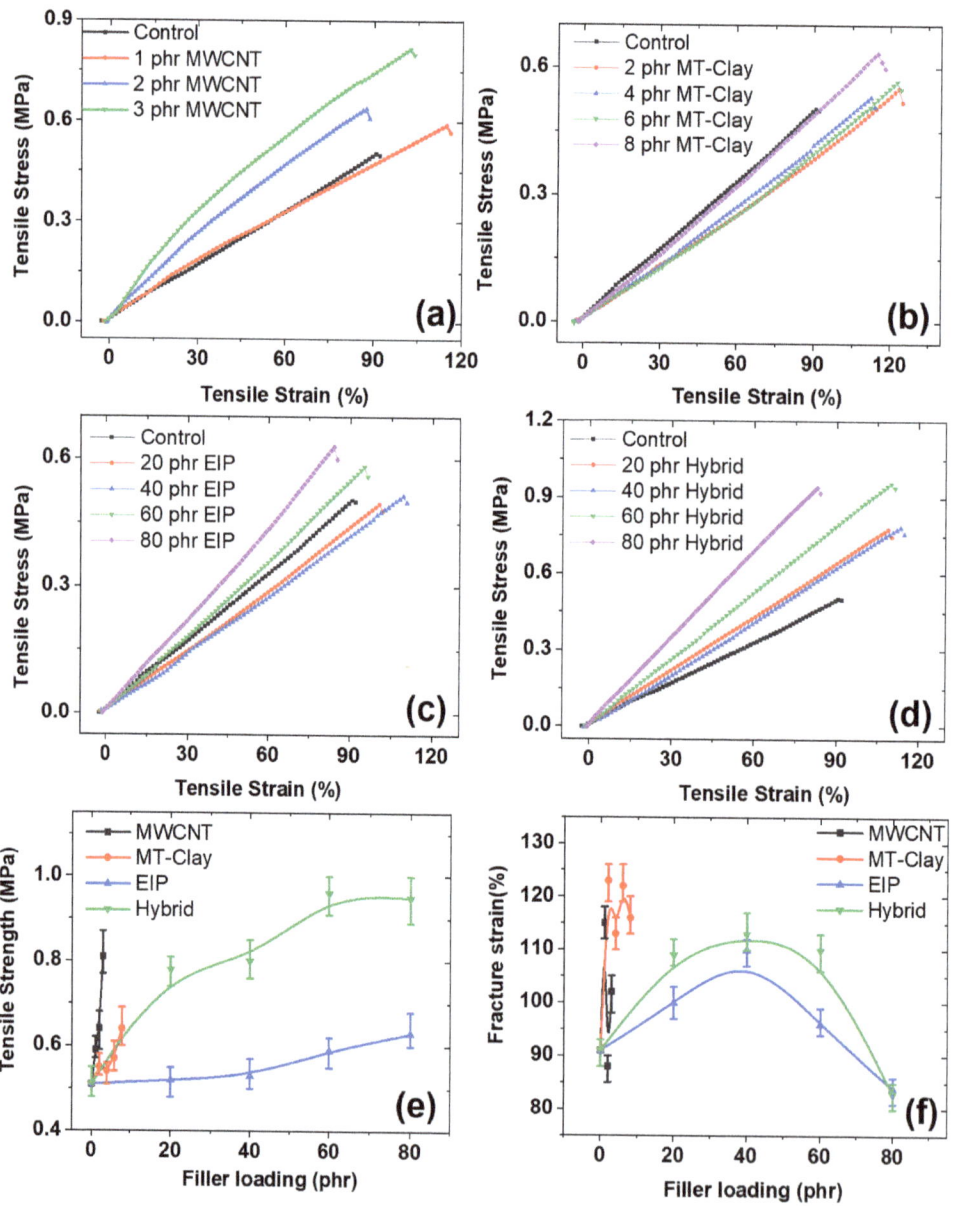

Figure 4. Stress–strain properties of the composites for (**a**) MWCNT; (**b**) MT-Clay; (**c**) EIP; (**d**) Hybrid filler; and (**e**) Tensile strength of different composites; (**f**) Fracture strain of composites.

The reinforcing ability of the fillers is dependent on their type and concentration in the rubber matrix [49,50]. Three types of fillers—MWCNT, EIP, and MT-Clay—were added to the rubber matrix in a single and hybrid state. All three fillers and their hybrid showed reinforcing abilities, with MWCNT being the most effective and EIP the least effective. Notably, MT-Clay improved the tensile strength moderately but significantly improved the fracture strain. However, due to the large particle size of EIP, a higher concentration is required, which is much higher than that of MWCNT and MT-Clay.

Figure 4e,f demonstrates the impact of filler concentration on tensile strength and fracture strain. MWCNT were found to be the most effective at reinforcing the rubber matrix due to their favorable characteristics, such as their tube-shaped morphology [51] These aspects aid in easy dispersion, a high aspect ratio that aids in forming robust filler–filler interconnections at a lower filler content, and a high surface area that facilitates higher interfacial interactions and leads to better properties [52–54]. Additionally, it is worth noting that the hybrid filler exhibits more robust mechanical properties than the three fillers used separately. Moreover, the hybrid filler displays a form of synergism in mechanical properties, with the tensile strength and fracture strain of the filled composites being higher in the hybrid filler than in MWCNT, MT-Clay, and EIP as single fillers. Therefore, it can be concluded that the hybrid filler system should be preferred over the single fillers used in this study.

3.5. Theoretical Modeling for Determining the Moduli of the MREs

The present study includes theoretical modeling to validate the experimental results using existing theoretical models. The Guth–Gold Smallwood model [55] and Halpin–Tsai theoretical model [56] are commonly used in literature for theoretical predictions, and their predictions strongly depend on morphological aspects of the filler such as aspect ratio, as well as the volume fraction of the filler [55,56]. The following equation was used for the Guth–Gold Smallwood prediction:

$$E_1 = E_o [(1 + 0.67 f_1 \phi_1)] \tag{1}$$

$$E_2 = E_o [(1 + 0.67 f_2 \phi_2] \tag{2}$$

$$E_3 = E_o [(1 + 0.67 f_3 \phi_3] \tag{3}$$

$$E_{1+2+3} = E_o [(1 + 0.67 f_1 \phi_1) + (1 + 0.67 f_2 \phi_2) + (1 + 0.67 f_3 \phi_3)] \times i \tag{4}$$

E_1, E_2, E_3, and E_{1+2+3} are the predicted theoretical moduli for MWCNT, MT-Clay, EIP, and their hybrid filler system, respectively. E_o is the experimental modulus of unfilled rubber. The f_1, f_2, and f_3 are the aspect ratios of the fillers. The ϕ_1, ϕ_2, and ϕ_3 are the volume fractions of the fillers. Moreover, the "i" is the interactive factor among the respective fillers in the hybrid system.

For the Halpin–Tsai theoretical model, the following equations are used—

$$E_1 = E_o [(1 + 2 f_1 \phi_1)/(1-\phi_1)] \tag{5}$$

$$E_2 = E_o [(1 + 2 f_2 \phi_2)/(1-\phi_2)] \tag{6}$$

$$E_3 = E_o [(1 + 2 f_3 \phi_3)/(1-\phi_3)] \tag{7}$$

$$E_{1+2+3} = E_o [(1 + 2 f_1 \phi_1)/(1 - \phi_1) + (1 + 2 f_2 \phi_2)/(1 - \phi_2) + (1 + 2 f_3 \phi_3)/(1 - \phi_3)] \times i \tag{8}$$

In this proposed theoretical model, the components have the same nomenclature as described in the Guth–Gold Smallwood equation. Figure 5a–d indicates that both models agree well with the experimental findings, further validating our results. However, in Figure 5c, the experimental data only agree up to 60 phr of EIP and then deviate. This behavior could be due to differences in assumptions made by the models, such as assuming perfect interfacial bonding between the filler–polymer interface [57] and perfect filler dispersion in the rubber matrix, which is difficult to achieve experimentally. Therefore, there is a deviation between the experimental data and the theoretical models. Additionally, it is worth noting that the hybrid filler system shows synergistic mechanical properties and is therefore more advantageous than using single-filled systems in the composites.

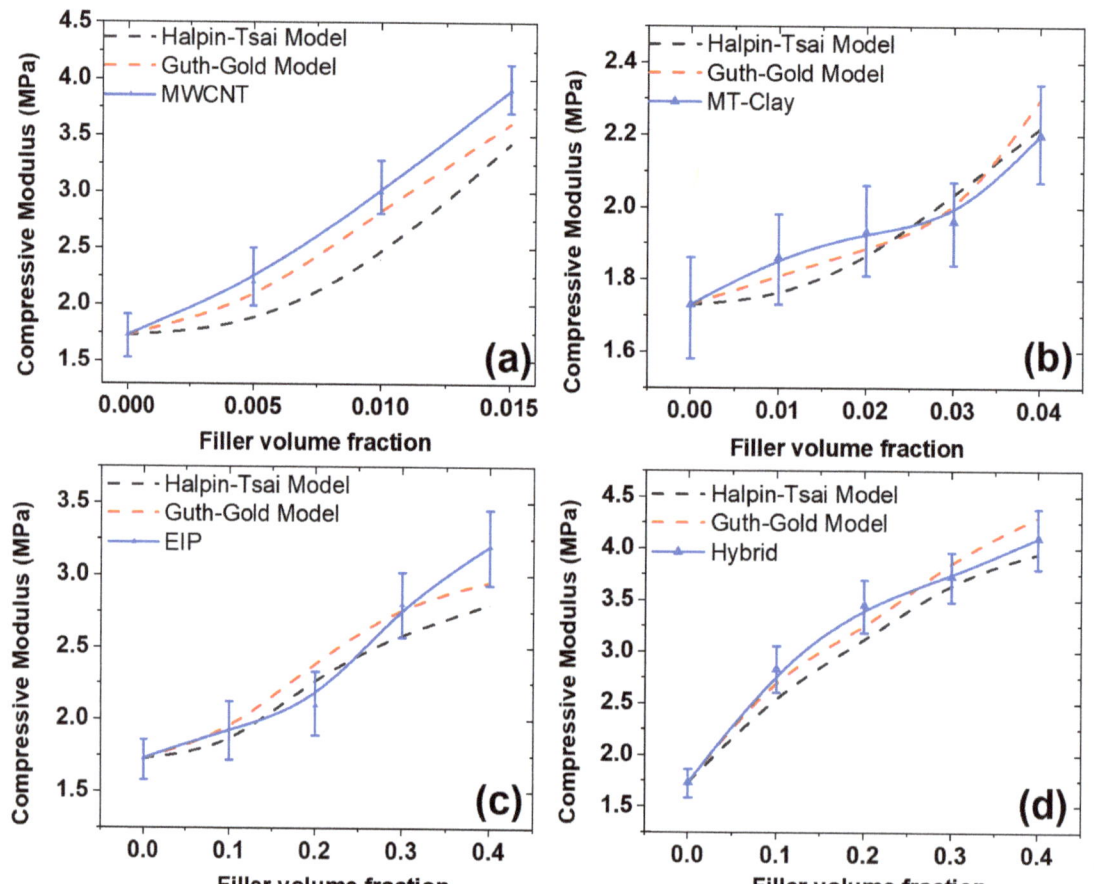

Figure 5. The theoretical and experimental fittings: (**a**) models for MWCNT; (**b**) models for MT-Clay; (**c**) models for EIP; (**d**) models for the hybrid filler system.

3.6. Experimental Deviation from Statistical Average in Hybrid Composites

The experimental behavior and its deviation from the theoretical values have been well-studied in the literature [57,58]. In this work, we use a simple theoretical model based on statistical averages to predict the mechanical properties of the hybrid composites [58]. The compressive behavior shown in Figure 6a can be derived from the following equation:

$$E_{1+2+3} = [0.1 \times E_1 + 0.4 \times E_2 + 1.5 \times E_3] \times i \qquad (9)$$

where "E_1" is the theoretical modulus of MWCNT, "E_2" is the theoretical modulus of MT-Clay, and "E_3" is the theoretical modulus of EIP-based composites. "E_{1+2+3}" is the theoretical modulus for hybrid composites containing all three components. In this theoretical model, the constants of 0.1 for 1 phr of MWCNT, 0.4 for 4 phr of MT-Clay, and 1.5 for 15 phr of EIP are related to the filler content in the sample and are used to predict mechanical properties through statistical averages for hybrid composites [58]. The interactive factor "i" considers the dispersion state and filler interactions in the composite. A low value of "i" (i = 0.1) indicates poor filler dispersion and interactions, while a high value of "i" (i ≥ 0.9) indicates good filler dispersion and interactions in the rubber matrix. For the determination of the compressive modulus, the value of "i" was found to be in the range of 0.7 to 0.8. It is worth noting that the literature has extensively studied experimental behavior and its deviation from theoretical values [57,58].

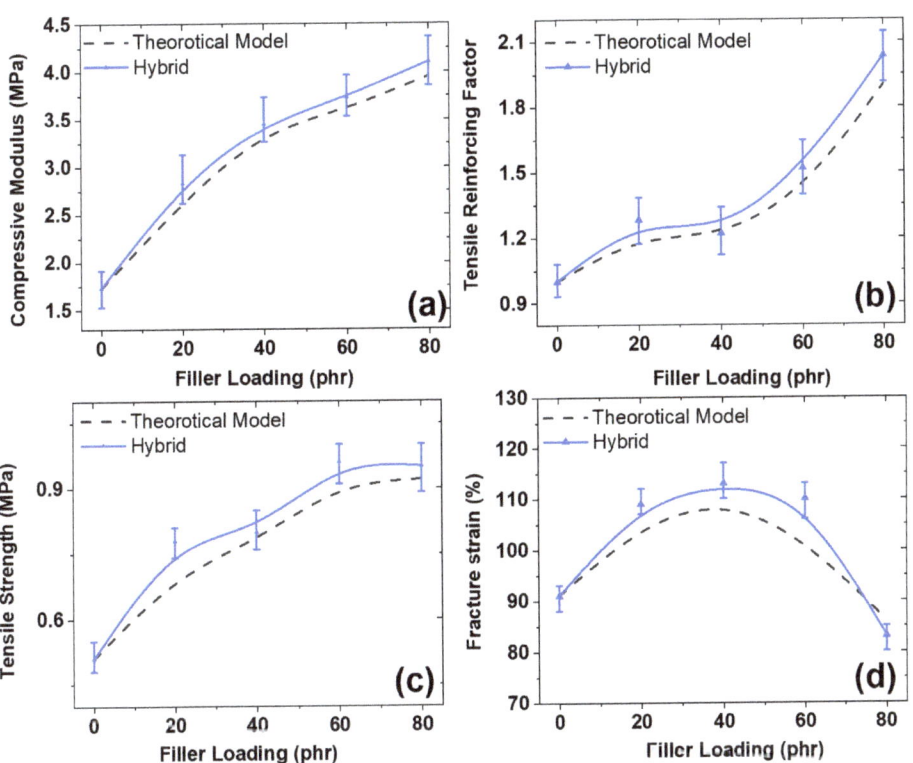

Figure 6. Experimental behavior against theoretical models: (a) for compressive modulus; (b) for tensile reinforcing factor; (c) tensile strength; (d) fracture strain.

The determination of the tensile reinforcing factor can be derived from the following equation—

$$R.F._{1+2+3} = [0.1 \times R.F._1 + 0.4 \times R.F._2 + 1.5 \times R.F._3] \times i \quad (10)$$

Here, "$R.F._{1+2+3}$" is the reinforcing factor for hybrid components. "$R.F._1$," "$R.F._2$," and "$R.F._3$" are the reinforcing factors of the individual components. Moreover, the interacting factor for determining theoretical R.F. was in the range of 0.5 to 0.8.

Similarly, the tensile strength in Figure 6c can be derived theoretically from the following equation—

$$T.S._{1+2+3} = 0.1 \times T.S._1 + 0.4 \times T.S._2 + 1.5 \times T.S._3 \times i \quad (11)$$

where "T.S.$_{1+2+3}$" is the theoretical tensile strength of the hybrid system, while "T.S.$_1$," "T.S.$_2$," and "T.S.$_3$" are the tensile strength of the individual components. Moreover, the interacting factor for determining the theoretical T.S. was in the range of 0.65 to 0.7.

Finally, the fracture strain in Figure 6d can be derived from the following equation—

$$F.S._{1+2+3} = 0.1 \times F.S._1 + 0.4 \times F.S._2 + 1.5 \times F.S._3 \times i \tag{12}$$

where "F.S.$_{1+2+3}$" is the theoretical fracture strain of the hybrid system, while "F.S.$_1$," "F.S.$_2$," and "F.S.$_3$" are the fracture strain of the individual components. Moreover, the interacting factor for determining the theoretical F.S. was around 0.5. From Figure 6a–d, it can be hypothesized that the theoretical models fit well with the experimental findings and are thus useful for further considerations in the literature.

3.7. Reinforcing Factor and Reinforcing Efficiency of the Fillers in MREs

Reinforcement via particulate filler in polymer composites is well documented [59]. It is known that fillers with small particle sizes produce higher reinforcement, so studying the reinforcing properties in rubber composites is important for understanding their mechanical properties, such as stiffness, stretchability, tensile strength, and modulus [60,61]. This study analyzed four categories of fillers with different concentrations for their reinforcing effect and efficiency, which is presented in Figure 7. The reinforcing factor of the composites can be calculated using the following equation:

$$R.F. = \frac{EF}{Eo} \tag{13}$$

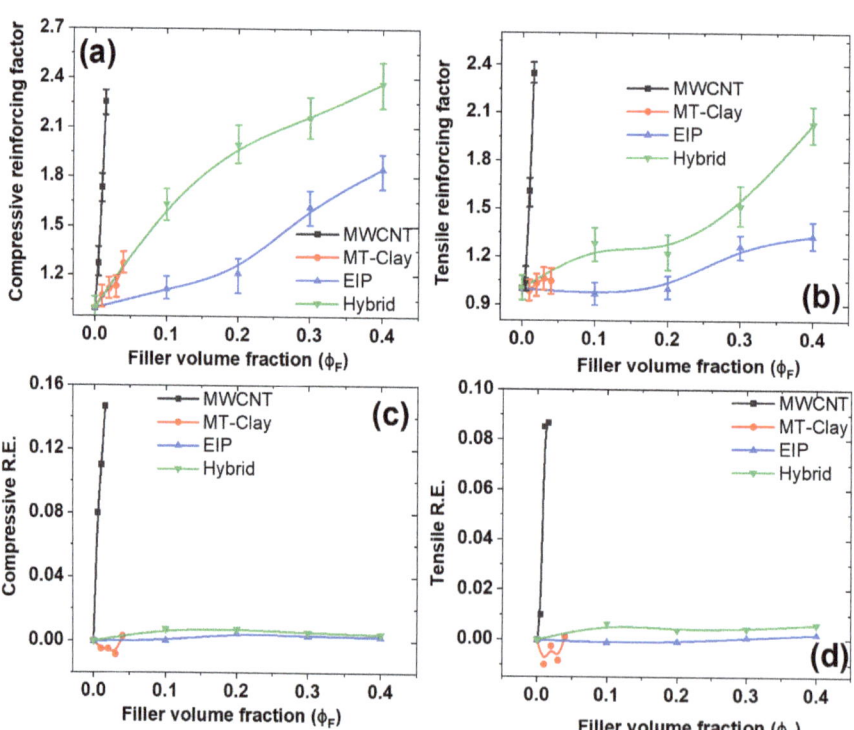

Figure 7. (a) R.F. under compressive strain; (b) R.F. under tensile strain; (c) R.E. under compressive strain; (d) R.E. under tensile strain.

Here, R.F. is the reinforcing factor, EF is the modulus of the filled composites, and E_o is the modulus of the unfilled composites. As shown in Figure 7a,b, the R.F. strongly depends on the type of filler. For example, MWCNT, with its small particle size and higher aspect ratio, was found to be the most promising source of reinforcement in the silicone rubber matrix. Other fillers such as MT-Clay provide medium reinforcement, while EIP, with its large particle size, shows poor reinforcing ability and is thus used in very high amounts compared to MWCNT to obtain optimum reinforcement [61]. Besides R.F., reinforcing efficiency (R.E.) is a significant parameter that affects the mechanical properties of the composites. It is also interesting to note that the R.E. is directly correlated with the concentration of the filler in the composites. The equation for calculating R.E. [62] is

$$\text{R.E. at compressive strain} = \frac{\sigma(35\%)\text{filled} - \sigma(35\%)\text{unfilled}}{\text{wt\% of filler}} \quad (14)$$

$$\text{R.E. at tensile strain} = \frac{\sigma(80\%)\text{filled} - \sigma(80\%)\text{unfilled}}{\text{wt\% of filler}} \quad (15)$$

where "σ" is the stress at a particular strain. The stress values used for calculating R.E. were 35% and 80% for compressive and tensile strain tests, respectively, as obtained from the stress–strain curves in Figures 3 and 4. Notably, MWCNT-based composites exhibited superior R.E. compared to MT-Clay and EIP particles, which can be attributed to their high aspect ratio, tube-shaped morphology, and higher interfacial area with the rubber matrix. These factors allowed for easy dispersion and stronger reinforcement, as seen in Figure 7.

4. Applications

4.1. Energy Harvesting Applications for the MREs

Energy harvesting using eco-friendly composites is a promising area of study for society. In this study, we fabricated an energy-harvesting device comprising conductive copper electrodes sandwiched with different substrates. The energy generated was due to the dielectric property of the elastomer used in the substrate against mechanical compressive loading, which was kept constant at 30% for all samples [63]. Although piezoelectric materials like PZT [64] or barium titanate [65] have shown promise for high-voltage generation, their use is limited due to their poisonous effects [65,66]. Recently, eco-friendly composite-based energy harvesting has been reported [67].

Figure 8 shows the different energy harvesting output voltages for the different substrates. From these measurements, we found that MWCNT-based substrates showed the highest output voltage while EIP-based substrates showed the lowest among all the substrates studied. However, the voltage stability was found to be less efficient in MWCNT-based substrates than in all other materials studied. Therefore, in conclusion, MWCNT-based substrates have higher voltage generation capabilities but the disadvantage of lower voltage stability. In addition to the type of substrate, electrode area is a critical factor affecting the output voltage. For instance, energy harvesting devices with larger electrode surface areas produce higher output voltages than those with smaller ones. We will explore this effect in our future work.

4.2. Magnetic Effect and Stress Relaxation Applications for the MREs

To magnetic sensitivity measured in this work has been optimized in our previous studies [5]. Figure 9a displays the magneto-mechanical response of the composites during the magnetic switching task. The measurements demonstrate that the compressive load increases when a magnetic field of 100 mT is applied and returns to normal when the magnetic field is turned off. This could be attributed to the orientation of EIP particles in the direction of the applied magnetic field, thereby enhancing the stiffness of the composites [5]. The increase in stiffness is correlated with a rise in compressive load, as shown in Figure 9a. These measurements establish that the composites containing magnetic fillers are sensitive to exposure to a magnetic field, as claimed in the objective of this research. Additionally, it is worth noting that a hybrid-filled composite provides higher sensitivity than using EIP as the only filler. Figure 9b shows the magnetic effect on the moduli of different composites. The magnetic effect was found to be higher for the hybrid composite than for EIP as the only filler. The higher magnetic sensitivity for the hybrid-filled composite could be attributed to the synergistic effect [5] between the MWCNT–EIP fillers, leading to greater sensitivity. Rubber composites reinforced with fillers often exhibit viscoelastic properties that affect their stress relaxation behavior [68]. The effect of magnetic switching and the type of mechanical reinforcement on stress relaxation in rubber composites is shown in Figure 9c. The results indicate that stress relaxation is higher when the magnetic field is on and is higher for hybrid composites than EIP-only-filled rubber composites. Additionally, the stress relaxation rate, as shown in Figure 9d, is influenced by the type of filler and magnetic switching. The stress relaxation rate is higher when the magnetic field is off and lower when it is on for both EIP-only-filled and hybrid-filled composites. The poor reinforcing and magnetic effect of EIP even at 60 phr filler content leads to a small change in magnetic effect and stress relaxation in composites. Moreover, these experiments were performed multiple times to make sure that the conclusions are convincing. Furthermore, hybrid-filled composites exhibit higher stress relaxation rates than EIP-only-filled composites. These results are consistent with the magnetic sensitivity tests shown in Figure 9a,b. The addition of reinforcing fillers, such as MWCNT, improves damping properties in MREs [69]. Therefore, the hybrid filler is the best candidate for achieving improved magnetic sensitivity and good damping in MREs.

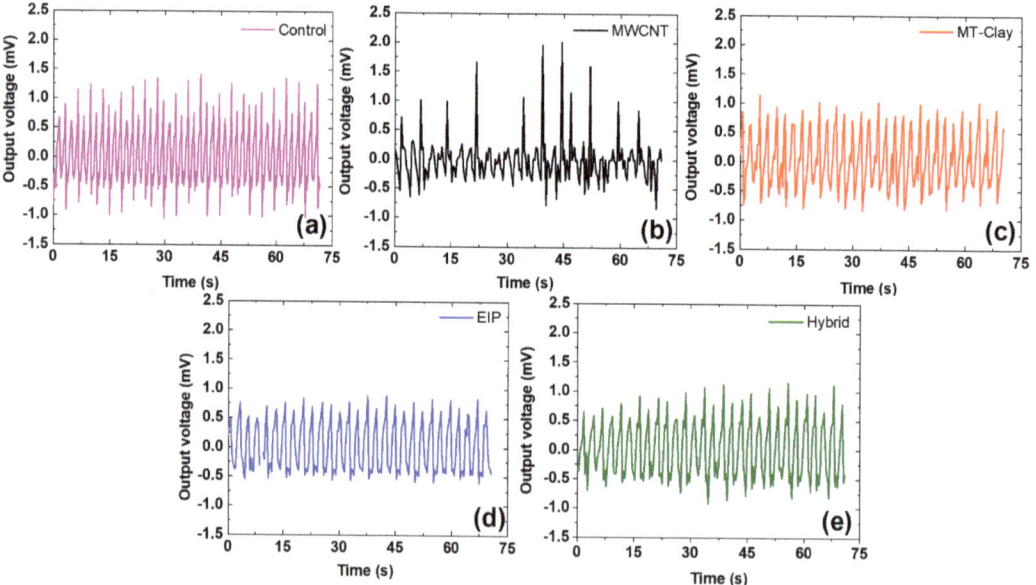

Figure 8. Energy harvesting using different substrates: (**a**) control sample; (**b**) 1 phr of the MWCNT-filled substrate; (**c**) 4 phr of the MT-Clay-filled substrate; (**d**) 40 phr of the EIP-filled substrate; (**e**) 40 phr of the hybrid substrate.

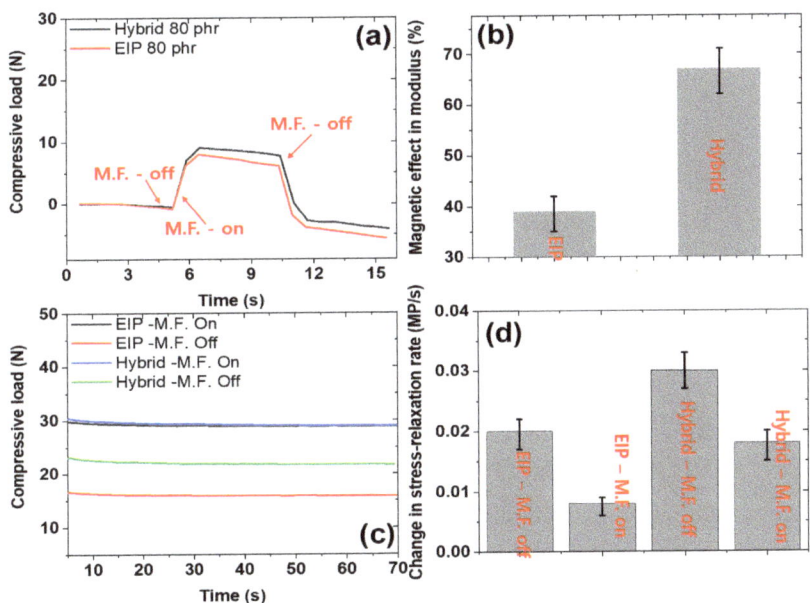

Figure 9. (a) Magnetic response of the composites under magnetic switching; (b) magnetic effect on moduli of different composites; (c) stress relaxation of different composites under magnetic switching; (d) change in stress relaxation rate of different composites under magnetic switching.

5. Conclusions

This study demonstrates that incorporating a triple-filler system into silicone rubber has the potential to yield superior mechanical performance, magnetic sensitivity, and energy generation. To this end, composite materials were prepared via solution mixing of MWCNT, MT-Clay, and EIP fillers in both single and hybrid states into the silicone rubber matrix. The improved mechanical performance of the resulting composites was then investigated and reported in this study. Specifically, mechanical stretchability was measured and found to be 91% (control), 102% (MWCNT composites, 3 phr), 116% (MT-Clay composites, 8 phr), 110% (EIP composites, 40 phr), and 113% (hybrid composites, 40 phr). The tensile strength was also analyzed and found to be 0.51 MPa (control), 0.81 MPa (MWCNT composites, 3 phr), 0.64 MPa (MT-Clay composites, 8 phr), 0.63 MPa (EIP composites, 80 phr), and 0.95 MPa (hybrid composites, 80 phr). Furthermore, the effect of the mechanical properties on magnetic sensitivity was explored, and it was found that EIP composites exhibited higher magnetic sensitivity than hybrid composites. However, the latter was identified as the most promising filler system due to its good reinforcement, optimum stiffness, and reasonable magnetic sensitivity. The key takeaway from this study is that selecting a hybrid filler system can result in balanced overall properties that are useful for different applications such as magnetic sensitivity or energy harvesting. For example, the triple-filler system was found to offer good reinforcement from MWCNT, stretchability from MT-Clay, and magnetic sensitivity from EIP in the composite material.

Author Contributions: Conceptualization, V.K. and M.N.A.; methodology, M.N.A., V.K. and M.A.Y.; validation, V.K., M.A.Y. and S.-S.P.; formal analysis, M.N.A., V.K. and M.A.Y.; investigation, V.K.; data curation, M.N.A. and V.K.; writing—original draft preparation, V.K.; writing—review and editing, M.N.A., V.K., M.A.Y. and S.-S.P.; visualization, V.K. and S.-S.P.; supervision, S.-S.P.; project administration, S.-S.P.; funding acquisition, S.-S.P. All authors have read and agreed to the published version of the manuscript.

Funding: This research is funded by the Korean Government (MOTIE, 2022) (P0002092, The Competency Development Program for Industry Specialists).

Institutional Review Board Statement: Not applicable.

Informed Consent Statement: Not applicable.

Data Availability Statement: The data presented in this study are available on request from the corresponding author.

Acknowledgments: This research was supported by the Korea Institute for Advancement of Technology (KIAT).

Conflicts of Interest: The authors declare no conflict of interest.

References

1. Chen, L.; Gong, X.L.; Jiang, W.Q.; Yao, J.J.; Deng, H.X.; Li, W.H. Investigation on magnetorheological elastomers based on natural rubber. *J. Mater. Sci.* **2007**, *42*, 5483–5489. [CrossRef]
2. Tagliabue, A.; Eblagon, F.; Clemens, F. Analysis of styrene-butadiene based thermoplastic magnetorheological elastomers with surface-treated iron particles. *Polymers* **2021**, *13*, 1597. [CrossRef] [PubMed]
3. Alam, M.N.; Kumar, V.; Lee, D.J.; Choi, J. Magnetically active response of acrylonitrile-butadiene-rubber-based magnetorheological elastomers with different types of iron fillers and their hybrid. *Compos. Commun.* **2021**, *24*, 100657. [CrossRef]
4. Guan, X.; Dong, X.; Ou, J. Magnetostrictive effect of magnetorheological elastomer. *J. Magn. Magn. Mater.* **2008**, *320*, 158–163. [CrossRef]
5. Shit, S.C.; Shah, P. A review on silicone rubber. *Natl. Acad. Sci. Lett.* **2013**, *36*, 355–365. [CrossRef]
6. Hamadi, S.H.K.; Isa, M.; Hashim, S.N.M.A.; Othman, M. Review on RTV silicone rubber coatings insulator for transmission lines. *IOP Conf. Ser. Mater. Sci. Eng.* **2020**, *864*, 012188. [CrossRef]
7. Hron, P. Hydrophilisation of silicone rubber for medical applications. *Polym. Int.* **2003**, *52*, 1531–1539. [CrossRef]
8. Borin, D.; Stepanov, G. Magneto-mechanical properties of elastic hybrid composites. *Phys. Sci. Rev.* **2022**, *7*, 1119–1140. [CrossRef]
9. Khayam, S.U.; Usman, M.; Umer, M.A.; Rafique, A. Development and characterization of a novel hybrid magnetorheological elastomer incorporating micro and nano size iron fillers. *Mater. Des.* **2020**, *192*, 108748. [CrossRef]
10. Bastola, A.; Hossain, M. Enhanced performance of core-shell hybrid magnetorheological elastomer with nanofillers. *Mater. Lett.* **2021**, *297*, 129944. [CrossRef]
11. Aloui, S.; Klüppel, M. Magneto-rheological response of elastomer composites with hybrid-magnetic fillers. *Smart Mater. Struct.* **2014**, *24*, 025016. [CrossRef]
12. Alam, M.N.; Kumar, V.; Jo, C.R.; Ryu, S.R.; Lee, D.J.; Park, S.S. Mechanical and magneto-mechanical properties of styrene-butadiene-rubber-based magnetorheological elastomers conferred by novel filler-polymer interactions. *Compos. Sci. Technol.* **2022**, *229*, 109669. [CrossRef]
13. Burgaz, E.; Goksuzoglu, M. Effects of magnetic particles and carbon black on structure and properties of magnetorheological elastomers. *Polym. Test.* **2020**, *81*, 106233. [CrossRef]
14. Yu, M.; Zhu, M.; Fu, J.; Yang, P.A.; Qi, S. A dimorphic magnetorheological elastomer incorporated with Fe nano-flakes modified carbonyl iron particles: Preparation and characterization. *Smart Mater. Struct.* **2015**, *24*, 115021. [CrossRef]
15. Lee, D.; Kwon, O.S.; Song, S.H. Tailoring the performance of magnetic elastomers containing Fe_2O_3 decorated carbon nanofiber. *RSC Adv.* **2017**, *7*, 45595–45600. [CrossRef]
16. Chen, L.; Gong, X.L.; Li, W.H. Effect of carbon black on the mechanical performances of magnetorheological elastomers. *Polym. Test.* **2008**, *27*, 340–345. [CrossRef]
17. Danafar, F.; Kalantari, M. A review of natural rubber nanocomposites based on carbon nanotubes. *J. Rubber Res.* **2018**, *21*, 293–310. [CrossRef]
18. Kumar, V.; Alam, M.N.; Park, S.S.; Lee, D.J. New Insight into Rubber Composites Based on Graphene Nanoplatelets, Electrolyte Iron Particles, and Their Hybrid for Stretchable Magnetic Materials. *Polymers* **2022**, *14*, 4826. [CrossRef]
19. Marins, J.A.; Mija, A.; Pin, J.M.; Giulieri, F.; Soares, B.G.; Sbirrazzuoli, N.; Bossis, G. Anisotropic reinforcement of epoxy-based nanocomposites with aligned magnetite–sepiolite hybrid nanofiller. *Compos. Sci. Technol.* **2015**, *112*, 34–41. [CrossRef]
20. Kumar, V.; Lee, D.J. Studies of nanocomposites based on carbon nanomaterials and RTV silicone rubber. *J. Appl. Polym. Sci.* **2017**, *134*, 44407. [CrossRef]
21. Srivastava, S.K.; Mishra, Y.K. Nanocarbon reinforced rubber nanocomposites: Detailed insights about mechanical, dynamical mechanical properties, payne, and mullin effects. *Nanomaterials* **2018**, *8*, 945. [CrossRef] [PubMed]
22. Poojary, U.R.; Hegde, S.; Gangadharan, K.V. Experimental investigation on the effect of carbon nanotube additive on the field-induced viscoelastic properties of magnetorheological elastomer. *J. Mater. Sci.* **2018**, *53*, 4229–4241. [CrossRef]
23. Zhao, D.; Cui, J.; Dai, X.; Liu, S.; Dong, L. Magneto-piezoresistive characteristics of graphene/room temperature vulcanized silicon rubber-silicon rubber magnetorheological elastomer. *J. Appl. Polym. Sci.* **2021**, *138*, 50051. [CrossRef]

4. Ahamed, R.; Choi, S.B.; Ferdaus, M.M. A state of art on magneto-rheological materials and their potential applications. *J. Intell. Mater. Syst. Struct.* **2018**, *29*, 2051–2095. [CrossRef]
5. Samal, S.; Škodová, M.; Abate, L.; Blanco, I. Magneto-rheological elastomer composites. A review. *Appl. Sci.* **2020**, *10*, 4899. [CrossRef]
6. Choi, S.B.; Li, W.; Yu, M.; Du, H.; Fu, J.; Do, P.X. State of the art of control schemes for smart systems featuring magneto-rheological materials. *Smart Mater. Struct.* **2016**, *25*, 043001. [CrossRef]
7. Kumar, V.; Lee, D.J. Iron particle and anisotropic effects on mechanical properties of magneto-sensitive elastomers. *J. Magn. Magn. Mater.* **2017**, *441*, 105–112. [CrossRef]
8. Diguet, G.; Sebald, G.; Nakano, M.; Lallart, M.; Cavaillé, J. Optimization of magneto-rheological elastomers for energy harvesting applications. *Smart Mater. Struct.* **2020**, *29*, 075017. [CrossRef]
9. Kwon, S.H.; Lee, J.H.; Choi, H.J. Magnetic particle filled elastomeric hybrid composites and their magnetorheological response. *Materials* **2018**, *11*, 1040. [CrossRef]
10. Moreno, M.A.; Gonzalez-Rico, J.; Lopez-Donaire, M.L.; Arias, A.; Garcia-Gonzalez, D. New experimental insights into magneto-mechanical rate dependences of magnetorheological elastomers. *Compos. Part B Eng.* **2021**, *224*, 109148. [CrossRef]
11. Molchanov, V.S.; Stepanov, G.V.; Vasiliev, V.G.; Kramarenko, E.Y.; Khokhlov, A.R.; Xu, Z.D.; Guo, Y.Q. Viscoelastic properties of magnetorheological elastomers for damping applications. *Macromol. Mater. Eng.* **2014**, *299*, 1116–1125. [CrossRef]
12. Lee, J.Y.; Kumar, V.; Tang, X.W.; Lee, D.J. Mechanical and electrical behavior of rubber nanocomposites under static and cyclic strain. *Compos. Sci. Technol.* **2017**, *142*, 1–9. [CrossRef]
13. Svoboda, P.; Zeng, C.; Wang, H.; Lee, L.J.; Tomasko, D.L. Morphology and mechanical properties of polypropylene/organoclay nanocomposites. *J. Appl. Polym. Sci.* **2002**, *85*, 1562–1570. [CrossRef]
14. Alter, H. Filler particle size and mechanical properties of polymers. *J. Appl. Polym. Sci.* **1965**, *9*, 1525–1531. [CrossRef]
15. Kobashi, K.; Nishino, H.; Yamada, T.; Futaba, D.N.; Yumura, M.; Hata, K. Epoxy composite sheets with a large interfacial area from a high surface area-supplying single-walled carbon nanotube scaffold filler. *Carbon* **2011**, *49*, 5090–5098. [CrossRef]
16. Karasek, L.; Sumita, M. Characterization of dispersion state of filler and polymer-filler interactions in rubber-carbon black composites. *J. Mater. Sci.* **1996**, *31*, 281–289. [CrossRef]
17. Taguet, A.; Cassagnau, P.; Lopez-Cuesta, J.M. Structuration, selective dispersion and compatibilizing effect of (nano) fillers in polymer blends. *Prog. Polym. Sci.* **2014**, *39*, 1526–1563. [CrossRef]
18. Bicerano, J.; Douglas, J.F.; Brune, D.A. Model for the viscosity of particle dispersions. *J. Macromol. Sci. Part C* **1999**, *39*, 561–642. [CrossRef]
19. Sápi, Z.; Butler, R.; Rhead, A. Filler materials in composite out-of-plane joints—A review. *Compos. Struct.* **2019**, *207*, 787–800.
20. Kumar, V.; Alam, M.N.; Manikkavel, A.; Song, M.; Lee, D.J.; Park, S.S. Silicone rubber composites reinforced by carbon nanofillers and their hybrids for various applications: A review. *Polymers* **2021**, *13*, 2322. [CrossRef]
21. Mohan, T.P.; Kuriakose, J.; Kanny, K. Effect of nanoclay reinforcement on structure, thermal and mechanical properties of natural rubber–styrene butadine rubber (NR–SBR). *J. Ind. Eng. Chem.* **2011**, *17*, 264–270. [CrossRef]
22. Park, S.; Vosguerichian, M.; Bao, Z. A review of fabrication and applications of carbon nanotube film-based flexible electronics. *Nanoscale* **2013**, *5*, 1727–1752. [CrossRef] [PubMed]
23. Zhou, W.; Wang, C.; An, Q.; Ou, H. Thermal properties of heat conductive silicone rubber filled with hybrid fillers. *J. Compos. Mater.* **2008**, *42*, 173–187. [CrossRef]
24. Warasitthinon, N.; Genix, A.C.; Sztucki, M.; Oberdisse, J.; Robertson, C.G. The Payne effect: Primarily polymer-related or filler-related phenomenon? *Rubber Chem. Technol.* **2019**, *92*, 599–611. [CrossRef]
25. Hentschke, R. The Payne effect revisited. *Express Polym. Lett.* **2017**, *11*, 278–292. [CrossRef]
26. He, X.; Ou, D.; Wu, S.; Luo, Y.; Ma, Y.; Sun, J. A mini review on factors affecting network in thermally enhanced polymer composites: Filler content, shape, size, and tailoring methods. *Adv. Compos. Hybrid Mater.* **2022**, *5*, 21–38. [CrossRef]
27. Zhang, X.; Wang, J.; Jia, H.; You, S.; Xiong, X.; Ding, L.; Xu, Z. Multifunctional nanocomposites between natural rubber and polyvinyl pyrrolidone modified graphene. *Compos. Part B Eng.* **2016**, *84*, 121–129. [CrossRef]
28. Chen, Y.; Sanoja, G.; Creton, C. Mechanochemistry unveils stress transfer during sacrificial bond fracture of tough multiple network elastomers. *Chem. Sci.* **2021**, *12*, 11098–11108. [CrossRef]
29. Bokobza, L. The reinforcement of elastomeric networks by fillers. *Macromol. Mater. Eng.* **2004**, *289*, 607–621. [CrossRef]
30. Zare, Y.; Rhee, K.Y. Effects of interphase regions and filler networks on the viscosity of PLA/PEO/carbon nanotubes biosensor. *Polym. Compos.* **2019**, *40*, 4135–4141. [CrossRef]
31. Wang, M.J. The role of filler networking in dynamic properties of filled rubber. *Rubber Chem. Technol.* **1999**, *72*, 430–448. [CrossRef]
32. Barus, S.; Zanetti, M.; Bracco, P.; Musso, S.; Chiodoni, A.; Tagliaferro, A. Influence of MWCNT morphology on dispersion and thermal properties of polyethylene nanocomposites. *Polym. Degrad. Stab.* **2010**, *95*, 756–762.
33. Guo, J.; Liu, Y.; Prada-Silvy, R.; Tan, Y.; Azad, S.; Krause, B.; Grady, B.P. Aspect ratio effects of multi-walled carbon nanotubes on electrical, mechanical, and thermal properties of polycarbonate/MWCNT composites. *J. Polym. Sci. Part B Polym. Phys.* **2014**, *52*, 73–83. [CrossRef]
34. Bronnikov, S.; Kostromin, S.; Asandulesa, M.; Pankin, D.; Podshivalov, A. Interfacial interactions and interfacial polarization in polyazomethine/MWCNTs nanocomposites. *Compos. Sci. Technol.* **2020**, *190*, 108049. [CrossRef]

55. Wolff, S.; Donnet, J.B. Characterization of fillers in vulcanizates according to the Einstein-Guth-Gold equation. *Rubber Chem. Technol.* **1990**, *63*, 32–45. [CrossRef]
56. Affdl, J.H.; Kardos, J.L. The Halpin-Tsai equations: A review. *Polym. Eng. Sci.* **1976**, *16*, 344–352.
57. Wu, Y.P.; Jia, Q.X.; Yu, D.S.; Zhang, L.Q. Modeling Young's modulus of rubber–clay nanocomposites using composite theories. *Polym. Test.* **2004**, *23*, 903–909. [CrossRef]
58. Kumar, V.; Alam, M.N.; Azam, S.; Manikkavel, A.; Park, S.S. The tough and multi-functional stretchable device based on silicone rubber composites. *Polym. Adv. Technol.* **2023**. Early view. [CrossRef]
59. Boonstra, B.B. Role of particulate fillers in elastomer reinforcement: A review. *Polymer* **1979**, *20*, 691–704.
60. Greenough, S.; Dumont, M.J.; Prasher, S. The physicochemical properties of biochar and its applicability as a filler in rubber composites: A review. *Mater. Today Commun.* **2021**, *29*, 102912. [CrossRef]
61. Hamed, G.R. Reinforcement of rubber. *Rubber Chem. Technol.* **2000**, *73*, 524–533. [CrossRef]
62. Das, C.; Bansod, N.D.; Kapgate, B.P.; Reuter, U.; Heinrich, G.; Das, A. Development of highly reinforced acrylonitrile butadiene rubber composites via controlled loading of sol-gel titania. *Polymer* **2017**, *109*, 25–37. [CrossRef]
63. Madsen, F.B.; Daugaard, A.E.; Hvilsted, S.; Skov, A.L. The current state of silicone-based dielectric elastomer transducers. *Macromol. Rapid Commun.* **2016**, *37*, 378–413. [CrossRef] [PubMed]
64. Kang, M.G.; Jung, W.S.; Kang, C.Y.; Yoon, S.J. Recent progress on PZT based piezoelectric energy harvesting technologies. *Actuators* **2016**, *5*, 5. [CrossRef]
65. Roscow, J.I.; Lewis RW, C.; Taylor, J.; Bowen, C.R. Modelling and fabrication of porous sandwich layer barium titanate with improved piezoelectric energy harvesting figures of merit. *Acta Mater.* **2017**, *128*, 207–217. [CrossRef]
66. Baek, C.; Yun, J.H.; Wang, J.E.; Jeong, C.K.; Lee, K.J.; Park, K.I.; Kim, D.K. A flexible energy harvester based on a lead-free and piezoelectric BCTZ nanoparticle–polymer composite. *Nanoscale* **2016**, *8*, 17632–17638. [CrossRef]
67. Liu, Y.; Zhao, L.; Wang, L.; Zheng, H.; Li, D.; Avila, R.; Yu, X. Skin-integrated graphene-embedded lead zirconate titanate rubber for energy harvesting and mechanical sensing. *Adv. Mater. Technol.* **2019**, *4*, 1900744. [CrossRef]
68. Wang, Y.X.; Ma, J.H.; Zhang, L.Q.; Wu, Y.P. Revisiting the correlations between wet skid resistance and viscoelasticity of rubber composites via comparing carbon black and silica fillers. *Polym. Test.* **2011**, *30*, 557–562. [CrossRef]
69. Kumar, V.; Alam, M.N.; Park, S.S. Soft Composites Filled with Iron Oxide and Graphite Nanoplatelets under Static and Cyclic Strain for Different Industrial Applications. *Polymers* **2022**, *14*, 2393. [CrossRef]

Disclaimer/Publisher's Note: The statements, opinions and data contained in all publications are solely those of the individual author(s) and contributor(s) and not of MDPI and/or the editor(s). MDPI and/or the editor(s) disclaim responsibility for any injury to people or property resulting from any ideas, methods, instructions or products referred to in the content.

Article

The Effect of Encapsulating a Prebiotic-Based Biopolymer Delivery System for Enhanced Probiotic Survival

Aida Kistaubayeva [1,*], Malika Abdulzhanova [1], Sirina Zhantlessova [1], Irina Savitskaya [1], Tatyana Karpenyuk [1], Alla Goncharova [1] and Yuriy Sinyavskiy [2]

1 Faculty of Biology and Biotechnology, Al-Farabi Kazakh National University, Almaty 050040, Kazakhstan
2 Kazakh Academy of Nutrition, Almaty 050008, Kazakhstan
* Correspondence: aida.kistaubaeva@kaznu.edu.kz

Abstract: Orally delivered probiotics must survive transit through harsh environments during gastrointestinal (GI) digestion and be delivered and released into the target site. The aim of this work was to evaluate the survivability and delivery of gel-encapsulated *Lactobacillus rhamnosus* GG (LGG) to the colon. New hybrid symbiotic beads alginate/prebiotic pullulan/probiotic LGG were obtained by the extrusion method. The average size of the developed beads was 3401 μm (wet), 921 μm (dry) and the bacterial titer was 10^9 CFU/g. The morphology of the beads was studied by a scanning electron microscope, demonstrating the structure of the bacterial cellulose shell and loading with probiotics. For the first time, we propose adding an enzymatic extract of feces to an artificial colon fluid, which mimics the total hydrolytic activity of the intestinal microbiota. The beads can be digested by fecalase with cellulase activity, indicating intestinal release. The encapsulation of LGG significantly enhanced their viability under simulated GI conditions. However, the beads, in combination with the prebiotic, provided greater protection of bacteria, enhancing their survival and even increasing cell numbers in the capsules. These data suggest the promising prospects of coencapsulation as an innovative delivery method based on the inclusion of probiotic bacteria in a symbiotic matrix.

Keywords: coencapsulation; symbiotic beads; *Lactobacillus rhamnosus* GG; bacterial cellulose; pullulan; simulated gastrointestinal conditions

1. Introduction

Microbial polysaccharides are becoming more and more attractive materials intended for use in encapsulation technology: the inclusion of biologically active substances, enzymes, vitamins, and probiotics in the matrix [1,2].

Probiotics are usually used in the form of biologically-active food additives, or even therapeutic drugs in the form of tablets, capsules, powders, and sachets. However, most often, probiotics are used as part of functional foods. Both these and other forms are applied orally and therefore enter the gastrointestinal (GI) tract, which they need to pass through safely to reach the colon where they function. The most critical points on this path are, first of all, the stomach, which has extreme pH values, and the upper parts of the intestine, where bile acids and digestive enzymes are present [3]. From this point of view, the microcapsule shell, serving as a physical barrier, protects the cells included in it and is therefore an excellent means of delivering probiotics to the lower intestine [4,5]. Hence, encapsulation ensures the viability of probiotics during their transportation to the place of action in the human body without negatively affecting their physiological properties.

On the other hand, even inside the capsule, there is not always a high viability of bacteria, which is one of the reasons for the low success of such protection of probiotics [6]. In this regard, in recent years, coencapsulation of a probiotic with a prebiotic has been proposed as an additional method, leading to an increase in the survival of probiotics in capsules [7]. A prebiotic is defined as "a substrate that is selectively utilized by host microorganisms, conferring a health benefit" [8]. The combination of a probiotic and a prebiotic is called a synbiotic, and can have either non-specific or selective effects. In the first case, a prebiotic is selected to enhance the local beneficial microbiota, i.e., it is a "universal agent". In the second case, the prebiotic is selected specifically to support the growth of the selected probiotic, regardless of the beneficial effect on the population of other bacteria. This particular method is used in the present study.

The model probiotic *Lactobacillus rhamnosus* GG (LGG) is one of the most extensively studied lactobacilli strains with established positive effects on human health [9].

The second component of the developed system, which this study proposes to introduce into the bead together with LGG, is a selective prebiotic. Evidence suggests that such a "nutrient component" may be polysaccharide pullulan (PUL), a neutral linear polysaccharide synthesized by *Aureobasidium pullulans*, consisting of α-1,6-linked maltotriose residues. Since PUL has several hydroxyl groups, if necessary, it can be easily modified. PUL is able not only to stimulate the growth of these bacteria in culture experiments but also increase their viability during encapsulation [10–13].

It should also be mentioned that according to some data, prebiotics can be destroyed to a certain extent in the small intestine [14]. Hence, prebiotics also need to be protected. With this in mind, the first objective of this study was to coat the beads in a durable biopolymer that is not directly subjected to enzymatic and physical destruction in the upper parts of the digestive system. For this purpose, bacterial cellulose (BC), an indigestible polysaccharide, also a product of microbial synthesis, was used. Since only cellulolytic bacteria live in the colon, cellulose and its derivatives can be decomposed there [15].

The second objective was to determine the protective properties of beads with prebiotic PUL coated with a cellulose shell as the proposed delivery system.

Human models are ideal for determining the functional effectiveness of beads; however, there are ethical limitations associated with this method. The closest to "intestinal nature" is an in vitro dynamic model that more accurately simulates the sequential kinetic conditions in the GI tract, including the active microbiota. To simulate the conditions of the cecum, Maathuis et al. [16] added fecal samples, which were used as an inoculum in a fed-batch fermenter. In our opinion, this is a rather complex system that requires hardware design. We suggested that a simplified modification of this system could be created, in which no intestinal microbes are added to the "intestinal fluid", except an enzymatic extract of feces. This is called "fecalase" and represents the enzymatic activity of intestinal microorganisms [17]. The results of these studies can provide predictive tools for determining the effectiveness of probiotic delivery systems in their ecological niche.

The purpose of this study was to create a symbiotic biopolymer capsule system, Alginate-Pullulan/Bacterial cellulose (Alg-PUL/BC), for targeted probiotic delivery.

2. Materials and Methods

2.1. Bacteria and Growth Conditions

Lactobacillus rhamnosus GG (ATCC® 53103™) strain was purchased from American Type Culture Collection. LGG was incubated in MRS medium (HiMedia, Mumbai, India) at 37 °C for 48 h to obtain a cell concentration of 10^{10} CFU/mL. Then, the cells were collected by a laboratory centrifuge RS-6MC (Dastan, Bishkek, Kyrgyzstan) (at $6000 \times g$, for 15 min) and washed twice with saline solution.

2.2. Preparation of BC

Komagataeibacter xylinus C3 strain was isolated at the Biotechnology department, Al-Farabi Kazakh National University; the strain was deposited in the Republic Collection of Microorganisms (Astana, Kazakhstan), with the Gen Bank accession number: KU598766. Inoculum of *K. xylinus* (1%, v/v) was added to the flasks with Hestrin–Shramm broth medium (Hi-Media, Mumbai, India) and incubated statically at 30 °C for 7 days. The cultivated films were first purified by washing with deionized water, treating with 1%

(w/v) NaOH at 35 °C for 24 h to remove microbial cells, and rinsed again with deionized water. The obtained films were dissolved in an aqueous solution of NaOH–urea–H_2O (w/v: 7%–12%–83%) at a low temperature (−12 °C) for 2 min.

2.3. Probiotics Encapsulation

Then, 10 mL of LGG cell culture ($\approx 10^9$–10^{10} CFU/mL) was carefully mixed with 40 mL of a granule-forming suspension of 2% sodium alginate (Sigma-Aldrich, Taufkirchen, Germany) (alginic acid sodium salt from brown algae, guluronic acid or glucuronic content ~65–70%; mannuronic acid content ~5–35%) solution. Beads were prepared by the extrusion method from a 10 mL plastic syringe through a 20-gauge needle (0.9 mm diameter) into a beaker, at a distance of 25 cm, containing calcium chloride (Fisher Scientific Inc., Ottawa, ON, Canada) solution (1%, w/v) with gentle agitation at room temperature. Armed MP-2003 syringe dispenser (Shanghai Leien Medical Equipment Co. Ltd., Shanghai, China) was used for extrusion. The formed beads were left to harden for 30 min and then washed with sterile distilled water.

To obtain Alg/BC samples, Alg beads were covered with an additional layer of 0.5–2% BC solution and incubated for 30–40 min on an orbital shaker incubator ES-20 (Biosan, Riga, Latvia) (130 rpm).

To obtain hybrid beads of PUL (Hayashibara Biochemical Laboratories, Okayama, Japan) with BC, a mixture of Alg (2% solution) and PUL (1–2% solution) was first prepared in a ratio of 1:1. 10 mL of probiotic culture was added to the resulting mixture at a cell concentration of at least 10^{10} CFU/mL. The mixture was then stirred on a homogenizer DG-360 (Stegler, Shanghai, China) at 3000 rpm for 30 min. The resulting mixture was passed through a syringe needle into a sterile calcium chloride solution (1%, w/v). After, the beads were kept in a solution of calcium chloride for 30 min for immobilization and washed twice with distilled water. The obtained beads Alg + PUL were put into a 0.5% BC solution, incubated on a shaker, and washed twice with sterile distilled water. The capsules were stored in sterile vials at 4 °C and used in further experiments.

2.4. Scanning Electron Microscopy (SEM)

SEM micrographs were obtained using a scanning electron microscope (Quanta 3D 200i, Hillsboro, OR, USA). To obtain samples, the beads were freeze-dried on a Lyoquest-80 lyophilizer (Telstar, Madrid, Spain). Dried samples were placed on strips of double-sided carbon tape attached to aluminum loops. SEM of all samples was carried out at the following parameters: accelerating voltage 15 kV; working distance $\approx$10 mm. All measurements were carried out in a high vacuum mode of 10^{-3} Pa.

2.5. Mechanical Characterization

The Instron bursting machine (model 3365, Norwood, MA, USA) was used to determine the tensile strength (MPa) in uniaxial mode. Randomly selected beads were placed on the lower platform of the bursting machine and subjected to pressure in the vertical direction from top to bottom until the bead was destroyed. The upper platform with a flat tip was connected to a force sensor, which, during compression, recorded the force values acting on the beads. The mechanical properties of each sample were the average values determined from fifty specimens.

2.6. Encapsulation Efficiency

The capsules containing LGG bacteria were decapsulated using cellulase from *Trichoderma* sp. (Sigma-Aldrich, Taufkirchen, Germany). Cellulase solution was made by dissolving 50 mg/mL enzyme in deionized water. The samples were added to cellulase solution (1:10), followed by shaking at 37 °C until bacteria were released from beads completely. The viability of released cells was determined by plating serial dilutions of the resulting suspension on MRS agar medium (Hi-Media, Mumbai, India). Colony-forming units were counted after 72 h of incubation at 37 °C.

The encapsulation efficiency (E_e) was determined by the equation:

$$Ee = \frac{N \times M}{N0} \times 100 \tag{1}$$

where N is the number of viable cells released from 1 g of beads, M is the total mass of the collected beads, and N_0 is the number of free cells before treatment.

2.7. Survival of LGG in Simulated Gastric Fluid (SGF) and Simulated Duodenum Fluid (SDF)

SGF was prepared by dissolving pepsin in sodium chloride solution (0.2%, w/v) to a final concentration of 3 g/L, and pH was adjusted to 2 with hydrochloric acid. SDF was prepared by dissolving pancreatin in sodium chloride solution (0.2%, w/v) to a final concentration of 1 g/L, with 4.5 g/L bile salts, and pH was adjusted to 6.8 with sodium hydroxide. Both solutions were filtered for sterilization through a 0.45 µm membrane. All reagents were purchased from Veld (Almaty, Kazakhstan).

The encapsulated samples (1 g) were placed in 10 mL of SGF. The tubes were incubated on an orbital shaker incubator ES-20 (Biosan, Riga, Latvia) (150 rpm) at 37 °C for 1–2 h. The samples were collected after 2 h in SGF, transferred into 10 mL of SDF, and incubated as described above for SGF.

At the end of the incubation period, each sample (1 g) was removed and rinsed with distilled water. The beads containing probiotic bacteria were disintegrated using cellulase. Surviving bacteria were enumerated by pour plate counts in MRS agar incubated at 37 °C for 72 h. The survival of probiotic LGG was presented as a number of viable cells (log CFU/g). The following equation was used to calculate the survival rate % of encapsulated bacteria cells.

$$\text{Survival rate \%} = \frac{\log \text{ CFU/g after treatment}}{\log \text{ CFU/g before treatment}} \times 100 \tag{2}$$

For the free cells, 1 mL of LGG suspension was inoculated into 9 mL of SGF. After incubation, 1 mL of suspension was collected and transferred into 9 mL of SDF solution. The incubation conditions for free bacteria were the same as for beads. At the end of the incubation period, the survival of free cells was determined in the way described above.

2.8. Preparation of Enzymatic Fecal Extracts (Fecalase) and Release of LGG into Simulated Colon Fluid (SCF)

Enzymatic fecal extracts were prepared from feces collected from three healthy donors on a regular diet (two females and one male, aged 22–30 years) who did not take pro- or prebiotics and antibiotics for at least 3 months before fecal sample donation. The fecal samples were suspended in 1 mL potassium phosphate buffer (0.01 M, pH 7.4) and homogenized for 1.5 min using a Mini-Beadbeater (BioSpec, Bartlesville, OK, USA). The suspension was centrifuged at $2000 \times g$ for 5 min using a Minispin centrifuge (Eppendorf, Hamburg, Germany), followed by another centrifugation of supernatant at $10,000 \times g$ for 20 min. The supernatant (fecalase) extracted after the second centrifugation was filtered for sterilization through a 0.45 µm membrane and used for the assay due to its ability to break down the cellulose shell of the capsules and release its content into SCF.

After 2 h incubation in SDF, beads (1 g) were placed in 9 mL of SCF (0.2 g/L of potassium chloride, 8 g/L of sodium chloride, 0.24 g/L of potassium phosphate monobasic, 1.44 g/L of sodium phosphate dibasic, pH 7.2). Then, 1 mL of fecalase was added to this solution. The free bacteria were treated similarly. The tubes were incubated on an orbital shaker incubator ES-20 (150 rpm) at 37 °C for 3–18 h. At the end of the incubation period, a 1 mL aliquot was removed and released bacteria were enumerated by pour plate counts in

MRS agar incubated at 37 °C for 72 h. The following equation was used to calculate the release rate % of bacteria cells.

$$\text{Release rate \%} = \frac{\log \text{CFU/g of released viable cells}}{\log \text{CFU/g before treatment}} \times 100 \quad (3)$$

2.9. Statistical Analysis

All analyses were conducted in triplicate and the results were presented as mean ± standard deviation unless otherwise stated. Data were analyzed using a one-way analysis of variance (ANOVA) with the Tukey test. Statistical analyses were performed using SPSS software (version 28.0, IBM Corp., Armonk, NY, USA). Significance was defined as $p < 0.05$.

3. Results and Discussion

3.1. Obtaining and Characterization of Probiotic Beads

Methods for encapsulating probiotics are very diverse and these techniques are well described in several reviews [18,19]. The preference for one or another method depends on the purpose of the study or the direction of application of the beads. The choice of the extrusion method for capsules obtaining was due to the fact that it does not require special expensive equipment, high temperatures, and the encapsulation efficiency is very high [20]. This method consists of mixing a suspension of bacteria and a hydrocolloid solution. Although PUL belongs to hydrocolloids, it is not capable of forming a gel [21], but it can be added to a solution of sodium alginate to obtain hybrid beads by the subsequent extrusion of a gel-forming agent (calcium chloride) through a nozzle into a solution. The principle of using calcium chloride as a carrier is due to the fact that Ca^{2+} forms a cross-link with guluronic acid, which is part of the Alg molecule, forming a G-G block. At the same time, the "Egg-box" model gelation mechanism is activated; guluronic acid multimers are "fixed" by ions, forming a pocket that balances negatively charged polymer chains [22].

Alg hydrogel can be called the "gold standard" for the encapsulation of probiotics since it has mucoadhesive properties and imitates the polysaccharide matrix of enterocytes, as well as the matrix of bacterial biofilms formed in the intestine by resident bacteria [23]. However, "simple" Alg capsules also have disadvantages: easy disintegration in an acidic environment, under the action of chelating agents, monovalent ions, as well as too high porosity, which can lead to rapid release of the active principle from the Alg gel [24].

To strengthen the Alg matrix, capsules are coated with additional and sometimes several layers (layer-by-layer) of other natural polymers; chitosan, gelatin, pectin, starch, and cellulose is also among them [13].

The design of the experiment for obtaining beads is shown in Figure 1.

The probiotic was included either in Alg alone or in Alg with PUL. Both versions of the capsules were placed from the gelling agent solution into regenerated cellulose gel to form an outer shell on the surface of the capsules.

Different ratios of polymers used may affect the properties of beads. The selection of the optimal composition was evaluated according to the parameters presented in Table 1, the key of which is mechanical strength. Since in a number of studies, the concentration of 2% Alg [25–27] is considered optimal, it was used for obtaining capsules by extrusion. The concentrations of PUL and BC varied in the range of 0.5–2%.

The size of the capsules has an important influence on the viability of probiotics and the sensory impact on food. There is an opinion that beads in the size range of 2000–5000 μm provide an optimal balance between these two requirements [28]. In our study, the bead size falls within this optimal range—3401 μm with PUL and 2820 μm without (Table 1). Due to the presence of PUL, the viscosity of the internal phase of the primary emulsion can increase, causing resistance to break down into smaller droplets and leading to an increase in the size of capsules [11]. Nevertheless, the size and diameter of the capsules almost coincided, i.e., no statistically significant differences ($p > 0.05$) were found between these parameters.

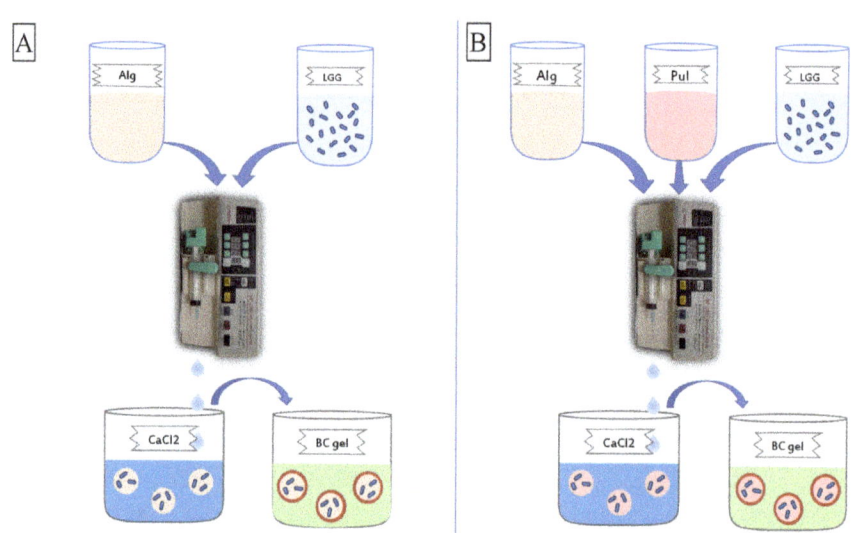

Figure 1. The scheme of beads with *L. rhamnosus* GG (LGG): (**A**) Alginate/Bacterial cellulose (Alg/BC), (**B**) Alginate-Pullulan/Bacterial cellulose (Alg-PUL/BC).

Table 1. Encapsulation efficiency (E_e), size, and mechanical strength of probiotic beads.

Type	Concentration, g/100 mL		E_e, %	Size, µm		Mechanical Strength, MPa
	PUL	BC		Dry	Wet	
Alg, 2%	-	-	78.8 ± 3.88	801 ± 55.1	2550 ± 127.1	26.6 ± 0.62
Alg, 2% + BC	-	0.5	81.2 ± 4.01	897 ± 60.1	2820 ± 143.0	28.8 ± 0.83 *
	-	1	80.2 ± 4.13	878 ± 59.9	2819 ± 140.1	26.9 ± 0.63
	-	2	77.2 ± 3.89	871 ± 71.4	2815 ± 168.9	24.9 ± 0.55
Alg, 2% + PUL/BC	1	0.5	88.3 ± 4.41 *	908 ± 61.1 *	3341 ± 233.4 *	36.8 ± 0.58 *
	1	1	87.6 ± 4.33 *	903 ± 72.7 *	3367 ± 167.1 *	35.2 ± 0.68 *
	1	2	87.1 ± 4.36 *	887 ± 97.3	3371 ± 235.9 *	35.8 ± 0.49 *
	2	0.5	89.1 ± 4.47 *	921 ± 61.0 *	3401 ± 204.0 *	37.1 ± 0.77 *
	2	1	87.8 ± 4.39 *	910 ± 85.4 *	3351 ± 134.4 *	34.1 ± 0.73 *
	2	2	87.6 ± 4.33 *	820 ± 61.3	3373 ± 168.1 *	34.6 ± 0.69 *

Alg—alginate, BC—bacterial cellulose, PUL—pullulan; * differences between alginate and hybrid beads were significant ($p < 0.05$).

The native beads obtained are translucent white spheres, and the dehydrated ones are "crumpled" irregularly shaped particles. In appearance, both types of capsules almost did not differ (Figure 2).

Figure 2. The appearance of beads: Alg (**A**), Alg/BC (**B**), Alg-PUL/BC (**C**), and dehydrated (**D**) beads.

Cellulose-coated Alg capsules have a higher mechanical strength. The improvement of compressive strength is associated with good interfacial interaction between cellulose and the matrix of Alg beads due to the structural similarity of these polysaccharides [29,30]. This contributes to the formation of multiple hydrogen bonds at the interface between BC and Alg, which leads to an increase in the strength of Alg capsules coated with cellulose [25].

However, capsules with PUL are even more rigid and strong. This may be due to the filler PUL reinforcing the hydrogen network and filling the voids [31,32]. Strength is a positive technological property, since fragile capsules are easily broken and destroyed, creating problems during handling, storage, and further processing.

The E_e is one of the most important parameters showing the effect of the encapsulation method, as well as the matrix of the core and the capsule wall on the "quantitative loading" of bacterial cells into it. The titer of cells in Alg/BC and Alg-PUL/BC capsules reaches 10^9 CFU/g. This is a fairly high indicator for capsules obtained by extrusion [33]. The high E_e obtained in our work indicates that the encapsulation process was adequate, and the wall materials were compatible with the probiotic strain.

Thus, the optimal composition of Alg-based capsules: 2% Alg + 2% PUL, 2% Alg + 0.5% BC, and 2% Alg + 2% PUL + 0.5% BC (Table 1).

The SEM study (Figure 3) demonstrated that bacteria was included in the matrix of the capsules (in the core).

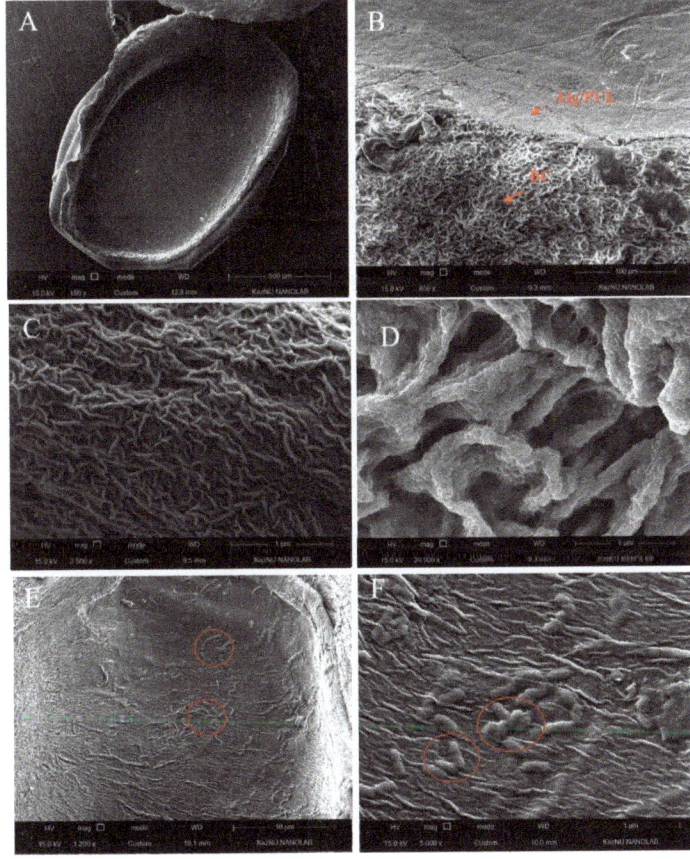

Figure 3. SEM images of Alg-PUL/BC beads: in cross section (**A**), in the shell section (**B**), bead shell (**C**), BC pores on the shell (**D**), LGG cells inside the bead (**E**,**F**).

The cellulose shell is also clearly visible. Moreover, it looks folded and dense and has a multi-layered structure covering the core. The formation of such a dense shell is associated with a unique three-dimensional BC network. The pore size is in a range that is insufficient for the release of encapsulated probiotics into the environment. Therefore, such a shell should be a strong framework to protect probiotics when passing through the digestive system. An experiment confirming this is presented in the next section.

3.2. Survival of Free and Encapsulated Bacteria in SGF and SDF

Protecting probiotic cells from exposure to a low pH gastric environment, bile salts, and hydrolytic enzymes is one of the primary objectives of encapsulation. Although the ultimate model for determining the functional effectiveness of capsules is a human organism, this "model" has ethical limitations. Therefore, in most such studies, an "artificial GI tract" system is used, simulating the physicochemical conditions of the main parts of the digestive system: stomach and the small and large intestines [11,12,31,34,35]. This is usually a buffer in which the pH value characteristic of a particular department is maintained and various digestive enzymes are added. In these departments, free and encapsulated cells are kept for a certain time, after which their number is determined.

To determine the effect of encapsulation, studies were conducted on the comparative survival of free and encapsulated LGG cells in SGF, SDF, and SCF, i.e., in the in vitro system [36]. In general, the design of this series of experiments is shown in Figure 4.

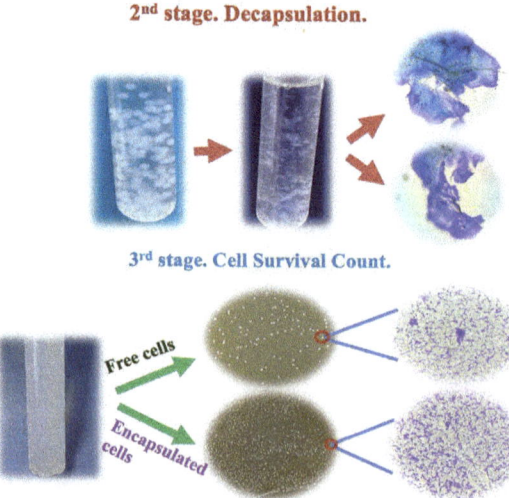

Figure 4. Experimental design of the simulating gastrointestinal (GI) tract.

In the 1st stage, the encapsulated cells were sequentially incubated for 2 h in SGF, SDF, and 18 h in SCF. The reason for this is to simulate the condition of the GI tract. To determine cell survival count after sequential incubation, LGG-loaded capsules were submitted to the decapsulation for cell survival count (2nd stage). According to some reports, capsules have been decapsulated using citrate and phosphate buffers [37]. In studies, where cellulose was a part of the encapsulation system, peptone water solution, phosphate buffer, and mixing were used to decapsulate beads [31,34,35]. In our research, capsules have not been able to disintegrate in these solutions. This could be due to the difference in the cellulose used. The fibrous structure of BC provides excellent mechanical properties [38], which probably made it impossible for capsules based on BC to disintegrate in citrate and phosphate solutions. Cellulose is mainly degraded by the cellulase enzyme [39]. Thus, in our study, to decapsulate the cellulose capsule shell, cellulase was used, followed by enumeration by pour plate counts in a nutrient medium (3rd stage).

The effect of capsule coating on LGG protection against SGF was studied by comparing the viability of free and coated cells over 1–2 h. They were incubated in SGF for 2 h because food is usually in the stomach for this period [40]. The viability of free and coated cells in SGF is shown in Figure 5.

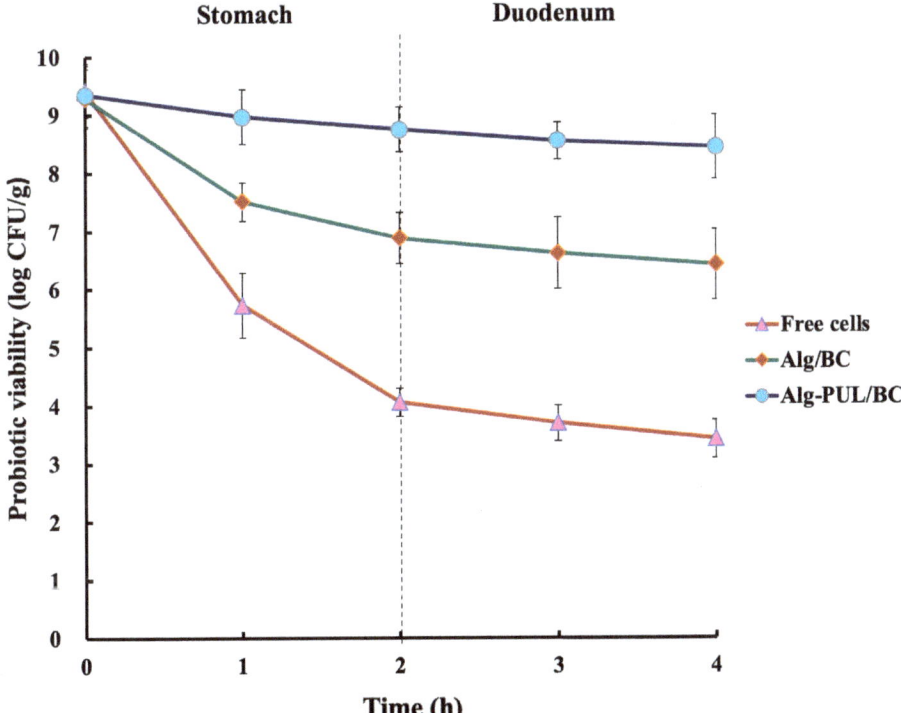

Figure 5. Survival of free and encapsulated LGG in upper GI conditions.

There was a noticeable decline in the viability of free probiotic cells in SGF at pH 2. The viable cell count of free cells dropped sharply in the first hour and continued to decline by 5.3 log units with a 43% survival rate. This is in agreement with other studies showing that LGG is acid sensitive [41–45].

Encapsulation of LGG into Alg/BC and Alg-PUL/BC beads offered more significant protection ($p < 0.05$). The viability of bacteria in beads based on BC exopolysaccharide was reduced by 2.41 log units, which provides a 74% survival rate. However, using PUL/BC as coating materials enables higher protection against low pH: 94% of cells in such capsules remained alive.

Studies where cellulose was a part of the encapsulation system demonstrated that the survivability of bacteria cells was around 83–91% [31,34]. A decrease in cell titer after incubation in SGF was also observed in capsules made of other polymers such as soy protein isolate, poly-L-lysine, or isomalto-oligosaccharide [45,46].

By comparing the loss of bacterial viability in the presence of BC and PUL/BC, it was found that the bacteria encapsulated with the prebiotic had a greater protective effect on bacteria against acidic conditions in SGF. This is in agreement with the findings reported by Çabuk et al. [11]. Their results demonstrated that probiotic cells were better protected in the presence of combined wall material PUL/whey protein after exposure to stomach conditions. PUL's ability to block the pores in the BC network led to the protective effect of the capsule. The large pores within a gel network led to a rapid release of encapsulated bioactive substances within the GI tract [47,48]. A similar effect was observed by Iyer et al. [49], who revealed that starch granules entrapped inside a Ca-alginate matrix might stop acid diffusion into the capsules.

The next stage of the GI tract, in which there are also bacteria-damaging factors (bile salts, hydrolytic enzymes of the pancreas), is the duodenum [3]. Capsules were extracted from SGF and transferred into SDF, followed by further incubation (Figure 2).

The viable cell counts for free cells dropped by 0.64 log CFU/g after 2 h incubation. According to other results, *L. rhamnosus* demonstrated strong bile resistance [50] and its viability decreased by approximately 0.5 log CFU/mL after 90 min of exposure to ox gall [51]. In contrast, Karu et al. [52] reported that the viability of LGG had reduced by about 4 and 5 log units after treatment with bile.

In the case of Alg/BC-coated capsules, the number of probiotic cells fell from 6.88 to 6.42, which was not statistically significant ($p > 0.05$). There was only a slight decrease in the viability of cells with Alg-PUL/BC coating by 0.31 log CFU/g after SDF incubation. The probiotic survival decreased proportionally by the time the cells were subjected to SGF and SDF solutions, which was in agreement with Morsy et al. [53], in which Alg 2% + anthocyanin 0.1% + whey protein 2% + PUL 2% + cocoa butter 1% were used as the encapsulation materials of LGG. Encapsulated bacteria survival level in Alg/BC and Alg-PUL/BC coatings after exposure to SDF for 2 h was 69% and 90%, respectively. In comparison, in a study by Afzaal et al. [34], where cellulose and chitosan were used as wall materials to encapsulate *L. plantarum*, the cell survivability reached 86% after exposure to bile conditions.

Alg-PUL/BC demonstrated effective protection against the damage of the bile salt solution. This can be explained by the reduced porosity and thicker structure that a double layer offers, which can prevent bile from entering the blended network [54]. According to Youssef et al. [55], the viability of *L. salivarius* encapsulated in Alg and coated with carboxymethyl cellulose was higher than the probiotics encapsulated in Alg alone under thermal treatment, storage, and simulated GI conditions. The combination of BC with prebiotic PUL not only increased the viability of probiotic bacteria but facilitated the formation of the integrated structure of beads. Capsules with larger size generally would not undergo the same rate of matrix degradation as the smaller capsules resulting in a longer release profile because they have a lower surface area to volume ratio [56,57]. According to these findings, Alg-PUL/BC beads retained more effectively under upper GI tract conditions. As a result, it was expected that the beads would reach the colon where they can exhibit their beneficial properties.

3.3. Release of LGG into SCF

In the traditional in vitro system simulating the colon conditions (pH 7.2 + electrolytes), cellulose-coated capsules were almost not destroyed, since, as has already been shown, cellulase treatment was needed for their destruction. It is well known that this enzyme is traditionally present in special sections of the GI tract of ruminants and other animals that feed exclusively on plant foods [58]. However, cellulolytic microorganisms have been found in human feces, in which all components of the microbial intestinal biofilm are present [59]. In this regard, it was hypothesized that cellulase should be present in the contents of the colon, namely in the enzyme fraction of feces: fecalase.

To detect cellulase activity, fecalase was placed in wells cut on agar with the substrate 1% carboxymethyl cellulose, which was then spilled with Congo red or Lugol solutions [60]. The presence of the enzyme was recorded by the appearance of hydrolysis zones, the diameter of which averaged 23–26 mm, which indicated significant cellulase activity (data not provided).

For the assay of the determination of encapsulated LGG release, beads were transferred into SCF with fecalase at pH 7.2 and incubated for a further 18 h. The results are shown in Figure 6.

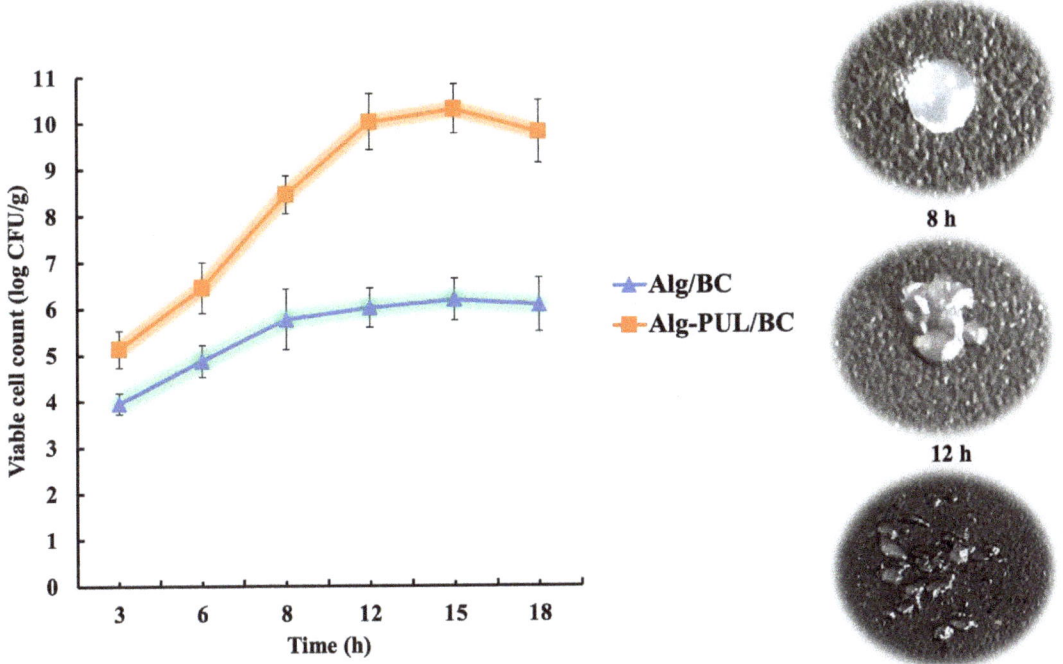

Figure 6. LGG release profile from beads into simulated colon fluid (SCF). All the samples were sequentially immersed in simulated gastric fluid (SGF) and simulated duodenum fluid (SDF) for 4 h before transferring into SCF.

Results indicated that cell counts of released bacteria increased with incubation exposure time. The surface of the capsules initially swells, then their porosity increases [61]. This leads to the release of bacteria from the core of capsules, the cellulose shell of which is gradually destroyed as the enzyme was exposed. Then, 60% (3.95 log CFU/g) of encapsulated probiotic bacteria from Alg/BC capsules were released after the first 3 h of incubation in SCF, and within 15 h this process was completed, reaching viable cell numbers of 6.18 log CFU/g. Then, the curve reached a plateau. The final number of viable cells released from the BC beads was 0.24 log CFU/g lower than the number after exposure to duodenum conditions (6.42 log CFU/g). The total loss of probiotic bacteria in Alg/BC beads by the end of the simulated GI conditions was 3.11 log CFU/g, while in the free state, only 10^3 CFU/mL cells reached the colon, i.e., the loss was 6 log units.

Under the same experimental conditions (SCF + fecalase), after 15 h, the highest number of viable bacteria released from Alg-PUL/BC capsules was recorded: 10.3 log CFU/g. It turned out that the number of cells in these capsules that reached the "terminal station" not only did not decrease but even increased by a log unit compared to the initial value (9.35 log CFU/g).

This phenomenon indicates that LGG in such a "container" not only survived under the harsh conditions of the human GI tract but also retained its ability to produce biomass inside the capsule. In its core, bacteria are distributed in the matrix of Alg and PUL, which is an effective prebiotic, i.e., a source of selective nutrition for this strain. Thus, coencapsulation of a probiotic with a suitable prebiotic and further coating with cellulose significantly increased the survival of bacteria compared to the delivery system without a prebiotic.

4. Conclusions

This study not only confirms that Alg-PUL/BC capsules can be used as a protective carrier for probiotic bacteria but also demonstrates the prebiotic effect of adding PUL, which leads to the reproduction of bacteria inside the capsule during intestinal transit. The intestinal microbiota contains microorganisms degrading cellulose. The combination of the peristaltic movement and the low amounts of cellulase presented in the lower GI tract is the result of bacteria release. In the presence of gut enzymes, the coating material was digested, resulting in the BC network's collapse. The degradation of the network led to the release of probiotic cells from the core. Delivery of the probiotic together with a nutrient source, providing its high cell number, will allow it to compete with the local microbiota, and, as a result, to colonize the colon, benefiting the host. Further research is needed on the selection of bacterial strains and carrier matrices, as well as the development of appropriate technologies that promote the survival of bacterial cells under other types of stress (heating, freezing, osmotic and oxygen stress, drying, and storage).

Author Contributions: Conceptualization, A.K., M.A. and I.S.; methodology, M.A., S.Z. and I.S.; software, A.K.; validation, A.K. and T.K.; formal analysis, A.K. and A.G.; investigation, A.K., M.A., I.S., S.Z., T.K., A.G. and Y.S.; resources, A.K.; data curation, Y.S.; writing—original draft preparation, A.K. and M.A.; writing—review and editing, M.A. and S.Z.; visualization, T.K., A.G. and Y.S.; supervision, A.K.; project administration, A.K.; funding acquisition, M.A. and I.S. All authors have read and agreed to the published version of the manuscript.

Funding: This work was supported by the grant AP09259491 "Biotechnology application in production of combined dairy products using polysaccharide matrix with probiotic biofilms" (2021–2023) funded by the Ministry of Education and Science of the Republic of Kazakhstan.

Institutional Review Board Statement: The study was conducted in accordance with the Declaration of Helsinki, and approved by the local ethical committee of Academy of Preventive Medicine and Kazakh Academy of Nutrition (No. 9; 7 November 2022).

Informed Consent Statement: Informed consent was obtained from all subjects involved in the study.

Data Availability Statement: Data are contained within the article.

Conflicts of Interest: The authors declare no conflict of interest.

References

1. Zabot, G.L.; Rodrigues, F.S.; Ody, L.P.; Tres, M.V.; Herrera, E.; Palacin, H.; Córdova-Ramos, J.S.; Best, I.; Olivera-Montenegro, L. Encapsulation of Bioactive Compounds for Food and Agricultural Applications. *Polymers* **2022**, *14*, 4194. [CrossRef]
2. Yao, M.; Xie, J.; Du, H.; McClements, D.J.; Xiao, H.; Li, L. Progress in microencapsulation of probiotics: A Review. *Compr. Rev. Food Sci. Food Saf.* **2020**, *19*, 857–874. [CrossRef]
3. Liu, W.J.; Chen, Y.F.; Kwok, L.Y.; Li, M.H.; Sun, T.; Sun, C.L.; Wang, X.N.; Dan, T.; Menghebilige; Zhang, H.P.; et al. Preliminary selection for potential probiotic *Bifidobacterium* isolated from subjects of different Chinese ethnic groups and evaluation of their fermentation and storage characteristics in bovine milk. *J. Dairy Sci.* **2013**, *96*, 6807–6817. [CrossRef]
4. Cook, M.T.; Tzortzis, G.; Charalampopoulos, D.; Khutoryanskiy, V.V. Microencapsulation of probiotics for gastrointestinal delivery. *J. Control. Release* **2012**, *162*, 56–67. [CrossRef]

5. Prakash, S.; Urbanska, A.M. Colon-targeted delivery of live bacterial cell biotherapeutics including microencapsulated live bacterial cells. *Biol. Targets Ther.* **2008**, *2*, 355–378. [CrossRef]
6. Mohd, G.K.; Vinod, G.; Chandel, H.S.; Asra, A.; Kasma, T. Development of Microencapsulation: A Review of Literature. *Int. J. Sci. Study* **2017**, *5*, 264–268. [CrossRef]
7. Kvakova, M.; Bertkova, I.; Stofilova, J.; Savidge, T.C. Co-Encapsulated Synbiotics and Immobilized Probiotics in Human Health and Gut Microbiota Modulation. *Foods* **2021**, *10*, 1297. [CrossRef]
8. Gibson, G.R.; Hutkins, R.; Sanders, M.E.; Prescott, S.L.; Reimer, R.A.; Salminen, S.J.; Scott, K.; Stanton, C.; Swanson, K.S.; Cani, P.D.; et al. Expert consensus document: The international scientific association for probiotics and prebiotics (ISAPP) consensus statement on the definition and scope of prebiotics. *Nat. Rev. Gastroenterol. Hepatol.* **2017**, *14*, 491–502. [CrossRef]
9. Capurso, L. Thirty Years of *Lactobacillus rhamnosus* GG. *J. Clin. Gastroenterol.* **2019**, *53*, S1–S41. [CrossRef]
10. Zhantlessova, S.; Savitskaya, I.; Kistaubayeva, A.; Ignatova, L.; Talipova, A.; Pogrebnjak, A.; Digel, I. Advanced "Green" Prebiotic Composite of Bacterial Cellulose/Pullulan Based on Synthetic Biology-Powered Microbial Coculture Strategy. *Polymers* **2022**, *14*, 3224. [CrossRef]
11. Çabuk, B.; Harsa, S.T. Protection of Lactobacillus acidophilus NRRL-B 4495 under in vitro gastrointestinal conditions with whey protein/pullulan microcapsules. *J. Biosci. Bioeng.* **2015**, *120*, 650–656. [CrossRef]
12. Zhang, M.; Cai, D.; Song, Q.; Wang, Y.; Sun, H.; Piao, C.; Yu, H.; Liu, J.; Liu, J.; Wang, Y. Effect on Viability of Microencapsulated *Lactobacillus rhamnosus* with the Whey Protein-pullulan Gels in Simulated Gastrointestinal Conditions and Properties of Gels. *Korean J. Food Sci. Anim. Resour.* **2019**, *39*, 459–473. [CrossRef]
13. Razavi, S.; Janfaza, S.; Tasnim, N.; Gibson, D.L.; Hoorfar, M. Microencapsulating Polymers for Probiotics Delivery Systems: Preparation, Characterization, and Applications. *Food Hydrocoll.* **2021**, *120*, 106882. [CrossRef]
14. Ferreira-Lazarte, A.; Gallego-Lobillo, P.; Moreno, F.J.; Villamiel, M.; Hernandez-Hernandez, O. In vitro digestibility of galactooligosaccharides: Effect of the structural features on their intestinal degradation. *J. Agric. Food Chem.* **2019**, *67*, 4662–4670. [CrossRef]
15. Russell, J.B.; Muck, R.E.; Weimer, P.J. Quantitative analysis of cellulose degradation and growth of cellulolytic bacteria in the rumen. *FEMS Microbiol. Ecol.* **2009**, *67*, 183–197. [CrossRef]
16. Maathuis, A.J.; Van den Heuvel, E.G.; Schoterman, M.H.; Venema, K. Galacto-oligosaccharides have prebiotic activity in a dynamic in vitro colon model using a ^{13}C-labeling technique. *J. Nutr.* **2012**, *142*, 1205–1212. [CrossRef]
17. Savitskaya, I.S.; Bondarenko, V.M. Suppression of mutagenic activity of intestinal metabolites in normobiocenosis. *Microbiol. J.* **2008**, *3*, 53–58. (In Russian)
18. Rodrigues, F.; Cedran, M.; Bicas, J.; Sato, H. Encapsulated probiotic cells: Relevant techniques, natural sources as encapsulating materials and food applications—A narrative review. *Food Res. Int.* **2020**, *137*, 109682. [CrossRef]
19. Rashidinejad, A.; Bahrami, A.; Rehman, A.; Rezaei, A.; Babazadeh, A.; Singh, H.; Jafari, S.M. Co-encapsulation of probiotics with prebiotics and their application in functional/synbiotic dairy products. *Crit. Rev. Food Sci. Nutr.* **2020**, *62*, 2470–2494. [CrossRef]
20. Lombardo, S.; Villares, A. Engineered multilayer microcapsules based on polysaccharides nanomaterials. *Molecules* **2020**, *25*, 4420. [CrossRef]
21. Tsujisaka, Y.; Mitsuhashi, M. Pullulan. In *Industrial Gums: Polysaccharides and Their Derivatives*; Academic Press: West Lafayette, IN, USA, 1993; pp. 447–460. ISBN 978-0-080-92654-4.
22. Cao, L.; Lu, W.; Mata, A.; Nishinari, K.; Fang, Y. Egg-box model-based gelation of alginate and pectin: A review. *Carbohydr. Polym.* **2020**, *242*, 116389. [CrossRef]
23. Mokarram, R.R.; Mortazavi, S.A.; Najafi, M.B.H.; Shahidi, F. The influence of multi stage alginate coating on survivability of potential probiotic bacteria in simulated gastric and intestinal juice. *Food Res. Int.* **2009**, *42*, 1040–1045. [CrossRef]
24. Dong, Q.Y.; Chen, M.Y.; Xin, Y.; Qin, X.Y.; Cheng, Z.; Shi, L.E. Alginate-based and protein-based materials for probiotics encapsulation: A review. *Int. J. Food Sci.* **2013**, *48*, 1339–1351. [CrossRef]
25. Zhang, S.; He, H.; Guan, S.; Cai, B.; Li, Q.; Rong, S. Bacterial cellulose-alginate composite beads as *Yarrowia lipolytica* cell carriers for lactone production. *Molecules* **2020**, *25*, 928. [CrossRef]
26. Jadhav, S.B.; Singhal, R.S. Pullulan-complexed α-amylase and glucosidase in alginate beads: Enhanced entrapment and stability. *Carbohydr. Polym.* **2014**, *105*, 49–56. [CrossRef]
27. Li, H.; Zhang, T.; Li, C.; Zheng, S.; Li, H.; Yu, J. Development of a microencapsulated synbiotic product and its application in yoghurt. *LWT* **2020**, *122*, 109033. [CrossRef]
28. Nag, A.; Han, K.S.; Singh, H. Microencapsulation of probiotic bacteria using pH-induced gelation of sodium caseinate and gellan gum. *Int. Dairy J.* **2011**, *21*, 247–253. [CrossRef]
29. Chang, P.R.; Jian, R.; Zheng, P.; Yu, J.; Ma, X. Preparation and properties of glycerol plasticized-starch (GPS)/cellulose nanoparticle (CN) composites. *Carbohydr. Polym.* **2010**, *79*, 301–305. [CrossRef]
30. Khan, A.; Khan, R.A.; Salmieri, S.; Le Tien, C.; Riedl, B.; Bouchard, J.; Lacroix, M. Mechanical and barrier properties of nanocrystalline cellulose reinforced chitosan based nanocomposite films. *Carbohydr. Polym.* **2012**, *90*, 1601–1608. [CrossRef]
31. Huq, T.; Fraschini, C.; Khan, A.; Riedl, B.; Bouchard, J.; Lacroix, M. Alginate Based Nanocomposite for Microencapsulation of Probiotic: Effect of Cellulose Nanocrystal (CNC) and Lecithin. *Carbohydr. Polym.* **2017**, *168*, 61–69. [CrossRef]

32. Pandey, S.; Shreshtha, I.; Sachan, S.G. Pullulan: Biosynthesis, Production and Applications. In *Microbial Exopolysaccharides as Novel and Significant Biomaterials*; Nadda, A.K., Sajna, K.V., Sharma, S., Eds.; Springer: Cham, Switzerland, 2021; pp. 121–141. ISBN 978-3-030-75289-7.
33. Fareez, I.M.; Lim, S.M.; Zulkefli, N.A.A.; Mishra, R.K.; Ramasamy, K. Cellulose derivatives enhanced stability of alginate-based beads loaded with *Lactobacillus plantarum LAB12* against low pH, high temperature and prolonged storage. *Probiotics Antimicrob. Proteins* **2018**, *10*, 543–557. [CrossRef]
34. Afzaal, M.; Saeed, F.; Ateeq, H.; Shah, Y.A.; Hussain, M.; Javed, A.; Ikram, A.; Raza, M.A.; Nayik, G.A.; Alfarraj, S.; et al. Effect of Cellulose–Chitosan Hybrid-Based Encapsulation on the Viability and Stability of Probiotics under Simulated Gastric Transit and in Kefir. *Biomimetics* **2022**, *7*, 109. [CrossRef]
35. Maleki, O.; Khaledabad, M.A.; Amiri, S.; Asl, A.K.; Makouie, S. Microencapsulation of *Lactobacillus rhamnosus* ATCC 7469 in whey protein isolate-crystalline nanocellulose-inulin composite enhanced gastrointestinal survivability. *LWT* **2020**, *126*, 109224. [CrossRef]
36. Voropaiev, M.; Nock, D. Onset of acid-neutralizing action of a calcium/magnesium carbonate-based antacid using an artificial stomach model: An in vitro evaluation. *BMC Gastroenterol.* **2021**, *21*, 112. [CrossRef]
37. Pupa, P.; Apiwatsiri, P.; Sirichokchatchawan, W.; Pirarat, N.; Muangsin, N.; Shah, A.; Prapasarakul, N. The efficacy of three double-microencapsulation methods for preservation of probiotic bacteria. *Sci. Rep.* **2021**, *11*, 13753. [CrossRef]
38. Potivara, K.; Phisalaphong, M. Development and Characterization of Bacterial Cellulose Reinforced with Natural Rubber. *Materials* **2019**, *12*, 2323. [CrossRef]
39. Brumm, P.J. Bacterial genomes: What they teach us about cellulose degradation. *Biofuels* **2013**, *4*, 669–681. [CrossRef]
40. Sensoy, I. A review on the food digestion in the digestive tract and the used in vitro models. *Cur. Res. Food Sci.* **2021**, *4*, 308–319. [CrossRef]
41. Corcoran, B.M.; Stanton, C.; Fitzgerald, G.F.; Ross, R.P. Survival of probiotic lactobacilli in acidic environments is enhanced in the presence of metabolizable sugars. *Appl. Environ. Microbiol.* **2005**, *71*, 3060–3067. [CrossRef]
42. Sohail, A.; Turner, M.S.; Coombes, A.; Bostrom, T.; Bhandari, B. Survivability of probiotics encapsulated in alginate gel microbeads using a novel impinging aerosols method. *Intl. J. Food Microbiol.* **2011**, *145*, 162–168. [CrossRef]
43. Qi, X.; Simsek, S.; Chen, B.; Rao, J. Alginate-based double-network hydrogel improves the viability of encapsulated probiotics during simulated sequential gastrointestinal digestion: Effect of biopolymer type and concentrations. *Int. J. Biol. Macromol.* **2020**, *165*, 1675–1685. [CrossRef]
44. Abbaszadeh, S.; Gandomi, H.; Misaghi, A.; Bokaei, S.; Noori, N. The effect of alginate and chitosan concentrations on some properties of chitosan-coated alginate beads and survivability of encapsulated *Lactobacillus rhamnosus* in simulated gastrointestinal conditions and during heat processing. *J. Sci. Food Agric.* **2014**, *94*, 2210–2216. [CrossRef] [PubMed]
45. Li, C.; Wang, C.L.; Sun, Y.; Li, A.L.; Liu, F.; Meng, X.C. Microencapsulation of *Lactobacillus rhamnosus* GG by Transglutaminase Cross-Linked Soy Protein Isolate to Improve Survival in Simulated Gastrointestinal Conditions and Yoghurt. *J. Food Sci.* **2016**, *81*, M1726–M1734. [CrossRef] [PubMed]
46. Siang, S.; Lai, K.; Kar, L.; Phing, L. Effect of added prebiotic (Isomalto-oligosaccharide) and Coating of Beads on the Survival of Microencapsulated *Lactobacillus rhamnosus* GG. *Food Sci. Technol.* **2019**, *39*, 601–609. [CrossRef]
47. Cordoba, A.L.; Deladino, L.; Martino, M. Effect of starch filler on calcium-alginate hydrogels loaded with yerba mate antioxidants. *Carbohydr. Polym.* **2013**, *95*, 315–323. [CrossRef]
48. Zhao, M.; Qu, F.; Wu, Z.; Nishinari, K.; Phillips, G.O.; Fang, Y. Protection mechanism of alginate microcapsules with different mechanical strength for *Lactobacillus plantarum* ST-III. *Food Hydrocoll.* **2017**, *66*, 396–402. [CrossRef]
49. Iyer, C.; Kasyphathy, K. Effect of Co-encapsulation of Probiotics with Prebiotics on Increasing the Viability of Encapsulated Bacteria under In Vitro Acidic and Bile Salt Conditions and in Yogurt. *J. Food Sci. Educ.* **2005**, *70*, 18–23. [CrossRef]
50. Succi, M.; Tremonte, P.; Reale, A.; Sorrentino, E.; Grazia, L.; Pacifico, S.; Coppola, R. Bile salt and acid tolerance of *Lactobacillus rhamnosus* strains isolated from Parmigiano Reggiano cheese. *FEMS Microbiol. Lett.* **2005**, *244*, 129–137. [CrossRef]
51. Kheadr, E. Impact of acid and ox gall on antibiotic susceptibility of probiotic Lactobacilli. *Afr. J. Agric. Res.* **2006**, *1*, 172–181.
52. Karu, R.; Sumeri, I. Survival of *Lactobacillus rhamnosus* GG during simulated gastrointestinal conditions depending on food matrix. *J. Food Res.* **2016**, *5*, 56. [CrossRef]
53. Morsy, M.; Morsy, O.; Abdelmonem, M.; Elsabagh, R. Anthocyanin-Colored Microencapsulation Effects on Survival Rate of *Lactobacillus rhamnosus* GG, Color Stability, and Sensory Parameters in Strawberry Nectar Model. *Food Bioprocess Technol.* **2022**, *15*, 352–367. [CrossRef]
54. Dafe, A.; Etemadi, H.; Zarredar, H.; Mahdavinia, G.R. Development of novel carboxymethyl cellulose/k-carrageenan blends as an enteric delivery vehicle for probiotic bacteria. *Int. J. Biol. Macromol.* **2017**, *97*, 299–307. [CrossRef]
55. Youssef, M.; Korin, A.; Zhan, F.; Hady, E.; Ahmed, H.Y.; Geng, F.; Chen, Y.; Li, B. Encapsulation of *Lactobacillus salivarius* in single and dual biopolymer. *J. Food Eng.* **2021**, *294*, 110398. [CrossRef]
56. Wang, J.; Korber, D.R.; Low, N.H.; Nickerson, M.T. Entrapment, survival and release of *Bifidobacterium adolescentis* within chickpea protein-based microcapsules. *Food Res. Int.* **2014**, *55*, 20–27. [CrossRef]
57. Klemmer, K.J.; Korber, D.R.; Low, N.H.; Nickerson, M.T. Pea protein-based capsules for probiotic and prebiotic delivery. *Int. J. Food Sci. Technol.* **2011**, *46*, 2248–2256. [CrossRef]

58. Tanimura, A.; Liu, W.; Yamada, K.; Kishida, T.; Toyohara, H. Animal cellulases with a focus on aquatic invertebrates. *Fish. Sci.* **2013**, *79*, 1–13. [CrossRef]
59. Piancone, E.; Fosso, B.; Marzano, M.; De Robertis, M.; Notario, E.; Oranger, A.; Manzari, C.; Bruno, S.; Visci, G.; Defazio, G.; et al. Natural and after colon washing fecal samples: The two sides of the coin for investigating the human gut microbiome. *Sci. Rep.* **2022**, *12*, 17909. [CrossRef]
60. Balla, A.; Silini, A.; Cherif-Silini, H.; Bouket, A.C.; Boudechicha, A.; Luptakova, L.; Alenezi, F.N.; Belbahri, L. Screening of Cellulolytic Bacteria from Various Ecosystems and Their Cellulases Production under Multi-Stress Conditions. *Catalysts* **2022**, *12*, 769. [CrossRef]
61. Omura, T.; Imagawa, K.; Kono, K.; Suzuki, T.; Minami, H. Encapsulation of Either Hydrophilic or Hydrophobic Substances in Spongy Cellulose Particles. *ACS Appl. Mater. Interfaces* **2017**, *9*, 944–949. [CrossRef]

Disclaimer/Publisher's Note: The statements, opinions and data contained in all publications are solely those of the individual author(s) and contributor(s) and not of MDPI and/or the editor(s). MDPI and/or the editor(s) disclaim responsibility for any injury to people or property resulting from any ideas, methods, instructions or products referred to in the content.

Article

In Vitro Comparison of Surface Roughness, Flexural, and Microtensile Strength of Various Glass-Ionomer-Based Materials and a New Alkasite Restorative Material

Alper Kaptan [1], Fatih Oznurhan [2],*, and Merve Candan [3]

1. Department of Restorative Dentistry, University of Cumhuriyet, Sivas 58140, Türkiye
2. Department of Pediatric Dentistry, University of Cumhuriyet, Sivas 58140, Türkiye
3. Department of Pediatric Dentistry, University of Osmangazi, Eskişehir 26040, Türkiye
* Correspondence: fatihozn@hotmail.com

Abstract: This study aims to evaluate the physical properties of Cention N and various glass-ionomer-based materials in vitro. The groups were obtained as follows: Group 1 (LC-Cent): light-cured Cention N; Group 2 (SC-Cent): self-cured Cention N; Group 3 (COMP): composite (3M Universal Restorative 200); Group 4 (DYRA): compomer (Dyract XP); Group 5 (LINER): Glass Liner; Group 6 (FUJI): FujiII LC Capsule; and Group 7 (NOVA): Nova Glass LC. For the microtensile bond strength (µTBS) test, 21 extracted human molar teeth were used. The enamel of the teeth was removed, and flat dentin surfaces were obtained. Materials were applied up to 3 mm, and sticks were obtained from the teeth. Additionally, specimens were prepared, and their flexural strength and surface roughness (Ra) were evaluated. Herein, data were recorded using SPSS 22.0, and the flexural strength, µTBS, and Ra were statistically analyzed. According to the surface roughness tests, the highest Ra values were observed in Group 6 (FUJI) (0.33 ± 0.1), whereas the lowest Ra values were observed in Group 2 (SC-Cent) (0.17 ± 0.04) ($p < 0.05$). The flexural strengths of the materials were compared, and the highest value was obtained in Group 2 (SC-Cent) (86.32 ± 15.37), whereas the lowest value was obtained in Group 5 (LINER) (41.75 ± 10.05) ($p < 0.05$). When the µTBS of materials to teeth was evaluated, the highest µTBS was observed in Group 3 (COMP) (16.50 ± 7.73) and Group 4 (DYRA) (16.36 ± 4.64), whereas the lowest µTBS was found in Group 7 (NOVA) (9.88 ± 1.87) ($p < 0.05$). According to the µTBS results of materials-to-materials bonding, both Group 2 (SC-Cent) and Group 1 (LC-Cent) made the best bonding with Group 3 (COMP) ($p < 0.05$). It can be concluded that self-cured Cention N had the highest flexural strength and lowest surface roughness of the seven materials tested. Although the bond strength was statistically lower than conventional composites and compomers, it was similar to resin-modified glass ionomer cements. Additionally, the best material-to-material bonding was found between self-cured Cention N and conventional composites.

Keywords: Cention N; alkasite restorative; self-cured restorative; resin-modified glass ionomer cement; microtensile bond strength

Citation: Kaptan, A.; Oznurhan, F.; Candan, M. In Vitro Comparison of Surface Roughness, Flexural, and Microtensile Strength of Various Glass-Ionomer-Based Materials and a New Alkasite Restorative Material. *Polymers* **2023**, *15*, 650. https://doi.org/10.3390/polym15030650

Academic Editors: Vineet Kumar and Xiaowu Tang

Received: 8 December 2022
Revised: 19 January 2023
Accepted: 23 January 2023
Published: 27 January 2023

Copyright: © 2023 by the authors. Licensee MDPI, Basel, Switzerland. This article is an open access article distributed under the terms and conditions of the Creative Commons Attribution (CC BY) license (https://creativecommons.org/licenses/by/4.0/).

1. Introduction

Masticatory forces, oral environment, oral habits, and physical properties of dental restorations play a crucial role in obtaining long-term restorations, and therefore, a healthy tooth structure. Dental amalgam, which has excellent physical properties, has been used as a dental restoration material for many years. However, it has many disadvantages, such as poor esthetics, leakage, postoperative sensitivity, plaque accumulation, tooth coloration, and difficulty in cavity preparation [1]. To overcome these disadvantages, various dental restorative materials are being developed. For example, composite resins are widely used by dental professionals. Nevertheless, they can cause many clinical problems, such as polymerization shrinkage, marginal leakage and discoloration, excessive surface loss, side effects due to monomer release, and bacterial adhesion [2,3].

Glass ionomer cements are also frequently preferred restorative materials in dentistry. Physicochemical bonding to enamel and dentin, biocompatibility with dental tissues, fluoride ion release properties, and low thermal expansion coefficients similar to tooth structure are its positive features. However, conventional glass ionomer cements also have negative properties, such as being affected by moisture, having an opaque color, low flexural strength, and high surface roughness. To overcome these negative properties, polymerizable resin-based glass ionomer materials have been developed. Thus, the setting time of glass ionomer cements is shortened, early moisture sensitivity is reduced, and better mechanical properties are exhibited [4].

Cention N, known as alkasite, was introduced as a new resin-based material containing alkacid fillers, such as fluoride, calcium, and hydroxide ions, to neutralize acids [5–7]. This material, which comprises powder and liquid phases, can be polymerized both by itself and by light after mixing the powder and liquid phases. The self-curing process is based on an initiator system consisting of a copper salt, a peroxide, and a thiocarbamide. The liquid part of Cention N contains hydroperoxide, and the standard filler in the powder part of the product is coated with the other initiator components. The copper salt accelerates the curing reaction. Additionally, Cention N contains the photoinitiator Ivocerin® and an acyl phosphine oxide initiator for optional light-curing, with a dental polymerization unit [8]. It is a restorative material based on urethane dimethacrylate (UDMA), which allows it to polymerize with light upon request, and it can also polymerize by itself. It is radio-opaque and contains alkaline glass fillers that can release fluoride, calcium, and hydroxide ions. Cention N is relatively more affordable and easier to use than the restorative materials available on the market [9]. Chole et al. [10] reported in their study that the flexural strength of Cention N is higher than that of light-cured composite resin and resin-modified glass ionomer. Due to the limited information available in the literature about Cention N more researches are needed on the use of this restoration material, particularly in primary teeth.

Physical properties are important factors for dental materials and also for restorative material selection in dental treatments. Since there are few studies about Cention N, this study aimed to compare the physical properties of Cention N with different glass-ionomer-based resin materials and a conventional composite. As Cention N is marketed as a posterior restorative, another aim of this study was to evaluate bonding strength when Cention N was used as a base material.

The null hypotheses tested were that (a) no difference exists between all restorative materials tested in terms of surface roughness, flexural strength, and microtensile bond strength, and that (b) no difference exists between the LC-CENT and SC-CENT bonding strength to other materials tested.

2. Materials and Methods

Ethical approval was obtained from the Clinical Research Ethics Committee of Sivas Cumhuriyet University (2020-01/42). Informed consent was obtained from all the patients for the collection of extracted teeth and their use in the in vitro study.

Our study consisted of 3 parts according to test methods (surface roughness test, flexural strength test, and microtensile bond strength (μTBS) test). The sample for each subgroup consisted of 10 specimens, and the power analysis revealed $p = 0.90145$ ($\alpha = 0.01$, $\beta = 0.10, 1 - \beta = 0.90$).

Seven experimental groups were obtained in this study as follows:
Group 1 (LC-Cent): light-cured Cention N (Ivoclar Vivadent, Schaan, Liechtenstein);
Group 2 (SC-Cent): self-cured Cention N (IvoclarVivadent, Schaan, Liechtenstein);
Group 3 (COMP): 3M Universal Restorative 200 (3M ESPE, St Paul, MN 55144, USA);
Group 4 (DYRA): Dyract XP (Dentsply, Konstanz, Germany);
Group 5 (LINER): Glass Liner (Willmann & Pein GmbH, Barmstedt, Germany);
Group 6 (FUJI): Fuji II LC Capsule (GC, Tokyo, Japan);
Group7 (NOVA): Nova Glass LC. (Imicryl, Konya, Türkiye)

The technical profiles and compositions of dental restorative materials are listed in Table 1.

Table 1. The technical profiles, codes, and compositions of the dental restorative materials.

Materials	Code	Composition	Manufacturer	LOT Number
Universal Restorative 200	COMP	Bis-GMA, Bis-EMA, UDMA, silica/zirconia, Filler 60% (volume)	3M ESPE, St Paul, MN 55144, USA	NA06972
Dyract XP	DYRA	UDMA, TCB resin, TEGDMA, trimethacrylate resin. 73 wt% Strontiumalumino-sodium-fluoro-phosphor-silicate	Dentsply, Konstanz, Germany	2003000720
Nova Glass LC	NOVA	*Powder:* Floro Alumino Silicate Glass, Pigments *Liquid:* Composite m resins 25–27% (Hema, dimethacrylates), Catalysts, Stabilisators	Imicryl, Konya, Türkiye	20031
Fuji II LC Capsule	FUJI	2-hydroxyethyl methacrylate, Polyacrylic acid, and water. 58 wt% Fluoro-aluminumsilicate	GC, Tokyo, Japan	1908281
Glass Liner	LINER	Glasionomerpulver, 1,6-Hexandioldimethacrylate, Bisphenol-A-bis (hydroxypropylmethacrylat), Isomere, 4-tert.-Butyl-N,N-dimethylaniline, Campherchinon	Willmann & Pein GmbH, Barmstedt, Germany	187576
Cention N	CENT	*Powder:* Barium aluminum silicate glass, ytterbium trifluoride, isofiller, calcium barium aluminum fluorosilicate glass, and calcium fluorosilicate glass *Liquid:* Urethane dimethacrylate, tricyclodecane dimethanol dimethacrylate, tetramethyl-xylylen diurethane dimethacrylate, polyethylene glycol 400 dimethacrylate, Ivocerin, and hydroxyperoxide	Ivoclar Vivadent, Schaan, Liechtenstein	Z0054T
AdperTM Easy One		2 HEMA, Bis-GMA, Methacrylated 85010 phosphoric esters, 1,6 hexaneddiol dimethacrylate, Methacrylate functionalized polyalkenoic acid (vitrebond copolymer), dispersed bonbed silica fillers 7 nm, ethanol, water, camphorquinone, stabilizers. pH = 2.4	3M ESPE, Seefeld, Germany	6744628

2.1. Surface Roughness Tests

Ten specimens were prepared for each group in a Teflon plastic mold (diameter and thickness of 8 and 2 mm, respectively). For each specimen, a plastic mold was placed on flat glass, and materials were applied by a single investigator (A.K.) according to the manufacturer's instructions. The excess material was removed using a mylar matrix strip. Except for the self-cured Cention N group, all the specimens were light-cured for 20 s with a power of 1200 mW/cm^2 using a second-generation LED device (Elipar S10 TM, 3M ESPE, St Paul, MN, USA) following the manufacturer's instructions. The light-curing device was positioned centrally 1 mm above the specimens. The light intensity was verified using a radiometer (Demetron LC, Kerr, Brea, CA, USA). The self-cured Cention N specimens were allowed to polymerize for 24 h in a dark chamber.

After the materials were polymerized, all the specimens were tested using a profilometer (Mitutoyo, Surftest SJ-301, Kawasaki, Japan). For each specimen, three measurements at randomly [11] different locations, with a cut-off length of 25 µm and 2 mm tracing length, and the average Ra values were recorded and analyzed using the SPSS program.

Scanning electron microscopy (SEM) images were taken from randomly chosen specimens from all groups. The specimens were coated with gold (Quorum Q150R ES, Quorum Technologies, Lewes, UK) and evaluated using SEM (Tescan MIRA3 XMU, Brno, Czech Republic). The entire sample surface was scanned and photographed at a magnification of 2000× with an accelerating voltage of 15 kV.

2.2. Flexural Strength

Ten specimens were prepared in a stainless-steel mold (2 × 2 × 25 mm^3) for each experimental group. Materials were applied by a single investigator (A.K.) according to the manufacturer's instructions and light-cured with an LED device (Elipar S10 TM, 3M ESPE, St Paul, MN, USA) as in the surface roughness test. Self-cured Cention N was removed from the mold after 24 h. The specimens were tested using a universal testing machine (LF Plus, LLOYD Instruments, Ametek, Inc., Bognor Regis, UK) with a crosshead speed of 0.5 mm/min, and the data were recorded using the SPSS program.

2.3. Microtensile Bond Strength (µTBS)

Twenty-one freshly extracted human third molars were used. The teeth were stored in saline solution at 4 °C and used within 1 month. All the root surfaces were cleaned to remove organic debris and deposits. One-third of the coronal teeth were removed using an Isomet low-speed diamond saw (Isomet, Buehler, Lake Bluff, IL, USA). A stereomicroscope was used to check the absence of enamel and pulp tissue on the resultant substrate. After grinding the residual occlusal enamel on wet #180 grit SiC paper, flat dentin surfaces were exposed. The exposed dentin surfaces were further polished with wet #600 grit SiC paper for 60 s to standardize the smear layer. Self-etch adhesive (AdperTM Easy One, 3M ESPE, Seefeld, Germany) was applied according to the manufacturer's instructions.

The prepared tooth samples were divided into seven groups randomly. Materials were applied by a single investigator (A.K.) according to the manufacturer's instructions. Dental restorative materials (up to 3 mm) were applied to the teeth with the help of a stainless-steel mold, and to standardize the application pressure, a 2 mm thick and 420 g circular glass slice was placed on top of the composite applied samples [12]. Afterward, in Group 1, 3–7 samples were light-cured immediately, and Group 2 samples waited in a dark chamber for self-polymerization. Polymerized samples were embedded in acrylic blocks. Samples of µTBS tests were prepared following the ISO/TS 11405:2015 guideline. Each tooth was sectioned in the x and y directions with a slow speed saw under water cooling, and 10 square-shaped (1 × 1 mm^2) sticks were obtained for each group (n = 10). Failed samples while obtaining the specimens before the µTBS test were not included in this study. The distribution of the failed samples per each group was as follows: Group 1=5, Group 2=4, Group 3=6, Group 4=5 Group 5=5, Group 6=6, and Group 7=4. A total of 10 obtained sticks per each group and a total of 70 sticks for 7 groups were stored in distilled water for 24 h and then fixed to a microtensile device with cyanoacrylate adhesive plus an accelerator (404 Super Cyanoplast, 404 Kimya, İstanbul, Turkey). The specimens were stressed under tension until failure using a microtensile testing machine (LF Plus, LLOYD Instruments, Ametek Inc., Bognor Regis, UK) with a crosshead speed of 0.5 mm/min; µTBS values were calculated and expressed in megapascals (MPa).

Additionally, microtensile tests were conducted to see the relationship with other materials in the use of Cention N as a base material. In this step, 4 × 4 × 4 mm^3 cubes were obtained, and all materials (from Groups 3 to 7) were restored on Group 1 and Group 2 samples with 4 × 4 × 4 mm³ dimensions. After obtaining 4 × 4 × 8 mm³ specimens, they were embedded in acrylic blocks. The specimens were sectioned perpendicularly to the bonding surface in the x and y directions with a low-speed saw under water cooling, and 10 sticks (1 × 1 × 8 mm³) were obtained for each group. The sticks were stressed in tension until failure using a microtensile testing machine.

2.4. Statistical Analysis

The data were processed using SPSS for Windows (version 22.0; SPSS Inc., Chicago, IL, USA). The mean and standard deviation of the flexural and µTBS and surface roughness was calculated for each group. Kolmogorov–Smirnov and Shapiro–Wilk tests were used to investigate the normality of data. Since the distribution of the data was normal, it was decided to use parametric tests. The physical properties of the dental restorations (flexural and microtensile bond strength and surface roughness) were analyzed using one-way ANOVA, and multiple comparisons were performed using Tukey's post hoc test. Statistical significance was set at $p < 0.05$.

3. Results

When the Ra values were compared, the highest value was observed in Group 6, whereas the lowest value was observed in Group 2. A statistically significant difference was observed between the Ra values of the materials ($p \leq 0.05$). The main Ra values and double comparisons of the materials are listed in Table 2. In addition, SEM photographs obtained at 2000× magnification from the surfaces of the materials are shown in Figure 1.

Table 2. Surface roughness and flexural strength values of the dental restorative materials.

Materials	Mean ± Standard Deviation (µm)	Mean ± Standard Deviation (MPa)
Group 1 (LC-CENT)	0.27 ± 0.04 [a]	58.17 ± 8.38 [a]
Group 2 (SC-CENT)	0.17 ± 0.04 [b]	86.32 ± 15.37 [b]
Group 3 (COMP)	0.18 ± 0.06 [b]	83.78 ± 16.65 [b]
Group 4 (DYRA)	0.22 ± 0.1 [b]	61.21 ± 9.82 [a]
Group 5 (LINER)	0.18 ± 0.07 [b]	41.75 ± 10.05 [c]
Group 6 (FUJI)	0.33 ± 0.1 [c]	48.17 ± 6.24 [d]
Group 7 (NOVA)	0.26 ± 0.07 [a]	57.83 ± 22.98 [a,d]

In each column, groups with the different lowercase superscripts are significantly different ($p < 0.05$).

When the flexural strength values of the materials were compared, the highest value was observed in Group 2, whereas the lowest value was observed in Group 5. A statistically significant difference was observed between the flexural strength values of the materials ($p \leq 0.05$). The main flexural strength values and double comparisons of the materials are listed in Table 2.

A statistically significant difference was found between the µTBS values of the materials to dentin bonding ($p \leq 0.05$). When µTBS test values were compared, the highest values were observed in Group 3 and Group 4, while the lowest was found in Group 7. The mean µTBS test values of the materials and their comparisons are presented in Table 3.

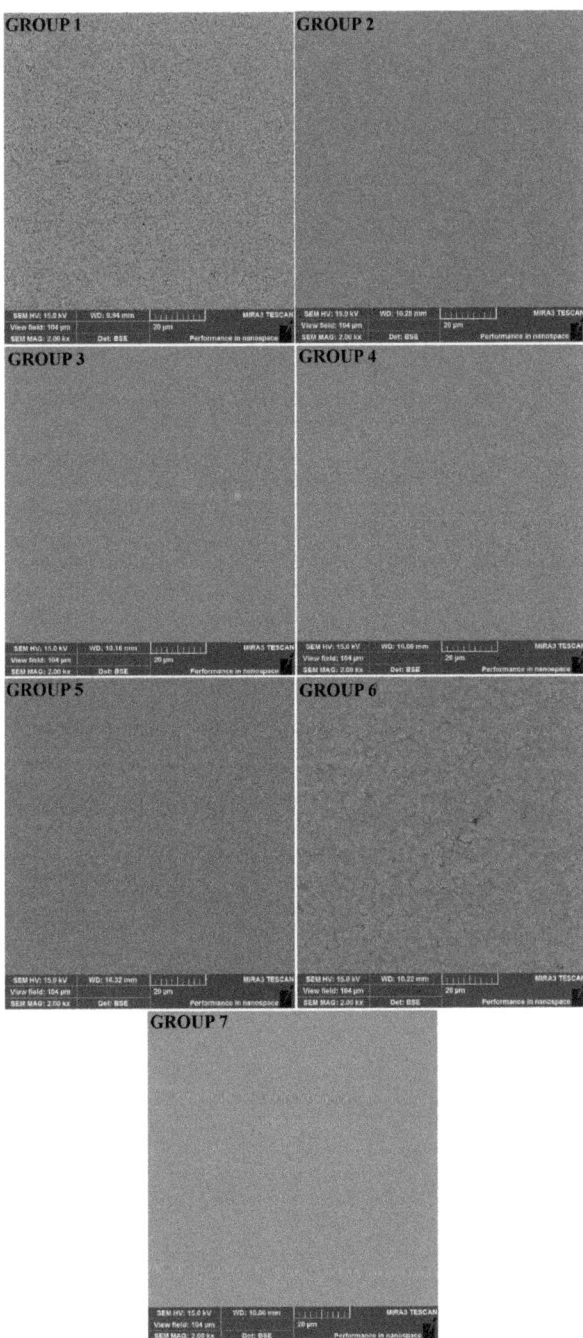

Figure 1. The SEM photographs obtained at 2000× magnification from the surfaces of the dental restorative materials. Group 1 (LC-Cent), Group 2 (SC-Cent), Group 3 (COMP), Group 4 (DYRA), Group 5 (LINER), Group 6 (FUJI), and Group 7 (NOVA).

Table 3. The microtensile bond strength (μTBS) values for each dental restorative material with teeth.

Materials	Mean ± Standard Deviation (MPa)	
Group 1 (LC-CENT)	13.25 ± 4.4 [B]	
Group 2 (SC-CENT)	12.50 ± 5.05 [B]	
Group 3 (COMP)	16.50 ± 7.73 [A]	F = 2.873
Group 4 (DYRA)	16.36 ± 4.64 [A]	p = 0.015 *
Group 5 (LINER)	11.98 ± 5.01 [B]	
Group 6 (FUJI)	11.17 ± 3.0 [B]	
Group 7 (NOVA)	9.88 ± 1.87 [C]	

The different uppercase letters represent the difference in the columns. * $p < 0.05$ was accepted as the significance level.

When comparing the bond strength of other restorative materials to Group 1 (LC-CENT), Group 3 (COMP) bonding to LC-CENT showed the highest μTBS values, and Group 6 (FUJI) and Group 7 (NOVA) bonding to LC-CENT showed the lowest μTBS values ($p < 0.05$). No statistically significant difference was found between the μTBS values to LC-CENT material between Group 5 (LINER) and Group 4 (DYRA).

The μTBS values between the SC-CENT material and all other materials were found to be higher than those between the LC-CENT material and all the other materials. When comparing the bond strength of other restorative materials to SC-CENT, the highest μTBS values were found in Group 3 (COMP) bonding to SC-CENT, while Group 6 (FUJI) bonding to SC-CENT showed the lowest μTBS values ($p < 0.05$). There was no statistically significant difference between Group 4 (DYRA), Group 5 (LINER), Group 6 (FUJI), and Group 7 (NOVA) to SC-CENT bonding ($p > 0.05$). The main μTBS values between the dental restorative materials and self and light-cured Cention N materials, and their double comparisons are shown in Table 4.

Table 4. Microtensile bond strength (μTBS) values of dental restorative materials bonding to light and self-cured Cention N specimens.

Materials	Group 1 (LC-CENT)	Group 2 (SC-CENT)	p Values
	Mean ± Standard Deviation (MPa)		
Group 3 (COMP)	20.00 ± 3.09 [a]	23.69 ± 6.68 [a]	p = 0.131
Group 4 (DYRA)	17.23 ± 2.53 [a,b]	20.09 ± 5.59 [a,b]	p = 0.156
Group 5 (LINER)	15.24 ± 2.20 [b,c]	17.39 ± 2.94 [a,b]	p = 0.081
Group 6 (FUJI)	11.52 ± 2.65 [c]	16.26 ± 5.09 [b]	p = 0.018 *
Group 7 (NOVA)	12.79 ± 4.99 [c]	21.72 ± 3.95 [a,b]	p = 0.001 *

* In each column, groups with the different lowercase superscripts are significantly different ($p < 0.05$).

4. Discussion

Both null hypotheses were rejected. The surface roughness, flexural strength, and μTBS values of the tested materials were found to be different. In addition, the bond strength of LC-CENT and SC-CENT to other tested restorative materials was found to be different.

4.1. Surface Roughness

Surface roughness is an important feature affecting biofilm formation in dental materials. Bacterial adhesion occurs over time with the formation of biofilms on dental surfaces. The number of microorganisms adhering to the restorative material depends on different factors, such as surface roughness, the hydrophobicity of the material surface, matrix type, electrostatic forces, material composition, filler size, and the configuration of fillers [13].

The surface roughness of dental restorative materials can be affected by internal factors, such as differences in the size, shape, volume, and distribution of inorganic fillers, as well as external factors, such as medications and liquids to which the materials are exposed [14–16]. As the filler size of the materials increases, the vertical surface roughness increases [17]. In addition, the inadequate polymerization of the materials may affect the average surface roughness values of dental restorative materials. In a previous study, the surface roughness value of the self-cured form of the glass ionomer material was lower than that of the light-cured form [18,19]. However, there was no statistically significant difference between the groups polymerized using different methods. Similarly, in the present study, the self-cured form of Cention N was smoother than the light-cured form [13].

Various methods can be used to polish the surfaces of dental restorative materials. The smoothest surface can be obtained using mylar strip bands [20,21]; therefore, in the present study, mylar strips were used to obtain a flat surface [22]. Setty et al. [23] compared composites' and Cention's surface roughnesses and found that the composites showed better results than light-cured cement. According to the results of this study, self-curing Cention and composites showed lower values, but light-cured Cention showed higher values than those reported by Setty et al. [23] In addition, acceptable Ra values should be less than 0.2 µm [24], and both self-cured Cention and composite showed this result. Further, self-cured Cention showed lower Ra values than light-cured Cention, probably because self-cured Centions are being cured slowly and for a long time period.

4.2. Flexural Strength

Flexural strength tests are important because they measure occlusal forces in the oral cavity. In this study, the flexural strengths of composite resin and self-cured Cention were 83.78 and 86.32 MPa, respectively. According to the International Organization for Standardization (ISO) 4049 standards, these are acceptable values [3,7,10,24]

Mishra et al. [5] tested the flexural strength of a composite, Cention N, GIC, and amalgam, and found that the best results were obtained for the composite. Kiran et al. [25] tested the mechanical properties of Cention N and type IX GIC, and Cention N showed higher flexural strength than type IX GIC. They suggested that this result was due to the filler content and monomers used in the materials. Sadananda et al. [7] compared the flexural strengths of Cention N, Fuji IX, Ketac-Molar, and Zirconomer and found that Cention N showed the highest flexural strength, whereas Fuji IX showed the lowest flexural strength. Panpisut et al. [6] tested two resin-modified GICs, Cention N, and a composite, and found that the composite displayed the highest flexural strength in the composite group, followed by that of Cention N, Fuji II, and Riva LC. The results were similar to those of the present study, and Cention N displayed higher results than Fuji II and other GICs. Chole et al. [10] tested the flexural strength of Cention N, a bulk-fill composite, nanocomposite, and resin-modified GIC. In their study, Cention N showed the highest flexural strength, whereas resin-modified GIC showed the lowest flexural strength, which is in accordance with the results of the proposed study. The authors attributed this highest flexural strength to the material's content and UDMA, and we agree with these authors [5–7,10,25]. Light-cured Cention N showed statistically lower results than self-cured Cention N, probably due to the light-cured Cention N's fast curing with LEDs.

4.3. Microtensile Bond Strength

µTBS has been used in several studies and is one of the most standardized and versatile bond strength tests. In this study, the materials were tested in two parts. One is to test bonding materials to teeth, and the other is bonding materials to another restorative material to test if Cention N could be used as a base material.

In this study, the composite (16.5 MPa) and compomer (16.36 MPa) materials showed higher µTBS values than the self-cured Cention N (12.5 MPa) and light-cured (13.25 MPa) Cention N; however, Cention N showed higher values than the other GIC groups. Yao et al. [26] tested the µTBS of Cention N, a bulk-fill composite, and Fuji II LC, and found the highest scores with Cention N, and the results were similar to those of this study. Both Cention N groups showed higher results than Fuji II LC. Naz et al. [27] tested the shear bond strength of composite and Cention N, and the highest results were shown by Cention N. Similarly, Eligeti et al. [28,29] reported that Cention N restorative material showed better bonding potential to dentin compared to the bonding potentials of resin-modified GIC, Zirconomer-enhanced, and Ketac-Molar. In microtensile or shear bond strength tests, many factors affect the results, such as tooth age, depth of dentine, dentine tubules, moisture, materials, and test conditions. In this study, Cention N showed lower µTBS values than the composite and compomer when compared to those obtained in other studies, and the possible explanation for this could be the conditions listed above. Cention N showed higher µTBS values than Group 5, Group 6, and Group 7, suggesting that Cention N has a high polymer network density and good bonding ability to dentin. This may also be explained by the fact that alkasites do not contain Bis-GMA, HEMA, or TEGDMA. UDMA was the main component of the monomer sequence. A combination of UDMA, DCP, aromatic and aliphatic UDMA, and PEG–400 DMA provides enhanced mechanical properties and good long-term stability during polymerization. PEG–400 DMA is a monomer liquid that increases the fluidity of the material, and its hydrophilic character supports the ability of the material to wet the substrate (enamel and dentin) and adapt to the smear layer [30].

To the best of our knowledge, this is the first study to use Cention N as a base material. Self-cured Cention N showed higher results than light-cured Cention N when bonded with other materials, and it was significantly different when bonded with Nova and Fuji II LC. This can primarily be attributed to the fact that while Cention N is self-curing, it provides stronger bond strengths.

Butera et al. [31] showed calcium and phosphorus ions' deposition on the surfaces of bulk-filled polymeric composite resins in the oral environment after one month of daily oral hygiene application with a toothpaste containing microRepair® (Zn-carbonate hydroxyapatite). Since Cention N is an ion-releasing restorative material, it may be a future goal to investigate whether the use of biomimetic hydroxyapatite will be effective in increasing ion deposition in this restorative material and reducing the incidence of secondary caries.

This study has some limitations. Firstly, only one material was used in the self-cure mode, which may have the potential effect to decrease the generalizability of findings. Further studies should examine the effect of different types of self-cured restorative materials. Another limitation was that this study was carried out under in vitro conditions. However, in the clinical use of this restorative material, the presence of saliva in the oral environment, visibility problems, transport of the material into the cavity, variable mixing rates, difficulties in the proximal cavities, and the need to complete the restoration in a short time may reduce the performance of this restorative material in oral conditions. In this context, there is a need for long-term clinical studies to evaluate the clinical performance of this material. Within the limitations of this study, Cention N showed superior results in all three tests when compared to conventional and resin-modified glass ionomer cements in in vitro conditions. Additionally, self-cured Cention N showed enhanced results when compared to those shown by light-cured Cention N.

5. Conclusions

Although self-cured Cention N had promising results with the highest flexural strength and lowest surface roughness of the seven materials tested, bond strength to dentin values were statistically lower than conventional composites and compomers, displaying similar strengths to resin-modified glass ionomer cements. These findings make the use of this material as a permanent restorative material questionable. However, considering that the best material–material bond in the present study is between self-cured Cention N and conventional composites, and when the ion-release (F^-, OH^- and Ca^{2+}) feature to induce the incidence of secondary caries of this material is taken into account, it can be considered that Cention N may be more suitable for use as a base material under composite materials. Further in vivo and in vitro studies are necessary to validate these findings. Because Cention N is a newly developed material, the present study is one of the few studies that investigate the mechanical properties of this material.

Author Contributions: Conceptualization, F.O. and A.K.; methodology, F.O.; software, M.C.; validation, A.K., M.C. and F.O.; formal analysis, A.K.; investigation, M.C.; resources, A.K.; data curation, M.C.; writing—original draft preparation, A.K.; writing—review and editing, A.K.; visualization, F.O.; supervision, A.K.; project administration, F.O.; funding acquisition, F.O. All authors have read and agreed to the published version of the manuscript.

Funding: This study was supported by the Scientific Research Project Fund Cumhuriyet University (Grant number DIS-243).

Institutional Review Board Statement: Clinical Research Ethics Committee of Sivas Cumhuriyet University (2020-01/42).

Informed Consent Statement: Informed consent was obtained from all the patients for the collection of extracted teeth and their use in the in vitro study.

Data Availability Statement: The data presented in this study are available within this article.

Conflicts of Interest: The authors declare no conflict of interest.

References

1. Kumar, S.A.; Ajitha, P. Evaluation of compressive strength between Cention N and high copper amalgam-An in vitro study. *Drug Invent. Today* **2019**, *12*, 255–257.
2. Bezerra, I.M.; Brito, A.C.M.; de Sousa, S.A.; Santiago, B.M.; Cavalcanti, Y.W.; de Almeida, L.F.D. Glass ionomer cements compared with composite resin in restoration of noncarious cervical lesions: A systematic review and meta-analysis. *Heliyon* **2020**, *6*, e03969. [CrossRef] [PubMed]
3. Zimmerli, B.; Strub, M.; Jeger, F.; Stadler, O.; Lussi, A. Composite materials: Composition, properties and clinical applications. A literature review. *Schweiz. Mon. Zahnmed. Rev. Mens. Suisse D'odonto-Stomatol. Riv. Mens. Svizz. Odontol. Stomatol.* **2010**, *120*, 972–986.
4. Hiremath, G.; Horati, P.; Naik, B. Evaluation and comparison of flexural strength of Cention N with resin-modified glass-ionomer cement and composite-An in vitro study. *J. Conserv. Dent.* **2022**, *25*, 288–291. [CrossRef] [PubMed]
5. Mishra, A.; Singh, G.; Singh, S.; Agarwal, M.; Qureshi, R.; Khurana, N. Comparative Evaluation of Mechanical Properties of Cention N with Conventionally used Restorative Materials—An In Vitro Study. *Int. J. Prosthodont. Restor. Dent.* **2018**, *8*, 120–124. [CrossRef]
6. Panpisut, P.; Toneluck, A. Monomer conversion, dimensional stability, biaxial flexural strength, and fluoride release of resin-based restorative material containing alkaline fillers. *Dent. Mater. J.* **2020**, *39*, 608–615. [CrossRef] [PubMed]
7. Sadananda, V.; Shetty, C.; Hegde, M.; Bhat, G. Alkasite restorative material: Flexural and compressive strength evaluation. *Res. J. Pharm. Biol. Chem. Sci.* **2018**, *9*, 2179.
8. Scientific Documentation of Cention N, Ivoclar Vivadent AG. Available online: https://downloadcenter.ivoclar.com/#search-text=cention&details=23015 (accessed on 26 January 2023).
9. Iftikhar, N.; Devashish, B.S.; Gupta, N.; Natasha Ghambir, R.-S. A Comparative Evaluation of Mechanical Properties of Four Different Restorative Materials: An In Vitro Study. *Int. J. Clin. Pediatr. Dent.* **2019**, *12*, 47. [CrossRef]
10. Chole, D.; Shah, H.; Kundoor, S.; Bakle, S.; Gandhi, N.; Hatte, N. In Vitro Comparision of Flexural Strength of Cention-N, BulkFill Composites, Light-Cure Nanocomposites And Resin-Modified Glass Ionomer Cement. *IOSR J. Dent. Med. Sci. (IOSR-JDMS)* **2018**, *17*, 79–82.
11. Loomans, B.A.; Cardoso, M.V.; Opdam, N.J.; Roeters, F.J.; De Munck, J.; Huysmans, M.C.; Van Meerbeek, B. Surface roughness of etched composite resin in light of composite repair. *J. Dent.* **2011**, *39*, 499–505. [CrossRef] [PubMed]
12. Scholz, K.J.; Bittner, A.; Cieplik, F.; Hiller, K.A.; Schmalz, G.; Buchalla, W.; Federlin, M. Micromorphology of the Adhesive Interface of Self-Adhesive Resin Cements to Enamel and Dentin. *Materials* **2021**, *14*, 492. [CrossRef] [PubMed]
13. Cazzaniga, G.; Ottobelli, M.; Ionescu, A.; Garcia-Godoy, F.; Brambilla, E. Surface properties of resin-based composite materials and biofilm formation: A review of the current literature. *Am. J. Dent.* **2015**, *28*, 311–320. [PubMed]
14. Candan, M.; Ünal, M. The effect of various asthma medications on surface roughness of pediatric dental restorative materials: An atomic force microscopy and scanning electron microscopy study. *Microsc. Res. Tech.* **2021**, *84*, 271–283. [CrossRef] [PubMed]
15. Gladys, S.; Van Meerbeek, B.; Braem, M.; Lambrechts, P.; Vanherle, G. Comparative physico-mechanical characterization of new hybrid restorative materials with conventional glass-ionomer and resin composite restorative materials. *J. Dent. Res.* **1997**, *76*, 883–894. [CrossRef] [PubMed]

16. Ünal, M.; Candan, M.; İpek, İ.; Küçükoflaz, M.; Özer, A. Evaluation of the microhardness of different resin-based dental restorative materials treated with gastric acid: Scanning electron microscopy-energy dispersive X-ray spectroscopy analysis. *Microsc. Res. Tech.* **2021**, *84*, 2140–2148. [CrossRef] [PubMed]
17. Marghalani, H.Y. Effect of filler particles on surface roughness of experimental composite series. *J. Appl. Oral Sci. Rev. FOB* **2010**, *18*, 59–67. [CrossRef]
18. de Sousa-Lima, R.X.; de Lima, J.F.M.; Silva de Azevedo, L.J.; de Freitas Chaves, L.V.; Alonso, R.C.B.; Borges, B.C.D. Surface morphological and physical characterizations of glass ionomer cements after sterilization processes. *Microsc. Res. Tech.* **2018**, *81*, 1208–1213. [CrossRef]
19. Maktabi, H.; Ibrahim, M.; Alkhubaizi, Q.; Weir, M.; Xu, H.; Strassler, H.; Fugolin, A.P.P.; Pfeifer, C.S.; Melo, M.A.S. Underperforming light curing procedures trigger detrimental irradiance-dependent biofilm response on incrementally placed dental composites. *J. Dent.* **2019**, *88*, 103110. [CrossRef]
20. Alfawaz, Y. Impact of Polishing Systems on the Surface Roughness and Microhardness of Nanocomposites. *J. Contemp. Dent. Pract.* **2017**, *18*, 647–651. [CrossRef]
21. Kaminedi, R.; Penumatsa, N.; Priya, T.; Baroudi, K. The influence of finishing/polishing time and cooling system on surface roughness and microhardness of two different types of composite resin restorations. *J. Int. Soc. Prev. Community Dent.* **2014**, *4*, S99–S104. [CrossRef]
22. Gumustas, B.; Sismanoglu, S. Effectiveness of different resin composite materials for repairing noncarious amalgam margin defects. *J. Conserv. Dent. JCD* **2018**, *21*, 627–631. [CrossRef] [PubMed]
23. Setty, A.; Nagesh, J.; Marigowda, J.; Shivanna, A.; Paluvary, S.; Ashwathappa, G. Comparative evaluation of surface roughness of novel resin composite Cention N with Filtek Z350 XT: In vitro study. *Int. J. Oral Care Res.* **2019**, *7*, 15. [CrossRef]
24. Bollen, C.M.; Lambrechts, P.; Quirynen, M. Comparison of surface roughness of oral hard materials to the threshold surface roughness for bacterial plaque retention: A review of the literature. *Dent. Mater. Off. Publ. Acad. Dent. Mater.* **1997**, *13*, 258–269. [CrossRef]
25. Kiran, N.K.; Chowdhary, N.; John, D.; Reddy, R.; Shidhara, A.; Pavana, M.P. Comparative Evaluation of Mechanical Properties of Cention-N And Type IX GIC-An In Vitro Study. *Int. J. Curr. Adv. Res.* **2019**, *8*, 20498–20501. [CrossRef]
26. Yao, C.; Ahmed, M.H.; Zhang, F.; Mercelis, B.; Van Landuyt, K.L.; Huang, C.; Van Meerbeek, B. Structural/Chemical Characterization and Bond Strength of a New Self-Adhesive Bulk-fill Restorative. *J. Adhes. Dent.* **2020**, *22*, 85–97. [CrossRef] [PubMed]
27. Naz, F.; Khan, A.; Kader, M.; Gelban, L.; Mousa, N.; Hakeem, A. Comparative evaluation of mechanical and physical properties of a new bulk-fill alkasite with conventional restorative materials. *Saudi Dent. J.* **2020**, *33*, 666–673. [CrossRef] [PubMed]
28. Eligetti, T.; Dola, B.; Kamishetty, S.; Gaddala, N.; Swetha, A.; Bandari, J. Comparative Evaluation of Shear Bond Strength of Cention N with Other Aesthetic Restorative Materials to Dentin: An in Vitro Study. *Ann. Rom. Soc. Cell Biol.* **2021**, *25*, 12707–12714.
29. Oznurhan, F.; Olmez, A. Morphological analysis of the resin-dentin interface in cavities prepared with Er,Cr:YSGG laser or bur in primary teeth. *Photomed. Laser Surg.* **2013**, *31*, 386–391. [CrossRef] [PubMed]
30. Valencia, J.; Felix, V. Alkasites, a New Alternative to Amalgam. Report of a Clinical Case. *Acta Sci. Dent. Sci.* **2019**, *3*, 11–19. [CrossRef]
31. Butera, A.; Pascadopoli, M.; Gallo, S.; Lelli, M.; Tarterini, F.; Giglia, F.; Scribante, A. SEM/EDS Evaluation of the Mineral Deposition on a Polymeric Composite Resin of a Toothpaste Containing Biomimetic Zn-Carbonate Hydroxyapatite (microRepair(®)) in Oral Environment: A Randomized Clinical Trial. *Polymers* **2021**, *13*, 2740. [CrossRef]

Disclaimer/Publisher's Note: The statements, opinions and data contained in all publications are solely those of the individual author(s) and contributor(s) and not of MDPI and/or the editor(s). MDPI and/or the editor(s) disclaim responsibility for any injury to people or property resulting from any ideas, methods, instructions or products referred to in the content.

Article

Experimental and Hybrid FEM/Peridynamic Study on the Fracture of Ultra-High-Performance Concretes Reinforced by Different Volume Fractions of Polyvinyl Alcohol Fibers

Kun Zhang [1,†], Tao Ni [2,3,*,†], Jin Zhang [4,5,*,†], Wen Wang [5], Xi Chen [1], Mirco Zaccariotto [3], Wei Yin [1], Shengxue Zhu [1] and Ugo Galvanetto [3]

1. Department of Transportation Engineering, Huaiyin Institute of Technology, Huaian 223003, China
2. State Key Laboratory of Geohazard Prevention and Geoenvironment Protection, Chengdu University of Technology, Chengdu 610059, China
3. Industrial Engineering Department, University of Padova, via Venezia 1, 35131 Padova, Italy
4. State Key Laboratory for GeoMechanics and Deep Underground Engineering, China University of Mining & Technology, Beijing 100083, China
5. College of Civil and Transportation Engineering, Hohai University, Nanjing 210098, China
* Correspondence: nitao_sklgp@cdut.edu.cn (T.N.); chelseazhangjin@163.com (J.Z.)
† These authors contributed equally to this work.

Abstract: In this study, a series of three-point bending tests were carried out with notched beam structures made of polyvinyl alcohol (PVA) fiber-reinforced ultra-high-performance concrete (UHPC) to study the effect of volume fractions of PVA fibers on the fracture characteristics of the UHPC-PVAs. Furthermore, in order to meet the increasing demand for time- and cost-saving design methods related to research and design experimentation for the UHPC structures, a relevant hybrid finite element and extended bond-based peridynamic numerical modeling approach is proposed to numerically analyze the fracture behaviors of the UHPC-PVA structures in 3D. In the proposed method, the random distribution of the fibers is considered according to their corresponding volume fractions. The predicted peak values of the applied force agree well with the experimental results, which validates the effectiveness and accuracy of the present method. Both the experimental and numerical results indicate that, increasing the PVA fiber volume fraction, the strength of the produced UHPC-PVAs will increase approximately linearly.

Keywords: ultra-high-performance concretes; polyvinyl alcohol fiber; fiber volume fraction; three-point bending test; extended peridynamics; finite element method

1. Introduction

Ultra-high performance concrete (UHPC) has developed as one of the most promising types of concrete in the last 25 years. This vanguard product presents both ultra-high compressive strength and remarkable durability, such as compressive strength of 150–200 MPa [1,2]. The superior performance is achieved by maximizing the packing density with very fine minerals and reactive powders. Unlike the steel bars in the reinforced concrete, the complicated design of the reinforcement layout is not necessary for UHPC elements.

It is generally accepted that the mechanical properties of UHPC can be remarkably improved with various types of fibers. The fibers made of steel, glass, polymer (such as PVA, PVC, PE), carbon, etc., mixed with high strength cement mortar could make the produced composites present quite different mechanical behaviours in the loading situation [3]. The performance is much influenced by a few parameters, e.g., the volume fraction and fiber distribution. Many authors have pointed out that steel fiber orientation can be influenced by flow patterns of mixture, rheological performance of mixture, casting methods, wall effect of formworks, extrusion of mixture and external electromagnetic field. Folgar [4] revealed that the distribution of steel fibers can be affected by the plastic viscosity and gradient of flow velocity of fresh mixture. Zhou and Uchida [5] reported that casting UHPC at the center of a slab with 1.2 m diameter can result in significant difference in steel fiber orientation value between the edge and center regions of the slab. Each of these factors could be eliminated or reduced in PVA fiber-reinforced UHPC (UHPC-PVA) material. The PVA fibers tend to develop very strong chemical bonding force with cement due to the presence of the hydroxyl group in its molecular chains [6], which causes the material to have more isotropic behaviour than other types of fiber-reinforced UHPCs. Although the PVA fiber reinforcement can increase the fracture toughness of concrete, there are still workability problems [7] to solve; an alcohol-based shrinkage reducing agent (ASRA) was first made by the authors, as reported in [8,9], with the ice-replaced mixing procedure in the production of UHPC and UHPC-PVA materials to reduce the shrinkage behavior and improve the workability.

As a fiber-reinforced composite material, it is essential to test the mechanical performance of the UHPC-PVA materials and the fracture behavior of the UHPC-PVA structures before applying a new type of UHPC-PVA to practical engineering [10,11]. Testing is the most commonly used method for revealing the mechanical properties of plain and fiber-reinforced concretes and studying their failure behaviors [12–14]. As reported by Yoo et al. [15], at low fiber volume fractions ($V_f \leqslant 1.0\%$), the twisted fibers provide the highest flexural strength, but they exhibit similar strength and poorer toughness than the straight fibers at a V_f equal to or higher than 1.5%. The three-point bending test on the notched beam structures is another alternative to study the mechanical performance of UHPC-PVA structures under flexure loading [9,11]. Critical stress intensity factor, tensile strength and fracture energy can be estimated from the test results [16,17]. Although the test method is visual and useful, it has its limitation in consuming a lot of material resources and time.

In addition to experiments, numerical simulation is another effective and cost-saving method to analyze the fracture mechanisms of the fiber-reinforced structures and to evaluate their mechanical properties. In recent decades, numerical studies have been carried on the fracture characteristics of plain and fiber-reinforced concrete. Most researchers used the general finite element (FE) software with some modifications to analyze the beams and slabs made of fiber-reinforced concretes. The earliest numerical study can be found in [18], where the authors reported the simulation techniques and input parameters required to accurately simulate the strengthened concrete structures. In [19], researchers also developed a meso-scale FE model to predict the de-bonding process in fiber-reinforced concrete using a fixed angle crack model. Chen et al. [20] investigated the effects of various modeling assumptions on the interfaces between concrete, steel fiber reinforcement and shear stirrups. These authors also stressed the importance of modeling the fibers' random distribution in the composite concrete to achieve good correlation with the measured experimental results. However, these models were only applicable in the simulations of two dimensional problems. In [21,22], ABAQUS, a general commercial FEA software, is used to perform 3D simulation of the failure of the fiber-reinforced UHPC structures under compression, flexural and tension loading by using the built-in concrete plasticity damage (CDP) model. In general, the existing relevant numerical studies did not present any techniques or recommendations to describe the crack growth process in fiber-reinforced concretes [23–25].

Peridynamics (PD), first proposed by Silling in 2000 [26], is a newborn non-local numerical theory, where integro-differential equations are used to describe the mechanical behavior of continuous media and discontinuities can be considered without singularities. As the earliest version of peridynamics, the bond-based peridynamics (BB-PD) theory defines the interaction by pairwise forces acting along the deformed bond, which has a limitation on the Poisson's ration of 1/3 for plane stress and 1/4 for plane strain and 3D problems. Then, state-based peridynamic models were introduced, including ordinary and non-ordinary versions (OSB-PD and NOSB-PD), to simulate the materials with any Poisson's ratio [27–29]. In recent years, PD-based computational methods have been widely used to investigate the toughening mechanisms of innovative materials [30–33]. Some relevant applications of PD-based tools to study the fracture mechanism of fiber-reinforced concretes can be found in [9,34–41]. However, in the existing literature, the fracture analysis on the fiber-reinforced concrete structures is only considered in plane stress or plane strain conditions. In addition, due to its natural non-locality, the PD-based models share the shortcoming of higher computing costs than those based on the local theory. To improve the computational efficiency and make use of the flexibilities of the PD approach in the simulation of fracture problems, coupling to the local models, such as the FE model [42–47], has become a popular and convenient choice.

The mechanical properties and fracture characteristics of the UHPCs and UHPC-PVAs produced following the manufacturing procedure reported in [8,9] have been investigated with the experimental and numerical tools. However, the cases with different PVA fiber volume fractions were not considered in the authors' previous works. In this paper, referring to [9], the three-point bending test is used to evaluate the fracture properties of the UHPC-PVA materials. Different PVA fiber volume fractions are considered to investigate the influence on the fracture process of the UHPC-PVA structures. To comprehensively analyze the fracture behaviors of the UHPC-PVAs, a 3D hybrid FE/PD modeling approach was developed. Different from that in [9], an extended bond-based peridynamic (XBB-PD) model [29,48] equipped with an energy-based failure criterion was adopted to overcome the limitation on the Poisson ratio of the classical BB-PD model.

The main contributions of this article with regard to numerical modeling are as follows:

- The XBB-PD model is adopted to describe the deformation and fracture behaviors of UHPC-PVA structures without the limitation on the Poisson ratio;
- The PD model is coupled to the FE model to decrease the overall computational costs and maintains its flexibility in simulating crack problems;
- The discrete-level modeling procedure of the UHPC-PVA materials and structures is illustrated in detail;
- Three-dimensional simulations are carried out and the numerical results are compared to the experimental results.

In addition to that, the experimental and numerical results will explain how the strength of the UHPC-PVAs changes in cases with different volume fractions of PVA fibers. The study is a supplement to those of [8,9]. The numerical modeling approach introduced in this paper is more advanced and capable of simulating 3D crack initialization and propagation with better computational efficiency.

2. Experimental Program
2.1. Preparation of the UHPC-PVA Materials

This study focuses on the effects of the volume fractions of the PVA fiber on the fracture properties of the UHPC-PVA structures and materials. The UHPCs were prepared following the same recipe as in [9]. As listed in Table 1, the main ingredients are as follows: ASTM Type-II Portland cement, sand (approximately 1000–1500 µm in diameter), fine quartz sand (approximately 150–500 µm in diameter), EBS-S silica fume (approximately 0.1–0.5 µm in diameter), Sika polycarboxylate superplasticizer (water reducing ratio ≥ 30%), sodium laurylsulfate and polyoxyethylene nonylphenolether compounded with alcohol-based shrinkage reducing agent (ASRA, weight ratio of 2%). Alcohol was used as a solvent to combine two additives (sodium laurylsulfate and polyoxyethylene nonylphenolether), which can reduce the existence of macro-pores in the hardening matrix [8]. More information on the ingredients can be found in [8,9]. Four cases with different volume fractions of the PVA fiber (V_f), 0.5%, 1%, 1.5% and 2%, were considered to produce the UHPC-PVA materials.

Table 1. Mixing ratios of main ingredients used in the production of the UHPC-PVA materials.

Items	Mixing Ratio
Cement	1
Sand	1
Quartz sand	0.3
Silica fume	0.25
Polycarboxylate Superplasticizer	2.5%
Shrinkage reducing agent	2%
Water	0.2
Ice cube	0.02

A strict procedure described in [9] was carried out in the production of the UHPC-PVA materials:

- Step 1: Mix the cement, quartz sand, manufactured sand and silica fume with a prescribed mixing ratio;
- Step 2: Add ice cube and 10% water with superplasticizer and ASRA and mix for 3 min;
- Step 3: Add the remaining 90% water (mixed with PVA fibers) and process the mixture unceasingly until smooth;
- Step 4: Pour the mixture into a selected mould and vibrate for 3 min on a vibrating table;
- Step 5: Cure the specimens at room temperature for 48 h before demoulding and then cure them in a fast curing box in hot water at 90°C for an additional 72 h.

In the mixing operation, the PVA fibers were mixed with water and then gradually added. Due to the excellent hydrophilicity, the PVA fibers can be uniformly dispersed into the hardened matrix.

2.2. Test Procedure

In this study, a series of three-point bending tests are carried out with a notched beam specimen to evaluate the fracture properties of the produced UHPC-PVA materials. The geometry of the beam specimen and loading conditions of the test is presented in Figure 1. The cuboid specimens were produced through the designed moulds with a size of 160 mm × 40 mm × 40 mm (length × width × thickness) and then cut into the designed beam specimens with a size of 160 mm × 40mm × 20mm (length × width × thickness). The notches, with geometric parameters of $C_l = 0$ mm, 20 mm and 40 mm, were fabricated by numerically controlled machine tools. All the produced notched beam specimens are shown in Figure 2a–d. As found in [8,9], the produced UHPCs and UHPC-PVAs have outstanding stable performance. Therefore, for the sake of saving material, we will use only one specimen in each case and a total of twelve specimens shown in Figure 2 will be involved in the experimental study.

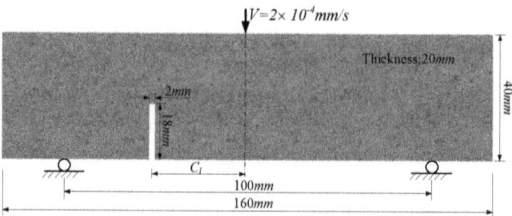

Figure 1. Geometry of the notched beam specimen and the loading conditions of the test.

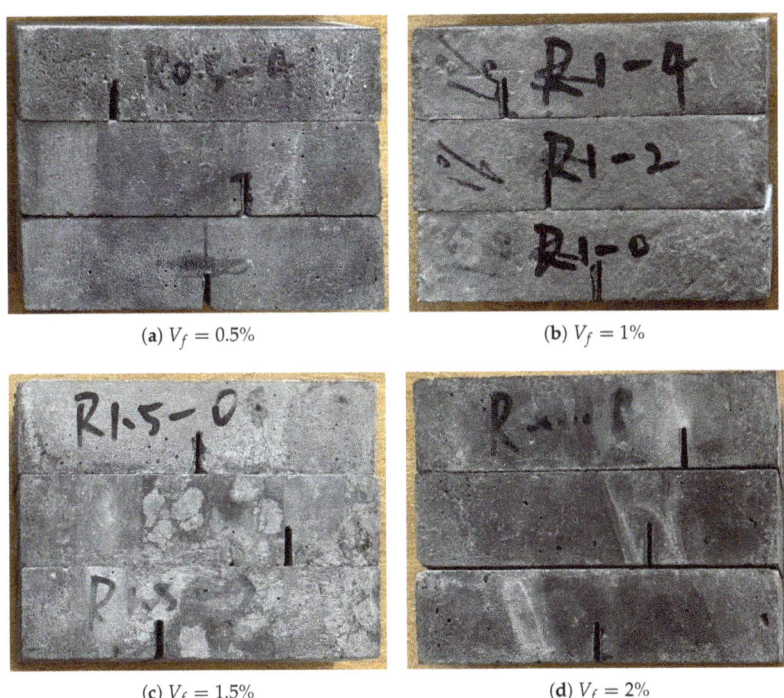

Figure 2. Photos of the notched UHPC-PVA beam specimens used for the three-point bending test.

The tests were carried out on a electromechanical compression testing machine (WAW1000) shown in Figure 3a. The loading conditions are described as in Figures 1 and 3a. The loading head forces the upper center of the beam specimens to gradually move downward at a rate of $\Delta v = 2 \times 10^{-4}$ mm/s until the crack propagates and penetrates the specimens.

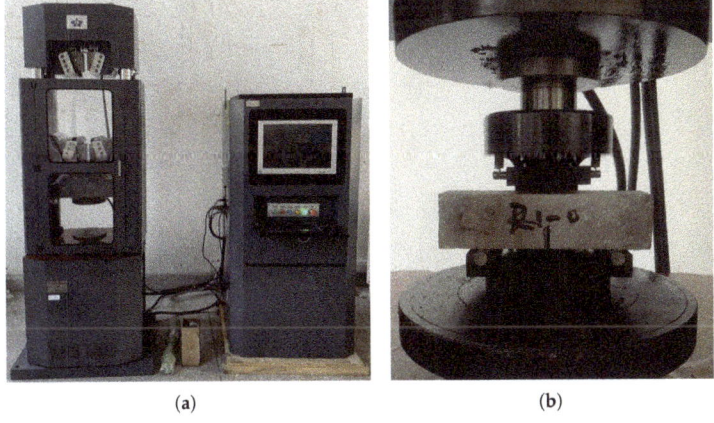

Figure 3. The mechanical testing system and the loading head for the three-point bending test. (**a**) Mechanical testing system, (**b**) Loading head for three-point bending test.

3. Numerical Model

In this section, a 3D hybrid finite element method (FEM) and extended bond-based peridynamic (XBB-PD) [29,48] modeling approach is introduced and applied to the numerical fracture analysis of UHPC-PVA materials and structures. Firstly, the governing equations of the local continuum model and the XBB-PD model are summarized. Subsequently, the model discretization and numerical implementation, including the discrete-level modeling procedure for the UHPC-PVAs, are described in detail.

3.1. Summary of the Mechanical Models

3.1.1. Governing Equations of the Local Continuum Model

In the classical continuum mechanics, the equation of motion can be expressed as:

$$\rho \ddot{u} = \nabla \cdot \sigma + b \tag{1}$$

where $\ddot{u}$ is the acceleration and σ is the stress tensor, b is the external force density. Under the assumption of small deformation, the stress tensor can be obtained as:

$$\sigma = C : \varepsilon \tag{2}$$

where C is the elasticity tensor, ε is the strain tensor. Considering the definition of strains, if the components of the continuous displacement field in the x, y and z directions are defined as u, v and w, respectively, the strain components can be given as:

$$\begin{cases} \varepsilon_{11} = \frac{\partial u}{\partial x}; & \varepsilon_{22} = \frac{\partial v}{\partial y}; & \varepsilon_{33} = \frac{\partial w}{\partial z} \\ \varepsilon_{21} = \varepsilon_{12} = \frac{\partial v}{\partial x} + \frac{\partial u}{\partial y}; & \varepsilon_{32} = \varepsilon_{23} = \frac{\partial w}{\partial y} + \frac{\partial v}{\partial z}; & \varepsilon_{31} = \varepsilon_{13} = \frac{\partial w}{\partial x} + \frac{\partial u}{\partial z} \end{cases} \tag{3}$$

3.1.2. Extended Bond-Based Peridynamic Model

As shown in Figure 4, a body $\mathcal{B}$, marked as $\mathcal{B}_0$ and $\mathcal{B}_t$ in the initial and deformed configurations, governed by the PD model, is usually seen to be composed of a series of material points. We can assume that x is a point in $\mathcal{B}$ interacting with all the other points over a prescribed domain $\mathcal{H}_x$. If point x' is a point within the domain $\mathcal{H}_x$, the relative position of x' to x in the initial configuration can be described as:

$$\xi = x' - x \tag{4}$$

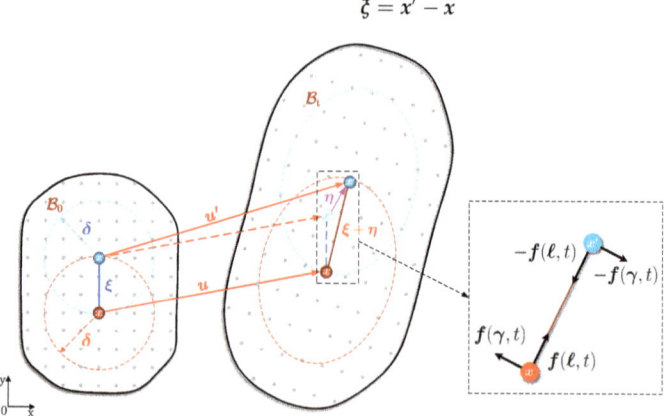

Figure 4. Schematic diagram of the extended bond-based peridynamic model.

Then, $\mathcal{H}_x$, the so-called neighbourhood, is usually a sphere space in 3D and a circle surface in 2D, which can be described as a radius of length δ (the *horizon* radius) and mathematically defined as:

$$\mathcal{H}_x = \mathcal{H}(x,\delta) = \{\|\xi\| \leq \delta : x' \in \mathcal{B}\} \tag{5}$$

where $\|\cdot\|$ denotes the Euclidean norm.

In the deformed configuration, the points x and x' will be displaced by u and u', respectively. Consequently, the relative displacement vector between the two points can be given as:

$$\eta = u' - u \tag{6}$$

and therefore the relative position vector in the deformed configuration can be given as $\xi + \eta$.

In the extended bond-based PD theory, the equation of motion at point x can be expressed as:

$$\rho \ddot{u}(x,t) = \int_{\mathcal{H}_x} f(\eta,\xi,t) dV_{x'} + b(x,t) \tag{7}$$

where ρ is the mass density, $\ddot{u}(x,t)$ is the acceleration of point x at time instant t, $dV_{x'}$ is the mass volume associated with point x', $b(x,t)$ is the body force density to point x applied by the external loads. $f(\eta,\xi,t)$ is the pairwise force density exerted to point x by the deformed bond, containing two contributions from the longitudinal and tangential deformations (see Figure 4), which can be expressed by [48]:

$$f(\eta,\xi,t) = c\ell(\eta,\xi,t)n + \kappa\gamma(\eta,\xi,t) \tag{8}$$

where c and κ are the normal and tangential micro moduli of the bond, $\ell(\eta,\xi,t)$ and $\gamma(\eta,\xi,t)$ are the longitudinal and tangential deformations of the bond. n is the unit directional vector along the deformed bond and its formulae can be given as:

$$n = \frac{\eta + \xi}{\|\eta + \xi\|} \tag{9}$$

The expressions of the normal and tangential micro moduli can be obtained from a comparison with the strain energy of local continuum mechanics for homogeneous deformation [29]. Their expressions in terms of the elastic constants of Young's modulus E and Poisson's ratio ν of the material can be obtained by:

$$\begin{cases} c = \frac{6E}{\pi\delta^4(1-2\nu)} \\ \kappa = \frac{6E(1-4\nu)}{\pi\delta^4(1+\nu)(1-2\nu)} \end{cases} \tag{10}$$

Referring to [29,48], based on the Cauchy–Born criterion, the relationships between the local deformations of the bond and the macroscopic strain can be constructed as:

$$\ell = n \cdot \varepsilon \cdot n \tag{11}$$

and

$$\gamma = n \cdot \varepsilon \cdot (I - n \otimes n) \tag{12}$$

where ε is the strain tensor and I is the second order unit tensor.

Furthermore, the longitudinal deformation can also be formulated based on the geometrical analyses [29]:

$$\ell = \frac{1}{\xi}\eta \cdot n \tag{13}$$

which is more efficient than Equation (11) and in this paper this formulae will be used to evaluate the longitudinal deformation.

To describe the material failure and crack propagation, a bond failure criterion is essential for the PD models. The critical bond-stretch criterion is the first introduced and most commonly used criterion to judge the bond breakage in the classical bond-based and state-based PD simulations. However, there are two deformation components in the micro-constitutive law and a failure criterion associated only with the bond stretch (longitudinal deformation) will not be able to reflect the effect of the tangential deformation on the failure behaviours. Thus, inspired by [49], an energy-based failure criterion will be adopted for the XBB-PD model to simulate the fracture problems.

The strain energy density stored in the deformed bond ξ can be computed by:

$$w(\xi) = w(\ell, \gamma) = \frac{1}{2}c\ell^2 \xi + \frac{1}{2}\kappa\gamma \cdot \gamma \xi \tag{14}$$

Following the derivations in [48,49], the critical strain energy density of the bond can be given as:

$$w_c = \frac{4G_c}{\pi \delta^4} \tag{15}$$

which means that the bond will be broken when its strain energy density $w(\xi)$ becomes greater than w_c and accordingly, a scalar variable is defined to indicate the connection state of the bond [50,51]:

$$\%(\xi) = \begin{cases} 1, & \text{if } w(\xi) < w_c \\ 0, & \text{otherwise} \end{cases} \tag{16}$$

Consequently, the damage level at point x can be defined as:

$$\varphi_x = 1 - \frac{\int_{\mathcal{H}_x} \%(\xi) dV_{x'}}{\int_{\mathcal{H}_x} dV_{x'}} \tag{17}$$

where $\varphi_x \in [0, 1]$ and the cracks are usually identified wherever $\varphi_x \geq 0.5$.

3.2. Discretization and Numerical Implementation

To obtain an acceptable numerical solution, a suitable discretization process is necessary. This section will introduce the numerical discretization of the FE and PD equations and their coupled modeling strategy for the UHPC-PVA materials and structures. In order to obtained a quasi-static solution of the coupled model and compare with the experimental observations, the adaptive dynamic relaxation algorithm is also briefly summarized.

3.2.1. FEM Discretization of the Governing Equations Based on Local Theory

The Galerkin finite element method [52] is adopted here to discretize the governing equations of the continuum mechanical model. The FE equation of motion can be written as the following matrix form:

$$M^{FEM}\ddot{U} + K^{FEM}U = F \tag{18}$$

where M^{FEM} and K^{FEM} are the mass and stiffness matrices of the FE domain. Given the shape function N_u for the displacement, the stiffness matrices in Equation (18) can be obtained by:

$$M^{FEM} = \int_\Omega N_u^T \rho N_u d\Omega \tag{19}$$

and

$$K^{FEM} = \int_\Omega (LN_u)^T D(LN_u) d\Omega \tag{20}$$

in which L is the differential operator defined as:

$$L = \begin{bmatrix} \frac{\partial}{\partial x} & 0 & 0 \\ 0 & \frac{\partial}{\partial y} & 0 \\ 0 & 0 & \frac{\partial}{\partial z} \\ \frac{\partial}{\partial y} & \frac{\partial}{\partial x} & 0 \\ 0 & \frac{\partial}{\partial z} & \frac{\partial}{\partial y} \\ \frac{\partial}{\partial z} & 0 & \frac{\partial}{\partial x} \end{bmatrix} \qquad (21)$$

and D is the elastic matrix given as:

$$D = \frac{E(1-v)}{(1+v)(1-2v)} \begin{bmatrix} 1 & \frac{v}{1-v} & \frac{v}{1-v} & 0 & 0 & 0 \\ \frac{v}{1-v} & 1 & \frac{v}{1-v} & 0 & 0 & 0 \\ \frac{v}{1-v} & \frac{v}{1-v} & 1 & 0 & 0 & 0 \\ 0 & 0 & 0 & \frac{1-2v}{2(1-v)} & 0 & 0 \\ 0 & 0 & 0 & 0 & \frac{1-2v}{2(1-v)} & 0 \\ 0 & 0 & 0 & 0 & 0 & \frac{1-2v}{2(1-v)} \end{bmatrix} \qquad (22)$$

where E and v are the Young's modulus and Poisson's ratio of the material.

3.2.2. Discretization of the XBB-PD Equations

After discretization, the spatial integrals in the XBB-PD equations will be written into forms of summation over nodes in the neighbourhood. Then, the equation of motion of node x_i at time t will be:

$$\rho \ddot{u}_i^t = \sum_{j=1}^{N_{\mathcal{H}_i}} f^t\left(\xi_{ij}\right) V_j + b_i^t \qquad (23)$$

where $N_{\mathcal{H}_i}$ is the number of family nodes in x_i's horizon. x_j represents x_i's family node and V_j is its volume. b_i^t is the body force density of node x_i. $f^t\left(\xi_{ij}\right)$ is the internal force density exerted to node x_i via the deformed bond ξ_{ij}, which can be computed by:

$$f^t\left(\xi_{ij}\right) = \left[f_{ij}\right] = \left[f_{ij}^{\ell}\right] + \left[f_{ij}^{\gamma}\right] = c\ell_{ij}\left[n_{ij}\right] + \kappa\left[\gamma_{ij}\right] \qquad (24)$$

in which ℓ_{ij} and $\left[\gamma_{ij}\right]$ are the longitudinal and tangential deformation components of the bond ξ_{ij} and $\left[n_{ij}\right]$ is the longitudinal unit vector. The two vectors can be defined as:

$$\left[n_{ij}\right] = \left[n_1 \ n_2 \ n_3\right]^T \quad \text{and} \quad \left[\gamma_{ij}\right] = \left[\gamma_1 \ \gamma_2 \ \gamma_3\right]^T \qquad (25)$$

if the displacement vectors of nodes x_i and x_j are given as $[U_i] = \left[U_i^1 \ U_i^2 \ U_i^3\right]^T$ and $[U_j] = \left[U_j^1 \ U_j^2 \ U_j^3\right]^T$. According to Equation (13), the stretch (longitudinal deformation) of the bond ξ_{ij} can be obtained by:

$$\ell_{ij} = \left[C_{ij}^{\ell}\right]\begin{bmatrix} U_i \\ U_j \end{bmatrix} \qquad (26)$$

where $\left[C_{ij}^{\ell}\right]$ can be given as:

$$\left[C_{ij}^{\ell}\right] = \frac{1}{\xi_{ij}}\left[\begin{array}{cccccc} -n_1 & -n_2 & -n_3 & n_1 & n_2 & n_3 \end{array}\right] \qquad (27)$$

Marking the strains at nodes x_i and x_j as $[\varepsilon_i]$ and $[\varepsilon_j]$, they can be written in vector forms as:

$$[\varepsilon_i] = [\varepsilon_i^1 \ \varepsilon_i^2 \ \varepsilon_i^3 \ \varepsilon_i^{12} \ \varepsilon_i^{13} \ \varepsilon_i^{23}]^T \quad \text{and} \quad [\varepsilon_j] = [\varepsilon_j^1 \ \varepsilon_j^2 \ \varepsilon_j^3 \ \varepsilon_j^{12} \ \varepsilon_j^{13} \ \varepsilon_j^{23}]^T, \tag{28}$$

respectively.

According to Equation (12), the tangential deformation vector of the bond ξ_{ij} can be obtained by:

$$[\gamma_{ij}] = [C_{ij}^\gamma][\varepsilon_{ij}] \tag{29}$$

where $[C_{ij}^\gamma]$ is given as:

$$[C_{ij}^\gamma] = \begin{bmatrix} n_1 - n_1^3 & -n_1 n_2^2 & -n_1 n_3^2 & n_2 - 2n_1^2 n_2 & n_3 - 2n_1^2 n_3 & -2n_1 n_2 n_3 \\ -n_1^2 n_2 & n_2 - n_2^3 & -n_2 n_3^2 & n_1 - 2n_1 n_2^2 & -2n_1 n_2 n_3 & n_3 - 2n_2^2 n_3 \\ -n_1^2 n_3 & -n_2^2 n_3 & n_3 - n_3^3 & -2n_1 n_2 n_3 & n_1 - 2n_1 n_3^2 & n_2 - 2n_2 n_3^2 \end{bmatrix} \tag{30}$$

and $[\varepsilon_{ij}]$ is the average strain of the bond ξ_{ij} defined as:

$$[\varepsilon_{ij}] = \frac{[\varepsilon_i] + [\varepsilon_j]}{2} \tag{31}$$

Based on the above notions, the PD force density exerted to node x_i by the bond ξ_{ij} can be obtained by:

$$[f_{ij}] = c[n_{ij}][C_{ij}^\ell]\begin{bmatrix} u_i \\ u_j \end{bmatrix} + \frac{1}{2}\kappa \begin{bmatrix} C_{ij}^\gamma & C_{ij}^\gamma \end{bmatrix}\begin{bmatrix} \varepsilon_i \\ \varepsilon_j \end{bmatrix} \tag{32}$$

Consequently, the equations of motion of the XBB-PD model can be assembled and written in the following matrix form:

$$M^{PD}\ddot{U} + cN^\ell C^\ell U + \kappa C^\gamma E = F \tag{33}$$

where M^{PD} is the diagonal mass matrix, N^ℓ, C^ℓ and C^γ are matrices assembled from the matrices of Equations (25), (27) and (30). $\ddot{U}$, U, E and F are the acceleration, displacement, strain and force vectors of the nodes, respectively.

As described in Equation (3), the strain components are the spatial partial derivatives of the displacement field. In the PD framework, the peridynamic differential operator (PDDO) proposed in [53] can be used to evaluate derivatives. Referring to [48], the global relationship between the displacement field and the strain field can be written in the following form:

$$E = GU \tag{34}$$

where G is the non-local strain coefficient matrix [48].

Therefore, substitution of Equation (34) into Equation (33) converts the PD equations of motion into the following concise form:

$$M^{PD}\ddot{U} + K^{PD}U = F \tag{35}$$

where $K^{PD} = cN^\ell C^\ell + \kappa C^\gamma G$ is the assembled stiffness matrix of the XBB-PD model.

3.2.3. Hybrid FEM and PD Modelling Approach for the UHPC-PVA Materials and Structures

The approach introduced in [9] is adopted here to model the UHPC-PVA materials and structures, where the interaction between the PVA fibers and the matrix is considered in the discrete level. The hybrid FEM/PD modeling procedure of UHPC-PVA materials and structures is described in Figure 5.

As shown in Figure 5a, the beam specimen is divided into FE and PD domains. Then, the hybrid model will be generated according to the following steps:

- Step 1: Discretize the FE and PD domains by using the FE mesh and PD grid with the same grid size; see Figure 5b;
- Step 2: Generate the PD bonds connecting all the FE and PD nodes; see Figure 5c;
- Step 3: Randomly select a certain number of bonds and set their parameters as the mechanical parameters of the PVA material; then, the rest are the matrix bonds with the mechanical parameters of the UHPC materials. The obtained model is shown in Figure 5d. The ratio of the total length of fiber bonds to the total length of all bonds, which is called the global numerical volume fraction of PVA fibers (V_f^g), is approximately equal to the volume fraction of fibers in the modeled UHPC-PVA material;
- Step 4: Determine the final FE/PD model, as shown in Figure 5e, where the reinforcement at the FE and coupling elements is considered based on the local numerical volume fraction of PVA fibers (V_f^l).

The global numerical volume fraction of PVA fibers can be calculated by:

$$V_f^g = \frac{\sum_{i=1}^{N_n} \sum_{j=1}^{N_i^f} \|\xi_{ij}\|}{\sum_{i=1}^{N_n} \sum_{j=1}^{N_i} \|\xi_{ij}\|} \tag{36}$$

where N_n is the number of nodes in the discrete model; $N_{\mathcal{H}_i}$ is the number of x_i's family nodes, while N_i^f is the number of x_i's family nodes connected by the fiber bonds. Consequently, the local numerical volume fraction of PVA fibers at node x_i can be obtained by:

$$V_f^{l_i} = \frac{\sum_{j=1}^{N_i^f} \|\xi_{ij}\|}{\sum_{j=1}^{N_i} \|\xi_{ij}\|} \tag{37}$$

Given the V_f^l value at each node, the reinforcement of the PVA fibers on the UHPC matrix will be expressed by:

$$P_i = P_m(1 - V_f^{l_i}) + P_f V_f^{l_i} \tag{38}$$

where P_i represents the mechanical parameters at node i; P_m and P_f are the parameters of the UHPC and PVA materials, respectively. Therefore, the reinforcement of the PVA fibers on the matrix will be considered in the calculation of the elastic matrix of Equation (22).

The system matrix of the hybrid FEM and PD model can be expressed by:

$$M^{Coup}\ddot{u} + K^{Coup}u = F \tag{39}$$

Note that, instead of the formation in Equation (19), the mass density matrix of the FE domain will use a diagonal form to maintain consistency with the PD domain.

3.2.4. Quasi-Static Solution Algorithm

The adaptive dynamic relaxation (ADR) algorithm was first proposed by Underwood in [54] to obtain the quasi-static solutions of non-linear problems. Later in [45,55–57], the ADR algorithm was successfully applied to solve the static or quasi-static solutions of PD models.

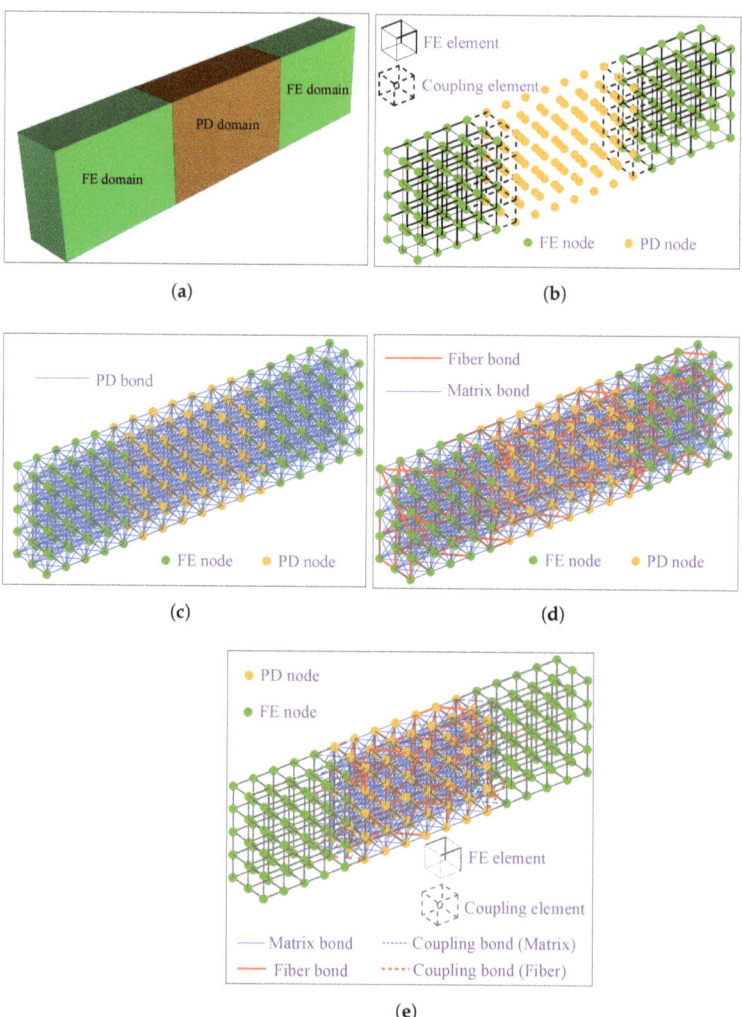

Figure 5. The diagrammatic sketch for the illustration of the hybrid FE/PD modeling procedure of the UHPC-PVA structures. Note that the size of the grid does not represent the real one used in the simulations. (**a**) Modelling schematic of a beam specimen; (**b**) Step 1: the hybrid FE/PD discretization; (**c**) Step 2: PD bond connections for all nodes; (**d**) Step 3: random selection of fiber bonds; (**e**) Step 4: the complete hybrid FE/PD model.

In accordance with our experience in [9], the tests described above adopted a quasi-static loading process. Therefore, the ADR algorithm will be equipped with the hybrid FEM/XBB-PD model to analyze the fracture process of the UHPC-PVA beams under the three-point bending load.

By introducing a damping term, the global governing equation of the hybrid model at the nth time increment can be written in the following form:

$$M\ddot{u}^n + C_d\dot{u}^n + Ku^n = F^n \tag{40}$$

where M, C_d and K are the fictitious mass, damping and stiffness matrices. F is the external force vector. Subsequently, the central time difference form will be adopted in the ADR algorithm and the displacement at the $(n+1)^{th}$ iteration can be obtained by:

$$u^{n+1} = u^n + \dot{u}^{n+1/2}\Delta t \qquad (41)$$

where the velocity at the $(n+1/2)^{th}$ iteration can be calculated by:

$$\dot{u}^{n+1/2} = \frac{M/\Delta t - \frac{1}{2}C_d}{M/\Delta t + \frac{1}{2}C_d}\dot{u}^{n-1/2} + \frac{[F - Ku^n]\Delta t}{M/\Delta t + \frac{1}{2}C_d} \qquad (42)$$

where Δt is the time increment.

In order to solve Equation (42) explicitly, a diagonal fictitious mass matrix is required. M_{ii} is the ith principal value of the fictitious mass matrix M, which needs to satisfy the following inequality:

$$M_{ii} \geq \frac{1}{4}\Delta t^2 \sum_j |K_{ij}| \qquad (43)$$

where K_{ij} are the elements of the global stiffness matrix K. To simplify the time integral process, the damping matrix is usually defined as multiples of the fictitious mass matrix:

$$C_d = c_d M \qquad (44)$$

where c_d is a system damping coefficient that needs to be updated during the iterations. The value of c_d at the nth iteration can be computed by:

$$c_d^n = 2\sqrt{\frac{(u^n)^T K_t^n u^n}{(u^n)^T M u^n}} \qquad (45)$$

where K_t^n is the "local" diagonal tangent stiffness matrix at the nth iteration and its diagonal entries are defined as:

$$(K_t^n)_{ii} = \frac{Ku^n - Ku^{n-1}}{\dot{u}^{n-1/2}\Delta t} \qquad (46)$$

Substituting Equation (44) into Equation (42), Equation (42) can be rewritten as follows:

$$\dot{u}^{n+1/2} = \frac{2 - c_d^n \Delta t}{2 + c_d^n \Delta t}\dot{u}^{n-1/2} + \frac{2[F - Ku^n]\Delta t}{(2 + c_d^n \Delta t)M} \qquad (47)$$

In addition, the iteration starts with:

$$\dot{u}^{1/2} = \frac{[F - Ku^0]\Delta t}{2M} \qquad (48)$$

3.2.5. The Model Parameters and Settings in the Simulations

In this section, the determination of the discretization and mechanical parameters needed in the numerical simulations is described.

According to the experimental results in [8,9] and the characteristics of the XBB-PD model, the mechanical parameters of the produced UHPC material adopted in the numerical simulations are taken as Young's modulus: $E_m = 34.5$ GPa; Poisson's ratio: $\nu_m = 0.1$; fracture energy density: $G_{cm} = 90$ J/m^2 (measured by the approach introduced in [58]). On the other hand, the mechanical parameters of the PVA materials provided by the manufacturer are given as Young's modulus: $E_f = 100$ GPa; Poisson's ratio: $\nu_f = 0.22$; fracture energy density: $G_{cf} = 8000$ J/m^2.

In [9], the produced UHPC-PVA materials were seen as a type of composite material. In order to keep the smoothness of the strain field in the PD simulation of such a composite material, the horizon radius used for the PD discretization should conform to the following inequality:

$$\delta \leqslant \sqrt{\frac{E_f h_f h_m^2}{3\mu_m \left(h_f + h_m\right)}} \tag{49}$$

where h_f and h_m represent the geometrically characteristic lengths of the matrix and fiber materials; μ_m is the shear modulus of the matrix material. As we stated in [9], the geometrically characteristic lengths can be taken as $h_f = 20$ mm and $h_m = 1.5$ mm, respectively. Given the Young's modulus of the PVAs $E_f = 100$ GPa and the shear modulus of the UHPCs $\mu_m = E_m/(1 + 2\nu_m) = 28.75$ GPa, using Equation (49), the horizon radius should satisfy $\delta \leqslant 1.558$ mm and $\delta = 1.5$ mm could be a convenient choice.

In [9], the m-ratio was taken as $m = 5$ for the 2D modeling of UHPC-PVA structures. However, the 3D condition is considered in this paper. For the purpose of compromise between accuracy and computational cost, the m ratio is adopted here as $m = 3$, then the grid size is obtained as $\Delta x = \delta/m = 0.5$ mm. The discrete models for the three cases in the experiments are shown in Figure 6a–c. The number of total nodes is 1,061,613 in the hybrid models. The discretization information is presented in detail in Table 2.

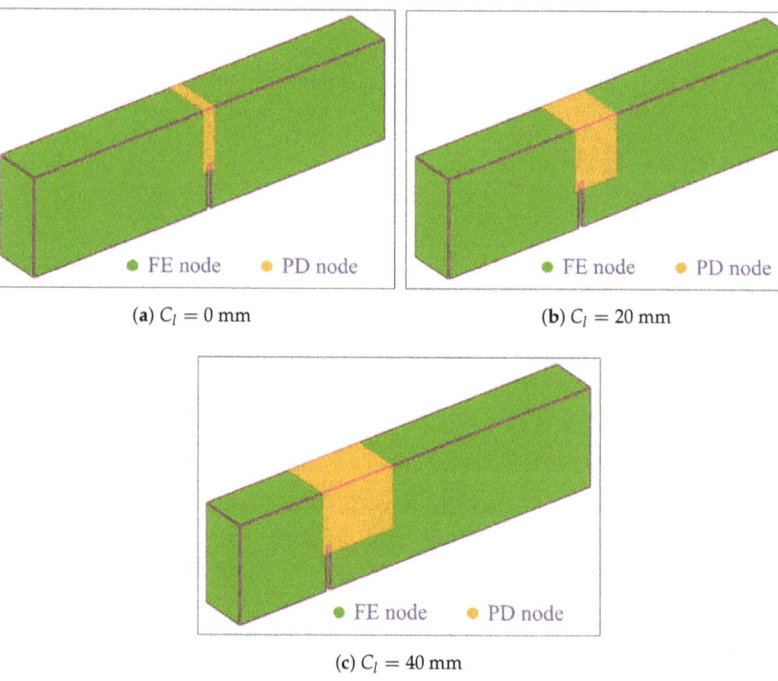

Figure 6. The discrete models used in the 3D simulations.

The ADR algorithm is sensitive to the value of Δt used in the time integration [9,45]. In [9], a numerical test was performed with a 2D model to determine the proper value of Δt to find the similar quasi-static characteristics for the experimental observations. Referring to that, $\Delta t = 5 \times 10^{-3}$s could be the most secure value and will be used in all the 3D simulations.

In order to compare with the experimental results, four different values of V_f (= 0.5%, 1%, 1.5% and 2%) are considered. Given the loading rate of $v = 2 \times 10^{-4}$ mm/s adopted in the experiments, the numbers of iterations in the simulations will be 350,000, 400,000 and 500,000 for cases 1, 2 and 3, respectively.

Table 2. Discretization information of the hybrid model for the notched beam specimens.

Case	Number of PD Nodes	Number of FE Elements	Number of Coupling Elements
1 ($C_l = 0$ mm)	22,263	994,880	4320
2 ($C_l = 20$ mm)	80,811	937,760	5440
3 ($C_l = 40$ mm)	135,177	884,720	6480

4. Experimental and Numerical Results

4.1. Experimental Results

The three-point bending tests on the beam specimens shown in Figure 2a–d were carried out. All the broken specimens are shown in Figures 7a–9a. The variations of the applied loads versus the central deflections are recorded and plotted in Figures 7b–9b. The shapes of the central deflection-force diagrams show different characteristics in the elastic, hardening, softening and failure stages in the cases with different PVA fiber volume fractions. In contrast, the results of the compression tests [8] and the three-point bending tests [9] performed with the produced plain UHPC materials showed a typical quasi-brittle behavior. It seems that, due to the addition of the PVA fibers and with the increase in the volume fraction, the UHPC-PVA materials gradually change from brittle to ductile. The significant toughening enhancement phenomena exist in the cases with greater PVA fiber volume fractions. Figure 10 shows the variations of the peak force values in the tests versus the volume fractions of PVA fibers mixed in the UHPC-PVAs, describing that the strength of the produced materials increases approximately linearly with PVA fiber volume fraction.

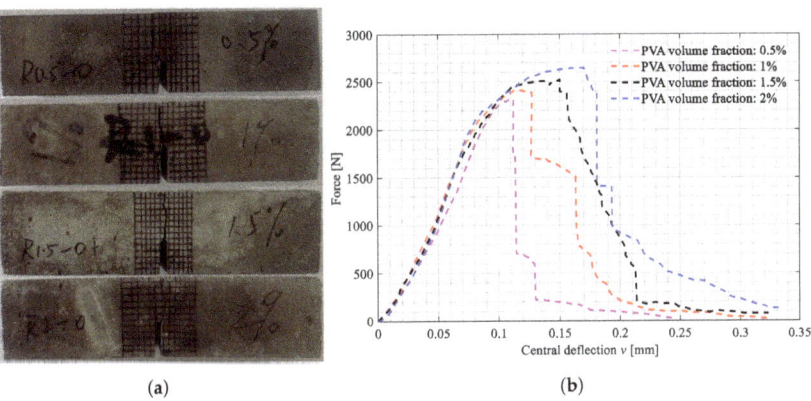

Figure 7. The broken beam specimens with $C_l = 0$ mm and the central deflection-force diagrams. (**a**) Broken specimens with $C_l = 0$ mm; (**b**) Central deflection-force diagrams.

The scanning electron microscope (SEM) shown in Figure 11a is used here to study the micro characteristics of the fracture surface in the UHPC-PVA beam specimen after tests. Figure 11b,c show two SEM micrographs near a PVA fiber and a quartz granule on the fracture surface. As shown in Figure 11b, during the fracture advancement, there is a granular peeling phenomenon near the quartzite–cement interface, but the PVA–cement interface is smooth (see Figure 11c). The difference suggests that the chemical bonding between the PVA fibers and cement matrix is much stronger than that of quartzite granules. This also explains why the PVA fibers can reinforce the produced UHPC materials. On the other hand, as stressed in [9], the PVA fibers were broken at the fracture surfaces and no pulling-out phenomena were observed, which justifies the proposed modeling approach for the produced UHPC-PVA materials and structures.

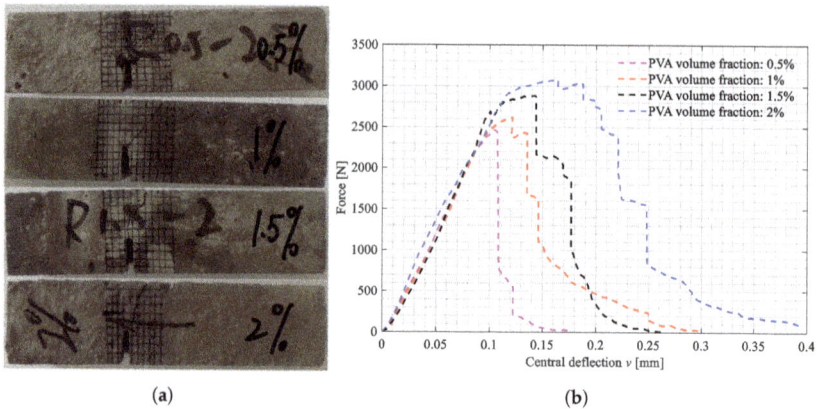

Figure 8. The broken beam specimens with $C_l = 20$ mm and the central deflection-force diagrams. (a) Broken specimens with $C_l = 20$ mm; (b) Central deflection-force diagrams.

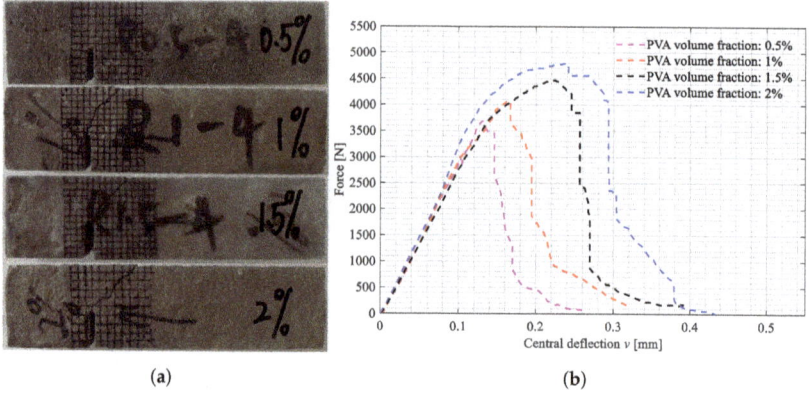

Figure 9. The broken beam specimens with $C_l = 40$ mm and the central deflection-force diagrams. (a) Broken specimens with $C_l = 40$ mm; (b) Central deflection-force diagrams.

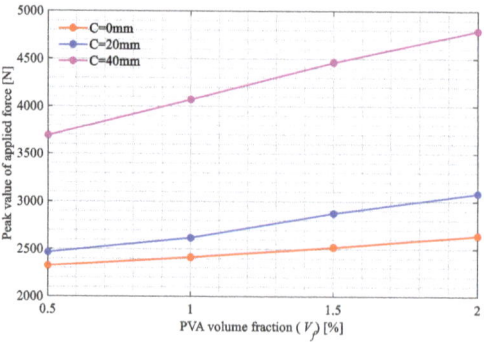

Figure 10. Variations in the peak force values in the tests versus the volume fractions of PVA fibers in the UHPC-PVAs.

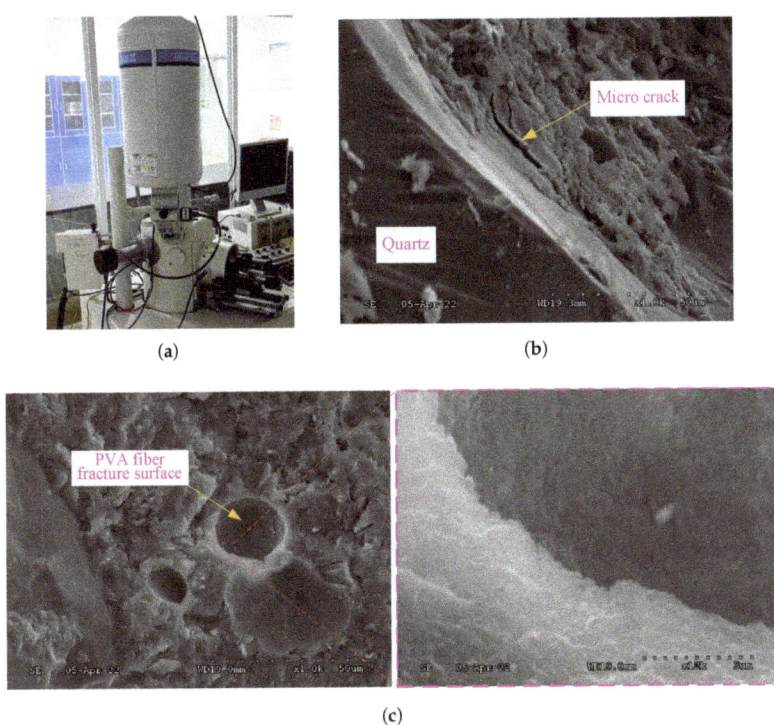

Figure 11. SEM micrographs of the fracture surfaces in the broken UHPC-PVA beams. (**a**) SEM; (**b**) A SEM micrograph near a quartz granule; (**c**) A SEM micrograph near a PVA fiber.

4.2. Numerical Results

The surface crack patterns obtained in the simulations are shown in Figures 12a–14a, the experimentally observed crack patterns (magenta curves) are also plotted for comparison. The 3D crack surfaces in the simulated specimens are shown in Figures A1–A3. The corresponding central deflection-force diagrams are plotted in Figures 12b–14b. Figure 15 shows the variations of the peak applied force versus the V_f values.

The differences between the surface crack patterns obtained in the experiments and numerical simulations may be caused by the random distribution of the PVA fibers in the UHPC-PVA structures. In addition, the damage zones, describing the crack patterns, are thicker in the cases with greater PVA fiber volume fractions, indicating that more bonds are broken in those cases. This is caused by the inconsistency between local deformation and force density due to the reinforcement of the fiber bonds, which is also the reason for the success of the proposed approach in modeling UHPC-PVA materials. Concerning the predicted fracture angles and the peak values of applied force, the simulation results agree well with the experimental results, explaining the adaptability of the proposed modeling approach in describing the failure and fracture behaviours of the produced UHPC-PVA materials and structures.

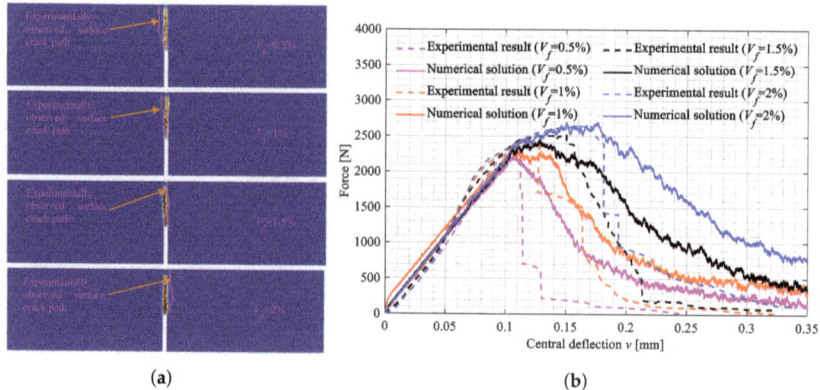

Figure 12. The crack patterns in the beam specimens with $C_l = 0$ mm and the central deflection-force diagrams. (**a**) Surface crack patterns; (**b**) Central deflection-force diagrams.

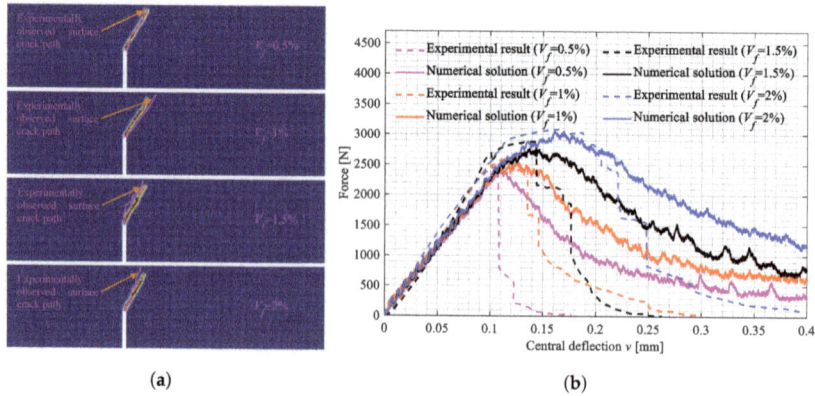

Figure 13. The crack patterns in the beam specimens with $C_l = 20$ mm and the central deflection-force diagrams. (**a**) Surface crack patterns; (**b**) Central deflection-force diagrams.

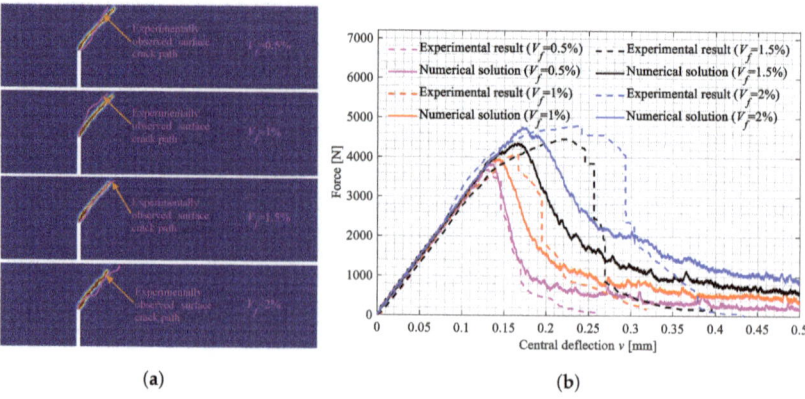

Figure 14. The crack patterns in the beam specimens with $C_l = 40$ mm and the central deflection-force diagrams. (**a**) Surface crack patterns; (**b**) Central deflection-force diagrams.

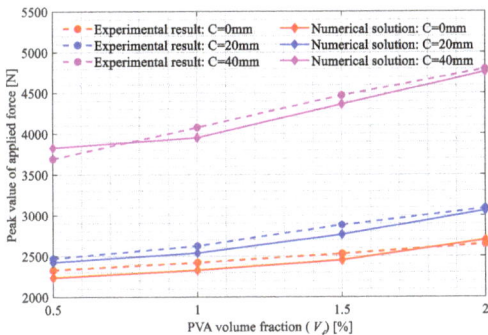

Figure 15. Variations in the peak force values in the simulations versus the volume fractions of PVA fibers in the UHPC-PVAs.

5. Conclusions and Discussions

In this paper, a series of three-point bending tests were carried out with the notched UHPC-PVA beam structures. Different cases were considered to study the effects of the PVA fiber volume fractions on the fracture behaviours of the UHPC-PVAs. Subsequently, in order to track the whole process of fracture advancement in the specimens, a 3D hybrid FE/PD modeling approach was proposed, where the XBB-PD model in conjunction with an energy-based failure criterion [48] was adopted to describe the deformation and failure behaviours of the UHPC-PVAs, removing the limitation on the Poisson ratio in the classical BB-PD model [9]. The comparison between the numerical solutions and the experimental results validates the proposed approach and further demonstrates the reliability of the experimental results.

Based on the above-presented results, we can conclude with the following points:

- With the increasing volume fractions, the PVA fibers show a significantly and linearly increased enhancement to the UHPC-PVAs (see Figure 10); as the PVA fiber volume fraction increased from 0.5% to 2%, the strength of the UHPC-PVA materials increased by 20.7%, 26.3% and 24.3%, respectively, in the cases with $C_l = 0$, 20 and 40 mm;
- In the experimental deflection-force diagrams of the specimens with a greater PVA fiber volume fraction, there exist non-negligible yield behaviors and residual strengths before and after the peak points, reflecting the brittle–ductile transition due to the PVA fiber reinforcement (see Figures 7–9);
- Due to the randomness of the fibers and initial defect distribution in the produced UHPC-PVA beam specimens, the crack patterns obtained by simulations show some differences to those from the experiments, which is reasonable;
- The obvious differences between the numerical crack patterns in the cases with the same initial cut position and similar structure strengths (shown in Figure 15) indicate that the proposed approach can reasonably describe the interaction between the fibers and matrix, as well as the reinforcement of the PVA fibers on the UHPC materials.

Remark: Although the peak values of applied force predicted by the proposed approach are very close to those obtained by the experiments, the behaviors described by the numerical deflection-force diagrams are different from all the experimental observations excepting the cases with $C_l = 40$ mm and $V_f = 0.5\%$. One of the reasons should be the use of the ADR algorithm. As reported in [9,45], the quasi-static solutions obtained by using the ADR algorithm will become closer to the static solutions but this involves greater computing costs. Another reason for the difference should be the linear microconstitutive relationship used to describe the bond behavior. In fact, such a constitutive relationship is for the prototype microelastic brittle materials. As discussed in [38,39], to accurately characterize the post-peak mechanical behaviors of ductile materials, the linear micro-constitutive relationship is not enough; bilinearity, trilinear or other more advanced non-liear constitutive relationships with more controlling parameters are needed.

Consequently, more efforts in the development of more appropriate constitutive relations and solution algorithms considering both the accuracy and computational efficiency should be made in the future to accurately simulate the mechanical and failure behaviors of the UHPC-PVA materials.

Author Contributions: Conceptualization, K.Z., T.N. and J.Z.; methodology, K.Z., T.N. and J.Z.; software, T.N.; validation, K.Z., T.N. and J.Z.; formal analysis, K.Z., T.N.; investigation, K.Z., T.N.; resources, W.Y. and S.Z.; data curation, K.Z. and T.N.; writing—original draft preparation, K.Z., T.N. and J.Z.; writing—review and editing, W.W., X.C., M.Z. and U.G.; visualization, T.N.; supervision, T.N.; project administration, K.Z., T.N. and U.G.; funding acquisition, K.Z. and T.N. All authors have read and agreed to the published version of the manuscript.

Funding: This research is financially supported by the Natural Science Foundation of the Higher Education Institutions of Jiangsu Province of China grant number 20KJB560006; National Natural Science Foundation of China, Grants 42207226 and 11902111; Funds for National Science Foundation for Outstanding Young Scholars, Grant 42125702; Funds for Creative Research Groups of China, Grant 41521002; Natural Science Foundation of Sichuan Province, Grant 2022NSFSC0003; State Key Laboratory of Geohazard Prevention and Geoenvironment Protection Independent Research Project SKLGP2021Z026; State Key Laboratory for GeoMechanics and Deep Underground Engineering, China University of Mining & Technology /China University of Mining & Technology, Beijing, grant No. SKLGDUEK2223.

Institutional Review Board Statement: Not applicable.

Informed Consent Statement: Not applicable.

Data Availability Statement: The data presented in this study are available on request from the corresponding author.

Acknowledgments: T. Ni would like to acknowledge the support he received from MIUR under the research project PRIN2017-DEVISU. U. Galvanetto and M. Zaccariotto would like to acknowledge the support they received from MIUR under the research project PRIN2017-DEVISU and from University of Padua under the research project BIRD2020 NR.202824/20.

Conflicts of Interest: The authors declare no conflict of interest.

Abbreviations

List of Symbols

C_l	distance of the initial crack to the middle of the beam specimen
V_f	PVA fiber volume fraction
ρ	mass density of the material
σ	stress tensor
ε	strain tensor
C	elasticity tensor
x, x'	location vector of material points
u	displacement vector
ξ	relative position vector of two material points
η	relative displacement vector of two material points
$\ddot{u}$	acceleration vector of material point
b	body force density
$\mathcal{H}_x$	neighborhood associated with the material point x
f	bond force density
c, κ	normal and tangential micro moduli of the bond
ℓ, γ	longitudinal and tangential deformations of the bond
n	unit directional vector along the deformed bond

δ	horizon radius
w	strain energy density stored in the deformed bond
w_c	critical strain energy density of the bond
G_c	critical energy release rate for mode I fracture
ϱ	characteristic function describing the connection status of bonds
φ_x	damage value at point x
M^{FE}, K^{FE}	mass and stiffness matrices of FE equations
N_u	shape functions for displacement
L	differential operator
D	elastic matrix
E, v	Young's modulus and Poisson's ratio
M^{PD}, K^{PD}	mass and stiffness matrices of PD equations
$\ddot{U}, \dot{U}, U, E, F$	acceleration, velocity, displacement, strain and force vectors
G	non-local strain coefficient matrix
V_f^g	global numerical volume fraction of PVA fibers in the discrete model
$V_f^{l_i}$	local numerical volume fraction of PVA fibers related to node i
P_m, P_f	parameters of the matrix and fiber materials
P_i	average parameter at node i
M^{Coup}, K^{Coup}	mass and stiffness matrices of coupled model
M, C_d, K	fictitious mass, damping and stiffness matrices of the system
Δt	time increment
M_{ii}	i^{th} principal value of the fictitious mass matrix
K_{ij}	elements of the global stiffness matrix
c_d	system damping coefficient
K_t^n	local diagonal tangent stiffness matrix at the nth iteration
E_m, v_m	Young's modulus and Poisson's ratio of the matrix materials
E_f, v_f	Young's modulus and Poisson's ratio of the fiber materials
h_m, h_f	geometrically characteristic lengths of the matrix and fiber materials
Δx	grid size
m	ratio of the PD horizon radius to the grid size

Appendix A. Numerical Simulation Results: Crack Surfaces in the Beam Specimens

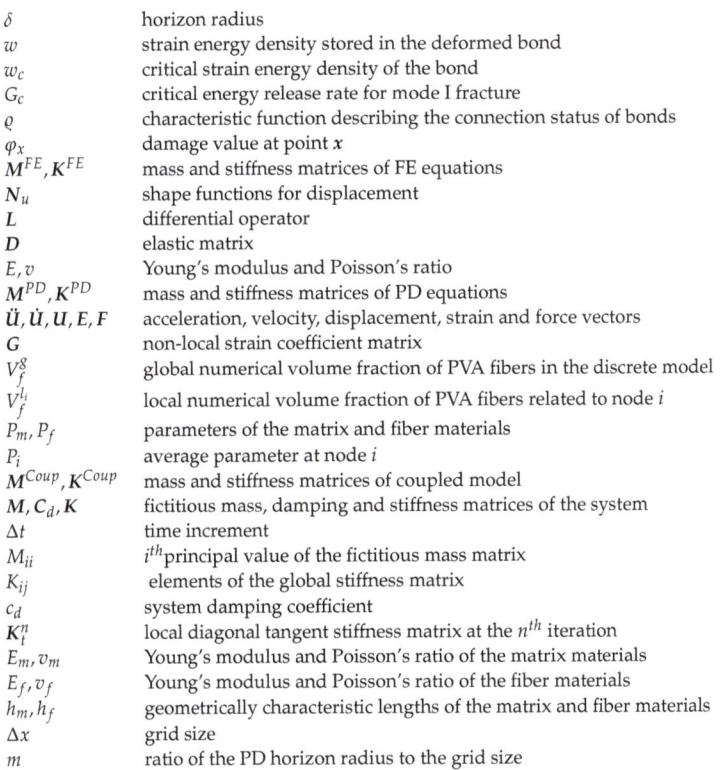

(a) $V_f = 0.5\%$ (b) $V_f = 1\%$

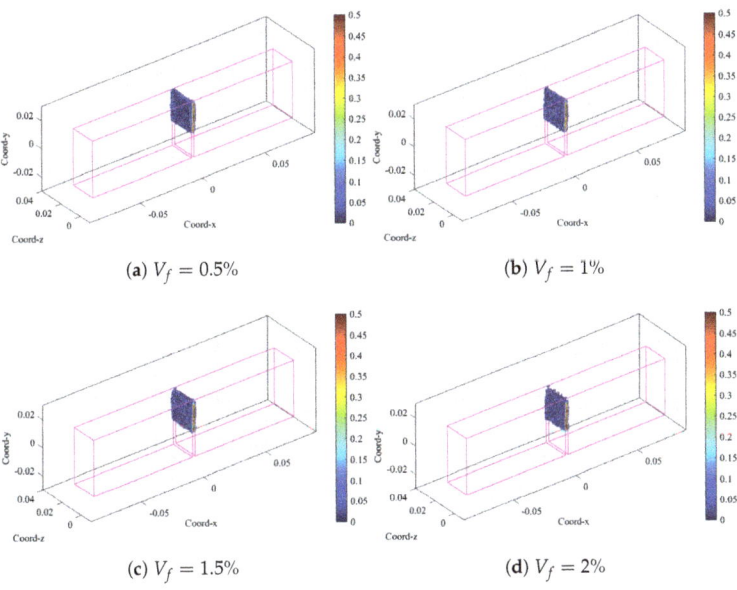

(c) $V_f = 1.5\%$ (d) $V_f = 2\%$

Figure A1. Damage levels in the simulated UHPC-PVA beam specimens with $C_l = 0$ mm.

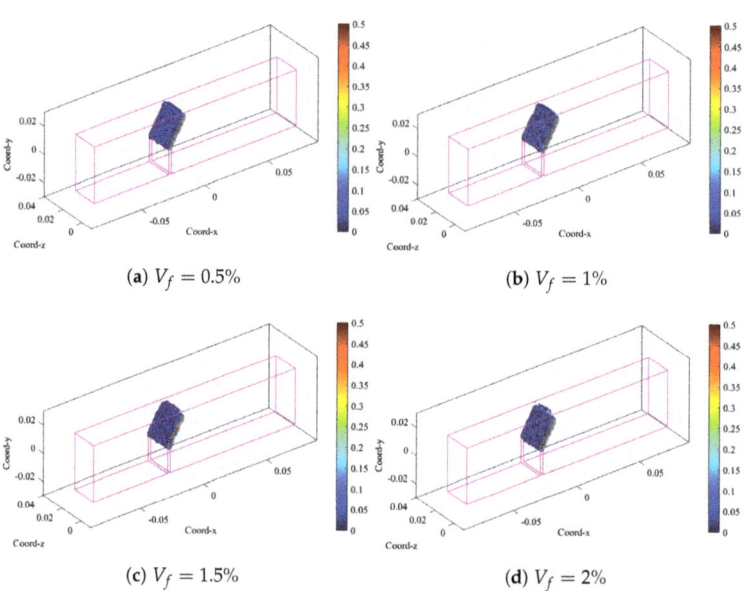

Figure A2. Damage levels in the simulated UHPC-PVA beam specimens with $C_l = 20$ mm.

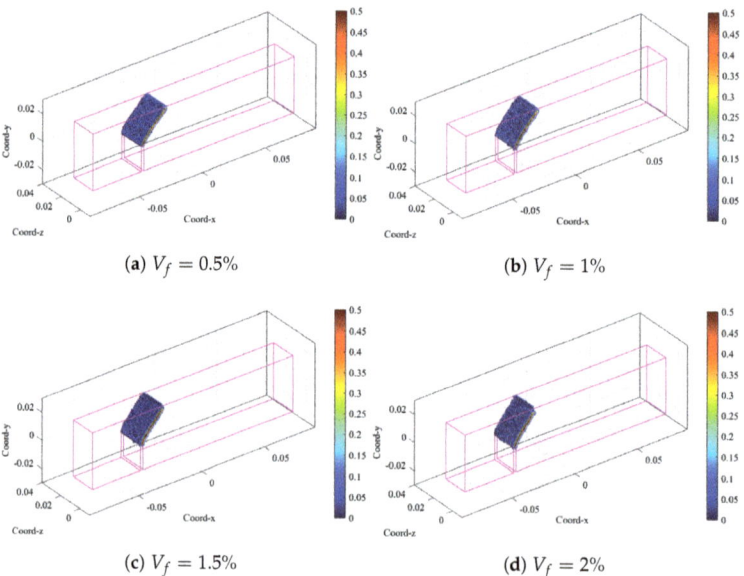

Figure A3. Damage levels in the simulated UHPC-PVA beam specimens with $C_l = 40$ mm.

References

1. Zhu, Y.; Hussein, H.; Kumar, A.; Chen, G. A review: Material and structural properties of uhpc at elevated temperatures or fire conditions. *Cem. Concrete Compos.* **2021**, *123*, 104212. [CrossRef]
2. Bajaber, M.; Hakeem, I. Uhpc evolution, development and utilization in construction: A review. *J. Mater. Res. Technol.* **2021**, *10*, 1058–1074. [CrossRef]
3. Shao, Y.; Billington, S.L. Impact of cyclic loading on longitudinally-reinforced uhpc flexural members with different fiber volumes and reinforcing ratios. *Eng. Struct.* **2021**, *241*, 112454. [CrossRef]

4. Folgar, F.; Tucker, C.L., III. Orientation behavior of fibers in concentrated suspensions. *J. Reinf. Plast. Compos.* **1984**, *3*, 98–119. [CrossRef]
5. Zhou, B.; Uchida, Y. Relationship between fiber orientation/distribution and post-cracking behaviour in ultra-high-performance fiber-reinforced concrete (uhpfrc). *Cem. Concr. Compos.* **2017**, *83*, 66–75. [CrossRef]
6. Kanda, T.; Li, V.C. Interface property and apparent strength of high-strength hydrophilic fiber in cement matrix. *J. Mater. Civ. Eng.* **1998**, *10*, 5–13. [CrossRef]
7. Noushini, A.; Vessalas, K.; Samali, B. Static mechanical properties of polyvinyl alcohol fibre reinforced concrete (pva-frc). *Mag. Concr. Res.* **2014**, *66*, 465–483. [CrossRef]
8. Zhang, K.; Zhao, L.-Y.; Ni, T.; Zhu, Q.-Z.; Shen, J.; Fan, Y.-H. Experimental investigation and multiscale modeling of reactive powder cement pastes subject to triaxial compressive stresses. *Constr. Build. Mater.* **2019**, *224*, 242–254. [CrossRef]
9. Zhang, K.; Ni, T.; Sarego, G.; Zaccariotto, M.; Zhu, Q.; Galvanetto, U. Experimental and numerical fracture analysis of the plain and polyvinyl alcohol fiber-reinforced ultra-high-performance concrete structures. *Theor. And Applied Fract. Mech.* **2020**, *108*, 102566. [CrossRef]
10. Yoo, D.-Y.; Banthia, N. Mechanical properties of ultra-high-performance fiber-reinforced concrete: A review. *Cem. Concr. Compos.* **2016**, *73*, 267–280. [CrossRef]
11. Gesoglu, M.; Güneyisi, E.; Muhyaddin, G.F.; Asaad, D.S. Strain hardening ultra-high performance fiber reinforced cementitious composites: Effect of fiber type and concentration. *Compos. Part Eng.* **2016**, *103*, 74–83. [CrossRef]
12. Voutetaki, M.E.; Naoum, M.C.; Papadopoulos, N.A.; Chalioris, C.E. Cracking diagnosis in fiber-reinforced concrete with synthetic fibers using piezoelectric transducers. *Fibers* **2022**, *10*, 5. [CrossRef]
13. Bhosale, A.B.; Prakash, S.S. Crack propagation analysis of synthetic vs. steel vs. hybrid fibre-reinforced concrete beams using digital image correlation technique. *Int. J. Concr. Struct. Mater.* **2020**, *14*, 1–19. [CrossRef]
14. Sahoo, S.; Selvaraju, A.K. Mechanical characterization of structural lightweight aggregate concrete made with sintered fly ash aggregates and synthetic fibres. *Cem. Concr. Compos.* **2020**, *113*, 103712. [CrossRef]
15. Yoo, D.-Y.; Kim, S.; Park, G.-J.; Park, J.-J.; Kim, S.-W. Effects of fiber shape, aspect ratio and volume fraction on flexural behavior of ultra-high-performance fiber-reinforced cement composites. *Compos. Struct.* **2017**, *174*, 375–388. [CrossRef]
16. Gontarz, J.; Podgórski, J. Analysis of crack propagation in a "pull-out" test. *Stud. Geotech. Mech.* **2019**, *41*, 160–170. [CrossRef]
17. Malvar, L.; Warren, G. Fracture energy for three-point-bend tests on single-edge-notched beams. *Exp. Mech.* **1988**, *28*, 266–272. [CrossRef]
18. Kachlakev, D.; Miller, T.R.; Yim, S.; Chansawat, K.; Potisuk, T. *Finite Element Modeling of Concrete Structures Strengthened with Frp Laminates*; Technical Report; Oregon Department of Transportation, Research Section : Salem, OR, USA, 2001.
19. Lu, X.; Ye, L.; Teng, J.; Jiang, J. Meso-scale finite element model for frp sheets/plates bonded to concrete. *Eng. Struct.* **2005**, *27*, 564–575. [CrossRef]
20. Chen, G.; Chen, J.; Teng, J. On the finite element modeling of rc beams shear-strengthened with frp. *Constr. Build. Mater.* **2012**, *32*, 13–26. [CrossRef]
21. Shafieifar, M.; Farzad, M.; Azizinamini, A. Experimental and numerical study on mechanical properties of ultra high performance concrete (uhpc). *Constr. Build. Mater.* **2017**, *156*, 402–411. [CrossRef]
22. Meng, W.; Khayat, K.H. Experimental and numerical studies on flexural behavior of ultrahigh-performance concrete panels reinforced with embedded glass fiber-reinforced polymer grids. *Transp. Res. Rec.* **2016**, *2592*, 38–44. [CrossRef]
23. Ingraffea, A.R.; Saouma, V. Numerical modeling of discrete crack propagation in reinforced and plain concrete. In *Fracture Mechanics of Concrete: Structural Application and Numerical Calculation*; Springer: Berlin/Heidelberg, Germany, 1985; pp. 171–225.
24. Rabczuk, T.; Akkermann, J.; Eibl, J. A numerical model for reinforced concrete structures. *Int. J. Solids Struct.* **2005**, *42*, 1327–1354. [CrossRef]
25. Kodur, V.; Dwaikat, M. A numerical model for predicting the fire resistance of reinforced concrete beams. *Cem. Concr. Compos.* **2008**, *30*, 431–443. [CrossRef]
26. Silling, S.A. Reformulation of elasticity theory for discontinuities and long-range forces. *J. Mech. Phys. Solids* **2000**, *48*, 175–209. [CrossRef]
27. Silling, S.A.; Epton, M.; Weckner, O.; Xu, J.; Askari, E. Peridynamic states and constitutive modeling. *J. Elast.* **2007**, *88*, 151–184. [CrossRef]
28. Chowdhury, S.R.; Rahaman, M.M.; Roy, D.; Sundaram, N. A micropolar peridynamic theory in linear elasticity. *Int. J. Solids Struct.* **2015**, *59*, 171–182. [CrossRef]
29. Zhu, Q.-z.; Ni, T. Peridynamic formulations enriched with bond rotation effects. *Int. J. Eng. Sci.* **2017**, *121*, 118–129. [CrossRef]
30. Oterkus, E.; Madenci, E.; Weckner, O.; Silling, S.; Bogert, P.; Tessler, A. Combined finite element and peridynamic analyses for predicting failure in a stiffened composite curved panel with a central slot. *Compos. Struct.* **2012**, *94*, 839–850. [CrossRef]
31. Rahimi, M.N.; Kefal, A.; Yildiz, M.; Oterkus, E. An ordinary state-based peridynamic model for toughness enhancement of brittle materials through drilling stop-holes. *Int. J. Mech. Sci.* **2020**, *182*, 105773. [CrossRef]
32. Basoglu, M.F.; Kefal, A.; Zerin, Z.; Oterkus, E. Peridynamic modeling of toughening enhancement in unidirectional fiber-reinforced composites with micro-cracks. *Compos. Struct.* **2022**, *297*, 115950. [CrossRef]
33. Ongaro, G.; Bertani, R.; Galvanetto, U.; Pontefisso, A.; Zaccariotto, M. A multiscale peridynamic framework for modeling mechanical properties of polymer-based nanocomposites. *Eng. Fract. Mech.* **2022**, *274*, 108751. [CrossRef]

34. Gerstle, W.; Sau, N.; Silling, S. Peridynamic modeling of concrete structures. *Nucl. Eng. Des.* **2007**, *237*, 1250–1258.
35. Yaghoobi, A.; Chorzepa, M.G. Meshless modeling framework for fiber reinforced concrete structures. *Comput. Struct.* **2015**, *161*, 43–54. [CrossRef]
36. Zhou, W.; Liu, D.; Liu, N. Analyzing dynamic fracture process in fiber-reinforced composite materials with a peridynamic model. *Eng. Fract. Mech.* **2017**, *178*, 60–76. [CrossRef]
37. Yaghoobi, A.; Chorzepa, M.G. Fracture analysis of fiber reinforced concrete structures in the micropolar peridynamic analysis framework. *Eng. Mech.* **2017**, *169*, 238–250. [CrossRef]
38. Zhang, Y.; Qiao, P. A fully-discrete peridynamic modeling approach for tensile fracture of fiber-reinforced cementitious composites. *Eng. Fract. Mech.* **2021**, *242*, 107454. [CrossRef]
39. Hattori, G.; Hobbs, M.; Orr, J. A review on the developments of peridynamics for reinforced concrete structures. *Arch. Comput. Methods Eng.* **2021**, *28*, 4655–4686. [CrossRef]
40. Chen, Z.; Chu, X. Numerical fracture analysis of fiber-reinforced concrete by using the cosserat peridynamic model. *J. Peridynamics Nonlocal Model.* **2022**, *4*, 88–111. [CrossRef]
41. Zhou, L.; Zhu, S.; Zhu, Z.; Xie, X. Simulations of fractures of heterogeneous orthotropic fiber-reinforced concrete with pre-existing flaws using an improved peridynamic model. *Materials* **2022**, *15*, 3977. [CrossRef]
42. Ni, T.; Zaccariotto, M.; Zhu, Q.-Z.; Galvanetto, U. Static solution of crack propagation problems in peridynamics. *Comput. Methods Appl. Mech. Eng.* **2019**, *346*, 126–151. [CrossRef]
43. Ni, T.; Pesavento, F.; Zaccariotto, M.; Galvanetto, U.; Zhu, Q.-Z.; Schrefler, B.A. Hybrid fem and peridynamic simulation of hydraulic fracture propagation in saturated porous media. *Comput. Methods Appl. Mech. Eng.* **2020**, *366*, 113101. [CrossRef]
44. Sun, W.; Fish, J. Coupling of non-ordinary state-based peridynamics and finite element method for fracture propagation in saturated porous media. *Int. J. Numer. Anal. Methods Geomech.* **2021**, *45*, 1260–1281. [CrossRef]
45. Ni, T.; Zaccariotto, M.; Zhu, Q.-Z.; Galvanetto, U. Coupling of fem and ordinary state-based peridynamics for brittle failure analysis in 3d. *Mech. Adv. Mater. Struct.* **2021**, *28*, 875–890. [CrossRef]
46. Ni, T.; Pesavento, F.; Zaccariotto, M.; Galvanetto, U.; Schrefler, B.A. Numerical simulation of forerunning fracture in saturated porous solids with hybrid fem/peridynamic model. *Comput. Geotech.* **2021**, *133*, 104024. [CrossRef]
47. Ni, T.; Sanavia, L.; Zaccariotto, M.; Galvanetto, U.; Schrefler, B.A. Fracturing dry and saturated porous media, peridynamics and dispersion. *Comput. Geotech.* **2022**, *151*, 104990. [CrossRef]
48. Ni, T.; Zaccariotto, M.; Fan, X.; Zhu, Q.; Schrefler, B.A.; Galvanetto, U. A peridynamic differential operator-based scheme for the extended bond-based peridynamics and its application to fracture problems of brittle solids. *Eur. J. Mech.-A/Solids* **2022**, *97*, 104853. [CrossRef]
49. Foster, J.T.; Silling, S.A.; Chen, W. An energy based failure criterion for use with peridynamic states. *Int. J. Multiscale Comput.* **2011**, *9*, 675–688. [CrossRef]
50. Zaccariotto, M.; Mudric, T.; Tomasi, D.; Shojaei, A.; Galvanetto, U. Coupling of fem meshes with peridynamic grids. *Comput. Methods Appl. Mech. Eng.* **2018**, *330*, 471–497. [CrossRef]
51. Ni, T.; Zhu, Q.-Z.; Zhao, L.-Y.; Li, P.-F. Peridynamic simulation of fracture in quasi brittle solids using irregular finite element mesh. *Eng. Fract. Mech.* **2018**, *188*, 320–343. [CrossRef]
52. Lewis, R.W.; Schrefler, B.A. *The Finite Element Method in the Static and Dynamic Deformation and Consolidation of Porous Media*; John Wiley: Hoboken, NJ, USA, 1998.
53. Madenci, E.; Barut, A.; Futch, M. Peridynamic differential operator and its applications. *Comput. Methods Appl. Mech. Eng.* **2016**, *304*, 408–451. [CrossRef]
54. Underwood, P. Dynamic relaxation. *Comput. Methods Transient Anal.* **1983**, *1*, 245–265.
55. Kilic, B.; Madenci, E. An adaptive dynamic relaxation method for quasi-static simulations using the peridynamic theory. *Theor. Appl. Fract.* **2010**, *53*, 194–204. [CrossRef]
56. Rabczuk, T.; Ren, H. A peridynamics formulation for quasi-static fracture and contact in rock. *Eng. Geol.* **2017**, *225*, 42–48. [CrossRef]
57. Wang, Y.; Zhou, X.; Wang, Y.; Shou, Y. A 3-d conjugated bond-pair-based peridynamic formulation for initiation and propagation of cracks in brittle solids. *Int. J. Solids Struct.* **2018**, *134*, 89–115. [CrossRef]
58. Rilem, D.R. Determination of the fracture energy of mortar and concrete by means of three-point bend tests on notched beams. *Mater. Struct.* **1985**, *18*, 285–290.

Disclaimer/Publisher's Note: The statements, opinions and data contained in all publications are solely those of the individual author(s) and contributor(s) and not of MDPI and/or the editor(s). MDPI and/or the editor(s) disclaim responsibility for any injury to people or property resulting from any ideas, methods, instructions or products referred to in the content.

Article

Fabrication of Nylon 6-Montmorillonite Clay Nanocomposites with Enhanced Structural and Mechanical Properties by Solution Compounding

Ahmed M. Abdel-Gawad [1], Adham R. Ramadan [2], Araceli Flores [3] and Amal M. K. Esawi [1,*]

[1] Department of Mechanical Engineering, The American University in Cairo, AUC Avenue, P.O. Box 74, New Cairo 11835, Egypt
[2] Department of Chemistry, The American University in Cairo, AUC Avenue, P.O. Box 74, New Cairo 11835, Egypt
[3] Department of Polymer Physics, Elastomers and Applications Energy, Institute of Polymer Science and Technology (ICTP), CSIC, Juan de la Cierva 3, 28006 Madrid, Spain
* Correspondence: a_esawi@aucegypt.edu; Tel.: +20-22-615-3102

Citation: Abdel-Gawad, A.M.; Ramadan, A.R.; Flores, A.; Esawi, A.M.K. Fabrication of Nylon 6-Montmorillonite Clay Nanocomposites with Enhanced Structural and Mechanical Properties by Solution Compounding. *Polymers* 2022, 14, 4471. https://doi.org/10.3390/polym14214471

Academic Editors: Vineet Kumar and Xiaowu Tang

Received: 1 October 2022
Accepted: 14 October 2022
Published: 22 October 2022

Publisher's Note: MDPI stays neutral with regard to jurisdictional claims in published maps and institutional affiliations.

Copyright: © 2022 by the authors. Licensee MDPI, Basel, Switzerland. This article is an open access article distributed under the terms and conditions of the Creative Commons Attribution (CC BY) license (https://creativecommons.org/licenses/by/4.0/).

Abstract: Melt compounding has been favored by researchers for producing nylon 6/montmorillonite clay nanocomposites. It was reported that high compatibility between the clay and the nylon6 matrix is essential for producing exfoliated and well-dispersed clay particles within the nylon6 matrix. Though solution compounding represents an alternative preparation method, reported research for its use for the preparation of nylon 6/montmorillonite clay is limited. In the present work, solution compounding was used to prepare nylon6/montmorillonite clays and was found to produce exfoliated nylon 6/montmorillonite nanocomposites, for both organically modified clays with known compatibility with nylon 6 (Cloisite 30B) and clays with low/no compatibility with nylon 6 (Cloisite 15A and Na$^+$-MMT), though to a lower extent. Additionally, solution compounding was found to produce the more stable α crystal structure for both blank nylon6 and nylon6/montmorillonite clays. The process was found to enhance the matrix crystallinity of blank nylon6 samples from 36 to 58%. The resulting composites were found to possess comparable mechanical properties to similar composites produced by melt blending.

Keywords: montmorillonite; nylon 6; nanocomposites; solution compounding; static melt annealing

1. Introduction

The production of polymer/layered silicate nanocomposites (PLSNs) with unique properties has been of interest to scientists and researchers. In 1989, researchers at the Toyota research center reported significant enhancements in the thermal and mechanical behaviour of nylon-6 after adding low contents of montmorillonite clay (MMT). This exposed an enormous research potential for this class of materials. Improvements in the mechanical, thermal, flame retardation, and gas separation properties as well as applications in biomedical and wastewater treatment fields have been frequently reported in the literature [1–8]. This strongly puts forward PLSNs as an attractive alternative to conventional micro-composites.

The mechanical and thermal properties of composites are generally dependent on the physico-chemical interaction between the matrix and the reinforcing phase. Due to the hydrophilic character of the silicate layers in the pristine clay (filler), and the organophilic character of most engineering polymers (matrix), interactions between the filler and the matrix are usually not favorable. This can be overcome by incorporating organic modifiers in the clay structure, which is usually achieved by ion exchange reactions where cations such as primary, secondary, tertiary, and quaternary alkylammonium or alkylphosphonium are used. Organic modification of the clay structure also leads to the increase in the distance between the silicate layers and inter-gallery spacing, which, in turn, facilitates the intercalation of the polymer matrix between the silicate layers [1–9].

It has been widely reported that significant improvements in properties can only be achieved for well-exfoliated and well-dispersed silicate layers [9]. Numerous studies addressed different preparation methods to improve the exfoliation and dispersion of the silicate layers [9–19]. The most common method for the preparation of PLSNs has been melt compounding. The key advantages are its speed and simplicity, as well as its compatibility with standard industrial techniques. This technique utilizes mechanical shearing forces, applied during extrusion or injection molding, to increase the inter-gallery spacing of the silicate layers, thus allowing the polymeric chains to diffuse into the clay galleries (intercalation). The technique can also lead to the separation of the silicate layers, resulting in the loss of their stacked form (exfoliation). However, complete exfoliation of the clay platelets was reported to be difficult to achieve [9]. The type of polymer affects the degree of intercalation or exfoliation, which are more easily achieved when organoclays are melt blended with polar polymers, such as polystyrene, polyamides, etc. but are a lot more challenging for apolar polyolefins such as polypropylene, polyethylene, etc. In addition, the change in melt properties, such as viscosity, upon the addition of the clay particles has been found to lead to polymer degradation under conditions of high shear rates, so careful selection of process parameters is crucial. Another challenge is that high melt temperatures can result in degradation of the organic modifier of the clay. Pre-exfoliation of clays prior to melt compounding by ultrasonication is common; however, as reported by Martinez-Colunga et al. [10], who prepared polyethylene–montmorillonite nanocomposites using different ultrasonic powers, restacking of the exfoliated nanolayers can occur after ultrasonication. To avoid the production of microcomposites, functionalization with polar monomers such as maleic anhydride (MA) or the addition of compatibilizers is usually utilized [16,20].

Some researchers have reported the possibility of the diffusion of the polymeric chains into clay galleries when samples are allowed to anneal above their melting temperature without being subjected to any shear forces, (sometimes referred to as unassisted exfoliation, melt intercalation or static melt annealing). For example, Vaia and Giannelis [21] indicated the possibility of the intercalation of organically-modified clays by polystyrene. In addition, Paci et al. [22] and Dennis et al. [23] reported full exfoliation for Cloisite30B/nylon6 using unassisted exfoliation. The degree of exfoliation was reported to depend on the annealing time needed for the polymer to diffuse into the clay galleries, which in turn depended on the molecular weight of the polymer and clay content. Investigations carried out with other clays presenting low compatibility between the polymer and the clay, such as Cloisite 25A, reported poor results. It was therefore suggested that polar interactions between the polymer and the clay are necessary for producing intercalated/exfoliated structures when shear forces are absent.

Solution compounding presents another approach for the preparation of PLSNs. The process entails the dispersion of the silicate layers in a solvent, which often results in swelling and then mixing the silicate solution with the dissolved polymer. Intercalation takes place when the polymer chains replace the solvent in the silicate layers galleries. Solvent removal is usually by precipitation in different non-solvents or by evaporation. Controlled removal of the solvent is crucial to the success of the process as any solvent residue can lead to polymer degradation during further processing. One of the benefits of solution compounding is that the agitation or stirring of the silicate solution prior to mixing with the dissolved polymer facilitates the separation and dispersion of the silicate layers in the final composite. Accordingly, the process can produce intercalated structures for polymers with little or no polarity. In addition, it has the advantage of not requiring high temperatures that may degrade the organic modifiers in the various clays. The process can be attractive for producing thin films with oriented intercalated clays. Selection of an appropriate solvent is a key requirement for the success of solution intercalation [9,24]. Many studies have focused on water-soluble polymers. For example, Strawhecker and

Manias [25] prepared polyvinyl alcohol (PVA)/montmorillonite (MMT) nanocomposites, whereas Aranda and Ruiz-Hitzky [26] prepared polyethylene oxide (PEO)/Na$^+$-MMT nanocomposites using a mixture of water and methanol as solvents. Similarly, Wu et al. [27] prepared intercalated PEO/Na$^+$-MMT and PEO/Na$^+$-hectorite nanocomposites. Some studies used non-aqueous solvents; for example, Ogata et al. [28] used chloroform to compound polylactic acid (PLA) with organically-modified MMT, but the process did not yield an intercalated structure.

Using solution compounding with engineering polymers has been reported in fewer studies. Yano et al. [29] prepared polyimide/MMT nanocomposites using a dimethylacetamide (DMAC) solution of polyamic acid and a DMAC dispersion of MMT modified with dodecylammonium cations. 12CH$_3$-MMT, 12COOH-MMT, and Cloisite 10A-MMT were used in their study. X-ray diffraction (XRD) analysis showed that the type of clay strongly influenced the obtained structure with uniform dispersion, and the exfoliated structure was only observed for the nanocomposites with 12CH$_3$-MMT. N-methyl-2-pyrrolidone was used as the solvent to prepare polyimide/MMT nanocomposites in another study [30]. A fully-exfoliated structure was reported when using low MMT contents, whereas nanocomposites with higher MMT contents were only partially exfoliated. High-density polyethylene (HDPE) and poly-dimethylsiloxane (PDMS) nanocomposites were also prepared by solution compounding [31,32]. Recently, Filippi et al. [20] reported that solution blending was unable to lead to intercalation in composites of ethylene copolymers and organically modified montmorillonite.

In addition to the study by Wu et al. [12], who prepared flame retardant polyamide 6/nanoclay/intumescent nanocomposite fibers through electrospinning using formic acid as a solvent, to the best of our knowledge, solution compounding was only employed in one other study of nylon6/layered silicate nanocomposites. In their work, Paci et al. [22] also prepared nylon6/Cloisite 30B nanocomposites using formic acid as a solvent. Their results showed that intercalation and even complete destruction of the silicate platelets stacking order is possible in cases of low filler content and small polymer/clay particles.

The current work focuses on the preparation of nylon-6/MMT nanocomposites by the solution compounding process. In addition to Cloisite 30B, which was investigated in the study by Paci as noted earlier, another MMT clay with a different organic modifier (Cloisite 15A) as well as unmodified clay (Na$^+$-MMT) were also used to prepare the composites. The structural morphologies of the resulting composites, their mechanical properties as well as their crystallization behavior were explored.

2. Experimental Procedure

2.1. Materials Used

Nylon6 (3 mm pellets) was purchased from Sigma-Aldrich, USA. Natural sodium-based montmorillonite (MMT) clay (Cloisite Na$^+$) and organically-modified sodium-based montmorillonite clays (Cloisite 15A and Cloisite 30B) were procured from Southern Clay Products, Inc., Austin, TX, USA. All of the clays used had an average particle size of 7 μm, as reported by the supplier. Table 1 presents the specifications of the clays used. Glacial acetic acid and 95% pure methanol were purchased from El Nasr Pharmaceutical Chemicals Co., Cairo, Egypt.

2.2. Sample Preparation

Composite samples (N6-Na$^+$, N6-15A, and N6-30B) were prepared by solution compounding of Cloisite Na$^+$, Cloisite 15A, and Cloisite 30B with nylon6. This entailed adding a suspension of each of the clays in 50 g glacial acetic acid to 50 g of nylon6 dissolved in 500 mL glacial acetic acid at 108 °C. This was followed by stirring until the mixture cools to room temperature. The amount of clay used was such that a final 5 wt% of clay in nylon6 was obtained after drying. Solvent evaporation was then carried out followed by flushing using methanol of any remaining acetic acid to give acetic acid-free composites. Those were dried at 90 °C until achieiving a constant weight. Blank samples of nylon6 without any clays (N6) were also prepared using the same routine to serve as reference samples.

Table 1. Specifications of MMT clays.

Clay	Compatibilizer	Gallery d-Spacing d_{001} (Å)	Organic Content (% Mass)
Cloisite Na$^+$	-	11.7	-
Cloisite 15A	$CH_3 - \overset{\overset{\displaystyle CH_3}{\displaystyle \mid +}}{\underset{\underset{\displaystyle HT}{\displaystyle \mid}}{N}} - HT$ T = (~65% C18; ~30% C16; ~5% C14)	31.5	43%
Cloisite 30B	$CH_3 - \overset{\overset{\displaystyle CH_2CH_2OH}{\displaystyle \mid +}}{\underset{\underset{\displaystyle CH_2CH_2OH}{\displaystyle \mid}}{N}} - T$ T = (~65% C18; ~30% C16; ~5% C14)	18.5	28%

The obtained powdered samples (composites as well as blanks) were then compression-molded at 240 °C for 5 min under a pressure of 65 MPa, to yield cylindrical samples 1 cm in diameter and 2 cm long. These were used for nano-indentation tests.

In order to investigate whether there are contributions to the attained morphologies from melt intercalation during the compression molding step, N6 samples were crushed into a fine powder and mixed with 5 wt% of the different MMT clays using Turbula® T2F mixer, at 96 rpm for 1 h. The mixtures were compression-molded using the same conditions specified above to produce additional composite samples (referred to as "mechanically-mixed" or "static melt annealing" samples).

2.3. XRD

A D8 Bruker X-ray diffractometer, operated at 40 kV and 30 mA, and using Cu Kα (λ = 0.1542 nm) was used to record diffraction patterns at middle angle, MAXS, (diffraction angle 2θ = 2° − 10°), in order to monitor the basal reflection (d_{001}). For each type of clay, four samples were analyzed: pristine clay (as-received); clay powder processed through the solution compounding routine without the addition of nylon6; nylon6 mixed with 5 wt% of Cloisite clay; and nylon6-Cloisite clay composite obtained by solution compounding. These four samples were analyzed in order to compare the effect of solution compounding on the basal reflection of the clays used. The Cloisite clay powder processed through the solution compounding routine without the addition of nylon6 served to divulge any effects of the solution compounding process on the clay basal reflection, and the blank N6 mixed with 5 wt% of Cloisite clay served to confirm that the significant decrease observed for the intensity of the d_{001} clay reflection was due to the exfoliation of the clay particles, and not merely to the lower clay content in the composite.

Two-dimensional wide-angle X-ray diffraction patterns, WAXS, (2θ = 8° − 35°) of pristine nylon 6 and the composites samples were obtained using a Micro Star rotating anode generator (Bruker, Karlsruhe, Germany) operating at 45 kV and 60 mA, and X-ray wavelength λ = 0.1542 nm. A Mar345 dtb image plate was used with a resolution of 3450 × 3450 pixels and 100 µm/pixel. The sample-to-detector distance was 250 mm. Diffraction patterns were analyzed using the FIT2D software [33]. All images showed isotropic diffraction rings indicating no preferred crystal orientation. The 2D diffraction patterns were azimuthally integrated to obtain intensity curves as a function of diffraction angle. The intensity profiles were fitted to several crystalline peaks and an amorphous halo as described in reference [34] and making use of the Peakfit program (Systat Software v4.12, San Jose, CA, USA). The degree of crystallinity, X_c, was calculated from the ratio of the area under the crystalline peaks to that of the total diffraction curve.

2.4. Nanoindentation

A Nanoindenter XP (Agilent, USA) was used for nanoindentation testing. Three arrays, each of twenty five indentations, were made at the top, middle, and bottom of the longitudinal cross-section of each sample. The distance between adjacent indentations was set to 100 µm in order to avoid the effect of interaction. A calibrated (Berkovich) tip was used under the continuous stiffness module (CSM). The indenter was loaded into the sample until a depth of 5000 nm. A constant strain rate of $0.05\ \text{s}^{-1}$ was maintained throughout the test. Thermal drift corrections were not performed in order to avoid possible creep of the tested polymer. However, the test was not started until the thermal drift was stabilized below 0.05 nm/s. In addition, the test was carried out overnight after leaving the samples for three hours inside the nanoindenter to thermally equilibrate. Minimum and maximum calculation depths were set to 2000 nm and 4000 nm, respectively.

2.5. Melt Flow Index

The melt flow index (MFI) in g/10 min (235 °C, 2.16 kg) was measured with RAY-RAN Melt Flow Indexer (Ray-Ran Test Equipment Ltd., Nuneaton, UK) according to ASTM D1238-04c. MFI values were considered as an indirect measure of the viscosity of the nanocomposite. MFI testing was conducted on nanocomposite materials prior to compression molding. Consistency of the flow of the molten polymer was ensured during testing by excluding any extrudate containing voids. The die was cleaned between successive runs to prevent contamination. Three MFI values were recorded for each sample.

2.6. TEM

Samples were covered with a protective sputter coating of Au-Pd, then subjected to Focused Ion Beam (FIB) milling, using the lift-out method, to obtain electron-thin foil sections. TEM analysis was conducted on a Tecnai F20 (200 kV) TEM.

3. Results and Discussion
3.1. X-ray Diffraction

Middle angle X-ray diffraction spectra are shown for $N6\text{-}Na^+$ (Figure 1), N6-15A (Figure 2), and N6-30B (Figure 3). Pristine Cloisite Na^+ (as-received) exhibited a peak corresponding to a basal spacing of 11.7 Å. This peak shifted to a lower angle when Cloisite Na^+ was subjected to the processing routine used for solution compounding without the addition of nylon6. The shift to a lower angle indicated a swelling of the clay structure as a result of stirring the clay in acetic acid. Mixed $N6/Na^+$ powder showed a noticeable decrease in the intensity of the basal reflection peak as compared to the pristine as-received clay. This was due to lower clay content in the sample. This peak however appeared at the same two-theta angle as the pristine as-received clay. Upon compounding Na^+ MMT with nylon-6, a significant decrease in the intensity of the basal reflection peak was observed. Furthermore, the peak appeared at a lower angle corresponding to a basal spacing of 12.8 Å. The decrease in the peak intensity of the compounded sample relative to the mixed sample (both containing 5 wt% clay) suggested partial exfoliation of the clay particles.

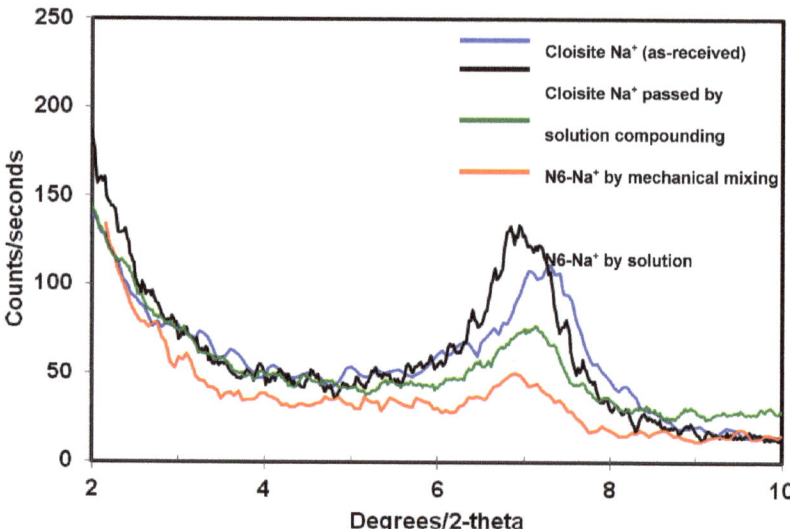

Figure 1. XRD diffraction patterns for the different Na⁺ and Na⁺ composite samples.

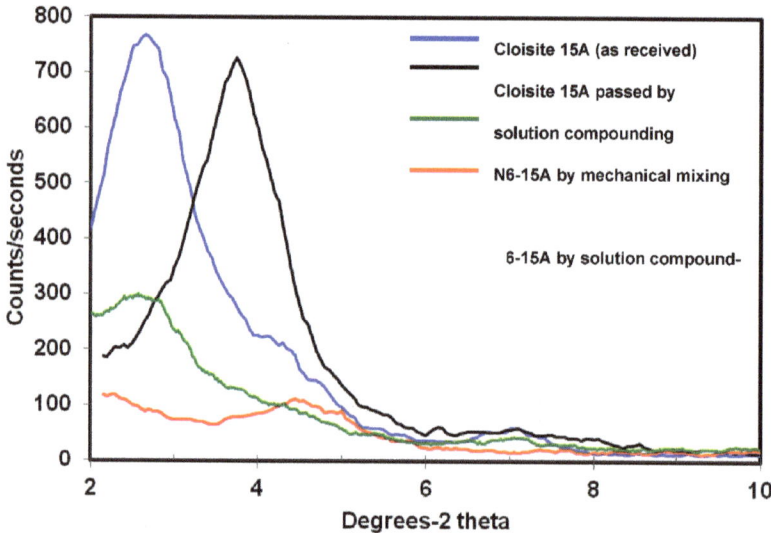

Figure 2. XRD diffraction patterns for the different 15A and 15A composite samples.

For Cloisite 15A (Figure 2), when subjecting the pristine clay to the solution compounding process without the addition of nylon6, the peak corresponding to the basal reflection shifted to a higher angle associated with the basal spacing of 23.6 Å. This decrease in basal spacing might be due to some destruction of the organic modifier. Lee and Char [35] argued that the organic modifier tails were more likely to bond to strongly acidic solvents, thus causing a collapse in the clay gallery spacing. Mixed N6/15A powder showed a noticeable decrease in the basal reflection peak intensity, due to the lower clay content in the sample, with the peak appearing at the same two-theta angle as the pristine as-received clay. Upon compounding with nylon6, the peak associated with basal reflection virtually disappeared, indicating significant clay exfoliation in the composite sample.

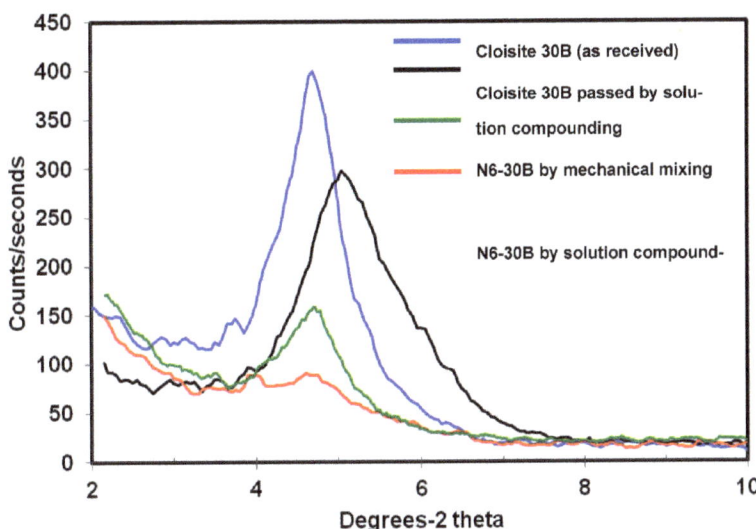

Figure 3. XRD diffraction patterns for the different 30B and 30B composite samples.

A similar trend was observed for Cloisite 30B samples, as shown in Figure 3 with the basal reflection peak shifting from a two-theta value, corresponding to 18.5 Å in the pristine as-received clay to a value corresponding to 17.5 Å for the clay subjected to the solution compounding routine, indicating a partial destruction of the organic modifier that could be due to rearrangement of its alkyl chains. Filippi et al. [36], who attributed a collapse in the interlayer spacing of Cloisite 30B to thermal degradation of the organic modifier after melt compounding 30B with polymer matrices, reported that the d-spacing collapse was reversed after dissolving the melt-compounded composite samples in appropriate solvents due to rearrangement of the alkyl chains of the clay modifier. Mixed nylon6-Cloisite 30B powders exhibited an XRD peak significantly lower in intensity due to the lower clay content and exhibiting a two-theta value similar to pristine as-received clay. Compounding Cloisite 30B with nylon6 led to the virtual disappearance of the basal reflection peak, indicating clay exfoliation in the composite.

Figure 4 illustrates the WAXS patterns of as-received and solution compounded nylon6, together with the solution compounded composites. Results show that nylon6 crystals mostly adopt the monoclinic α-form, which is characterized by two strongly diffracting peaks at $2\theta = 24°$ and $2\theta = 20.5°$. Only a very weak reflection at $2\theta = 21.5°$ associated to the γ crystal structure can be discerned in the WAXS pattern of the nylon6 original sample. It is well known that nylon6 exhibits polymorphic structures, α and γ being the most common [37]. The α-form appears to be more stable than the γ structure partly due to enhanced hydrogen bonding interaction. Several authors have reported that melt compounding silicate layers into nylon6 induced the formation of the γ form [38]. Researchers have reported that slow crystallization or crystallization at high temperatures favors the α form whereas rapid crystallization or crystallization at low temperatures favors the γ form [39].

It is interesting to note that preparing samples using the solution compounding technique did not significantly alter the type of crystal structure obtained, with α crystal structure continuing to prevail. Furthermore, the addition of layered silicates did not cause a change in the crystal structure, in contrast to other nylon-layered silicates composites reported in the literature [37]. Another relevant aspect is that no preferential crystal orientation has been observed on any of the composites or neat nylon6. Indeed, the 2D WAXS images show isotropic diffraction rings for all samples and also the relative intensity of the two main reflections at 2θ = 24° and 2θ = 20.5° was found to remain constant. This behavior is different from that reported in the literature for nylon6/30B prepared by melt compounding in which the reflection at 2θ = 20.5° disappears upon the addition of clay, indicating a strong orientation of both the clay platelets and the polymer crystallites due to the shear stresses involved [22].

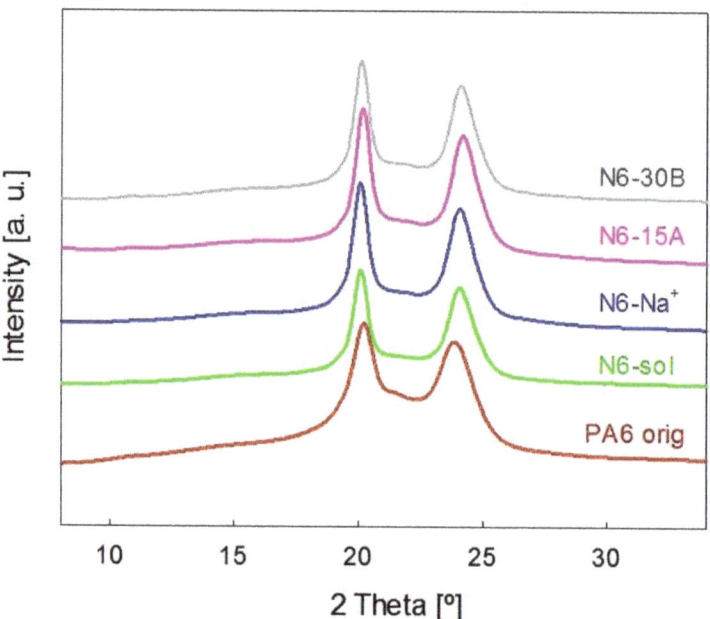

Figure 4. WAXS patterns of as-received and solution compounded nylon6, together with the solution compounded composites.

Concerning the amount of crystalline material, Table 2 shows similar X_c values for all the solution compounded materials (approx. 0.58), including the neat polymer and regardless of the type of clay filler in the composite. However, PA6orig shows significantly lower values (0.36). It is also apparent that the crystalline peaks are wider in the case of PA6orig. This suggests that the crystal size is approximately the same for all the samples except for PA6 orig, which shows more limited crystal size.

Table 2. Degree of crystallinity by WAXS, X_c of the various samples.

Sample	Degree of Crystallinity by WAXS, X_c
PA6 orig	0.36
N6-sol	0.58
N6-Na$^+$	0.57
N6-15A	0.58
N6-30B	0.58

The crystallization behavior of nylon6-clay nanocomposites processed by solution compounding has not been explored previously and therefore the current observations are reported for the first time. Results appear to indicate that the solution compounding process has a higher influence on the matrix crystallization behavior than does the polymer-silicate layers interaction.

3.2. Nanoindentation

Nanoindentation results, presented in Figures 5 and 6, show an improvement in the mechanical behaviour (modulus and hardness) of the nylon-6 blank sample prepared by solution compounding, as compared to the as-received polymer. The increase in modulus of the blank polymer can be attributed to the higher amount of crystals with larger crystal sizes that develop during solution compounding, as indicated earlier based on the WAXS results. Dissolution of nylon-6 followed by solvent evaporation resulted in finer particles compared to the as-received pellets that melt at a faster rate. Since compression molding was carried out at 240 °C for 5 min, solution-compounded samples were annealed at this temperature for a longer period, which affected their crystallization behaviour and which is known to depend on melt annealing time, as observed in [40]. With regards to composite samples, all samples of nylon-6 compounded with the different clays showed improved modulus and hardness compared to the N6 blank samples as a consequence of the intrinsic higher modulus of the clays. On the other hand, static melt annealing produced composites with much lower mechanical properties compared to solution compounded ones, thus confirming that the observed enhancements in mechanical properties are due to the solution compounding process. Table 3 presents enhancements in nanoindentation modulus and hardness.

Table 3. Nanoindentation modulus and hardness enhancements for the different composite samples.

	Modulus Enhancement		Hardness Enhancement	
	I ◊	II *	I ◊	II *
N6-Na⁺	12%	11%	10%	8%
N6-15A	17%	15%	19%	15%
N6-30B	18%	16%	14%	11%

◊ Percent enhancement values determined relative to the pristine polymer. * Percent enhancement values determined relative to the N6 blank.

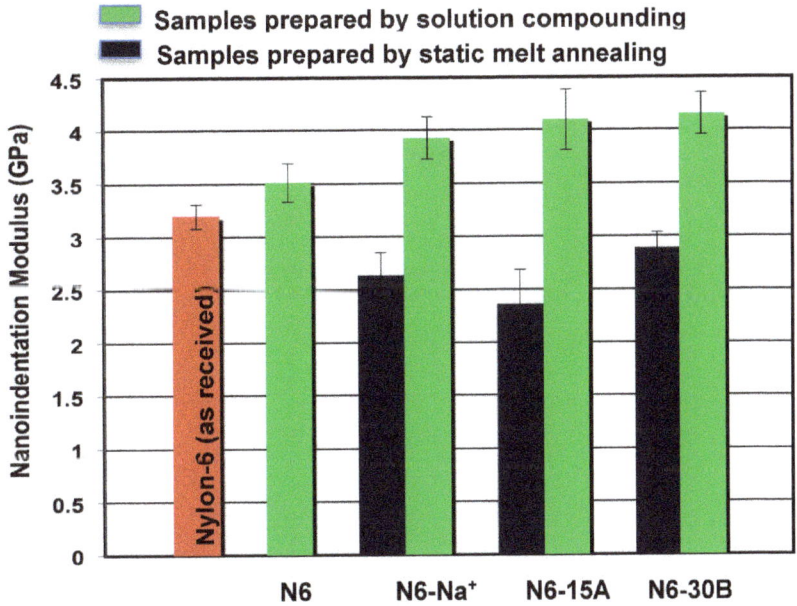

Figure 5. Average nanoindentation modulus values for the different composite samples.

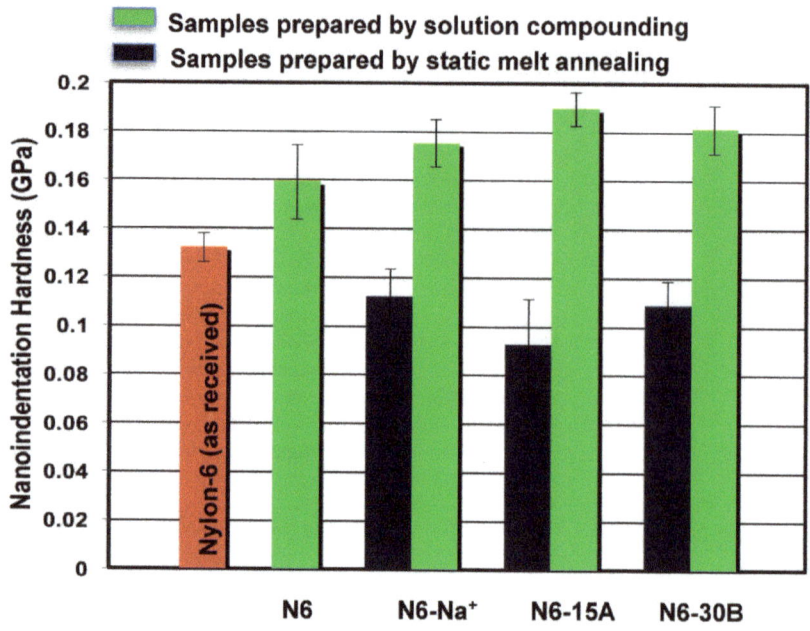

Figure 6. Average nanoindentation hardness values for the various samples.

3.3. Melt Flow Index

The MFI results in Figure 7 show a significant decrease in MFI for the N6 blank sample compared to the as-received polymer. This might be due to the effect of melt annealing for longer time on the molecular weight of the polymer [40]. All composite samples exhibited lower MFI values with the N6-30B sample exhibiting the lowest value, i.e., highest melt viscosity which can be attributed to the better compatibility between the clay's organic modifier and the polymer matrix. Similarly, the less pronounced decrease in MFI of the N6-Na$^+$ and N6-15A samples is believed to be due to the lower compatibility between nylon6 and the clay layers (Na$^+$ having no organic modifier and 15A having an organic modifier with lower compatibility with nylon6).

3.4. TEM

For the N6-Na$^+$ composite, TEM images at low and high magnifications are presented in Figure 8a,b, respectively. The images reveal a composite with a mixture of intercalated and delaminated silicate layers. However, upon analyzing several areas of the sample, it became apparent that there are areas with a low density of silicate layers. The non-uniform dispersion is believed to be due to the absence of organic modifier in Cloisite Na$^+$. This result agrees with the MAXS finding that showed a peak for the solution compounded composite that was quite similar in intensity and width to that of the mechanically-mixed composite.

The TEM images of the N6-15A composite are presented in Figure 9 at low and high magnifications (Figure 9a,b, respectively) and show individual clay layers uniformly dispersed within the polymer matrix. However, similar to Na$^+$, the examination of various areas across the sample revealed non-uniform clay distribution. A low intensity MAXS peak was detected, as presented in Figure 2, in agreement with the TEM observations.

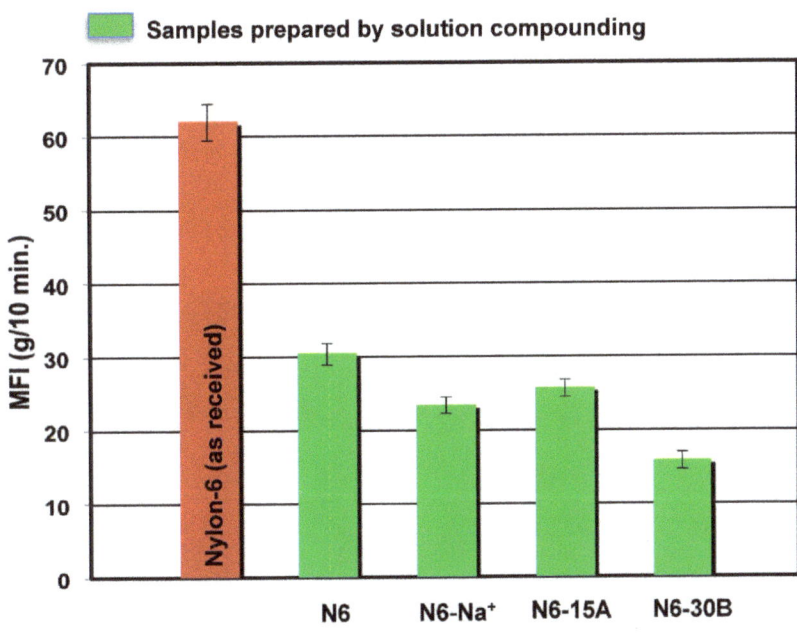

Figure 7. Melt flow index of nylon6/MMT nanocomposites.

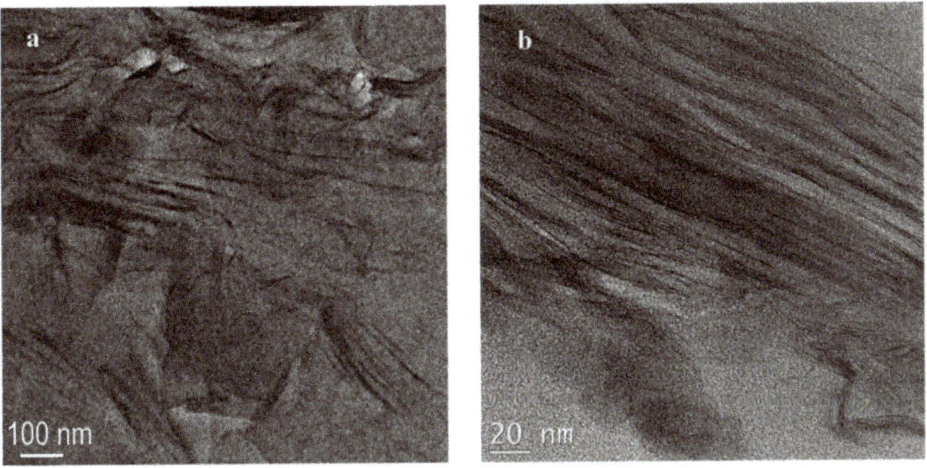

Figure 8. TEM micrographs for N6-Na⁺ prepared by solution compounding at low (a) and high (b) magnifications.

TEM images of the N6-30B samples are presented in Figure 10. Both low and high magnification images (Figure 10a,b, respectively) reveal a structure primarily composed of uniformly dispersed individual clay layers. This supports the MAXS results presented in Figure 3. The exfoliated layers coupled with the uniform dispersion—not observed in the other samples—elucidate the role of the type of organic modifier in influencing the morphology of the nanocomposites. We have reported in a previous study that the alkylammonium ions preserve the structure of the silicate layers in Cloisite 30B [41]. A similar observation was reported by Mani et al. for Cloisite 15A [42]. This challenges the common understanding that the increase in the gallery d-spacing due to the organic modifier makes the intercalation of the polymer chains easier. The organic modifier can therefore be considered to act in a twofold manner. On the one hand, it would bond with the polymer, facilitating its intercalation and the increase in the disorder of the clay structure, and on the other, it would stabilize the clay structure by maintaining the order of the layers. The polar organic modifier in Cloisite 30B is more likely to bond with nylon6 and thus increase the disorder in the N6-30B nanocomposite structure. This eventually leads to full exfoliation. For the case of the N6-15A composite with its non-polar organic modifier, the second factor is more dominant and thus exfoliation and a uniform dispersion of the clay layers is more difficult.

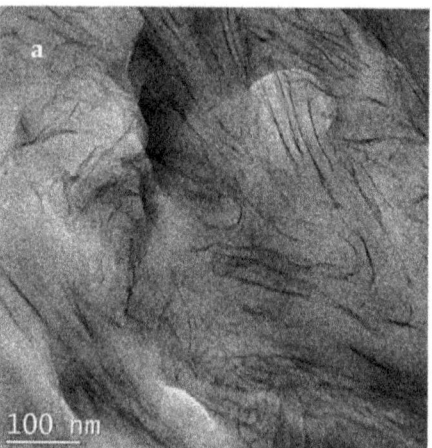 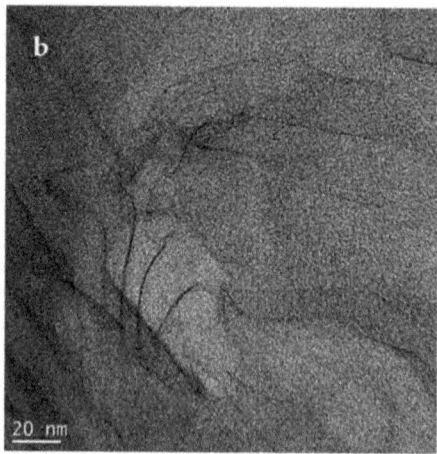

Figure 9. TEM micrographs for N6-15A prepared by solution compounding at low (**a**) and high (**b**) magnifications.

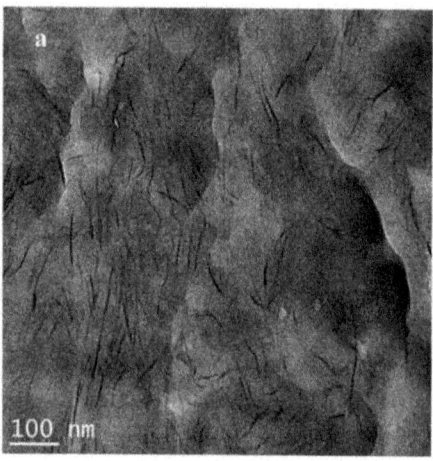 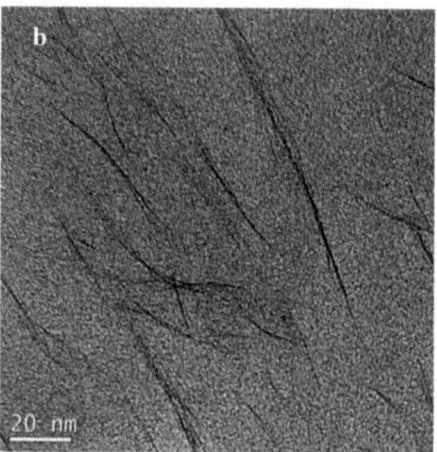

Figure 10. TEM micrographs for N6-30B prepared by solution compounding at low (**a**) and high (**b**) magnifications.

N6-clay samples prepared by the mechanical mixing of clay with N6 (i.e., static melt annealed samples), which were also analyzed by TEM. Results of N6-Na$^+$ (Figure 11a,b) show unevenly dispersed clay particles and no isolated silicate layers. In addition, large areas of the examined sample did not have any clay particles, while in a few areas high densities of clay particles could be found. Composites of N6-15A showed some intercalation as well as some individual silicate layers (Figure 12a,b). However, similar to the N6-Na$^+$ composites, large areas of the examined samples did not contain any clay particles, indicating poor dispersion. Composites of N6-30B (Figure 13a,b) showed better results with some areas having a high density of mostly evenly distributed silicate layers as well as other areas with very few layers. These observations are in line with the nanoindentation results that confirmed low mechanical properties for samples prepared by static melt annealing. This corroborates the fact that the intercalated/exfoliated structures of solution-compounded composite samples result from the solution mixing process.

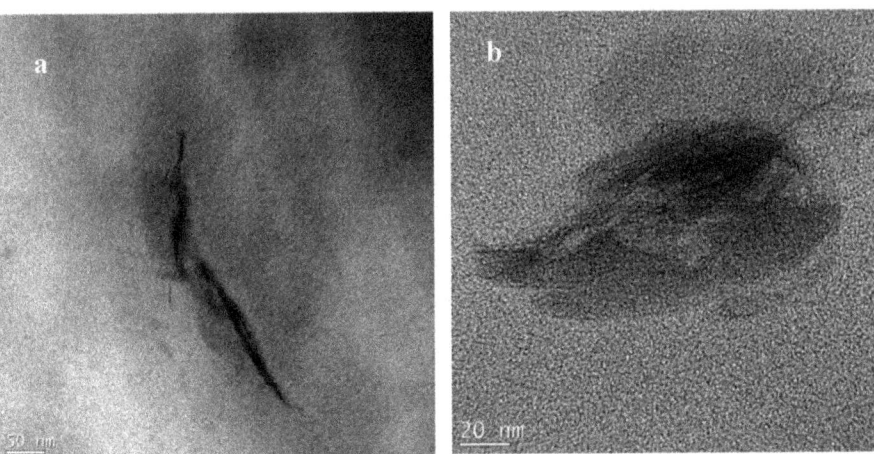

Figure 11. (**a**,**b**) TEM micrographs for N6-Na$^+$ prepared by static melt annealing showing unevenly dispersed clay particles in different regions in the sample.

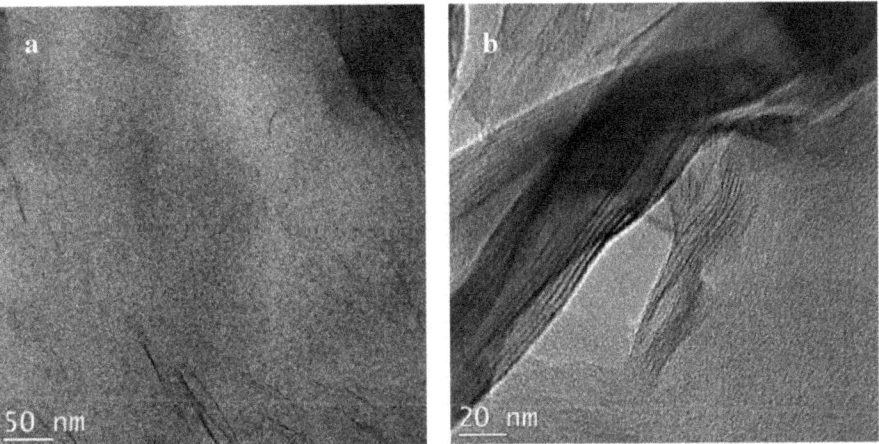

Figure 12. TEM micrographs for N6-15A prepared by static melt annealing showing some individual silicate layers (**a**) and intercalated particles (**b**).

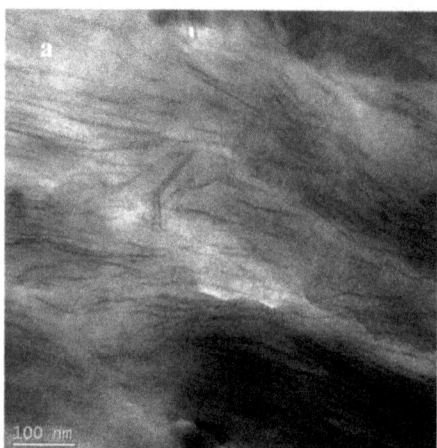 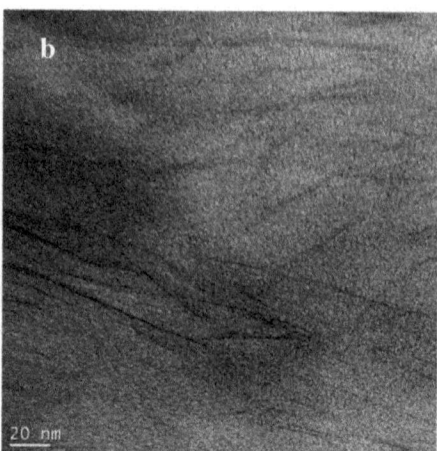

Figure 13. TEM micrographs for N6-30B prepared by static melt annealing showing evenly distributed silicate layers at low (**a**) and high (**b**) magnifications.

With the exception of one study using solution compounding for the preparation of Cloisite 30B-nylon6 PLSNs [22], studies published on nylon6 PLSNs, in which testing of mechanical properties is reported, have predominantly used melt compounding [43–54]. Due to the limited use of solution compounding, it is important at this stage to compare the results obtained in the current investigation to those published in the literature pertaining to melt compounding. A comparison of our results to results from the literature is presented in Table 4. The search for those results has revealed that although many studies have been published on PLSNs, the general focus/theme has been the preparation method and how it influences the composite morphology. Testing of mechanical properties of nylon6-montmorillonite has been reported in only a few studies. Since enhancement of Young's modulus has been the objective in these studies, we will focus on this property to give us a guide on how well our composites compare to others.

All results pertain to composites of nylon6 with 5 wt% clay. $E_{polymer}$ is for nylon6 with the same process history as the composite.

Results in the table show that reported enhancements in Young's modulus vary significantly from one study to the other, with one study reporting up to 88% enhancement upon the addition of only 5 wt% clay, whereas others report values as low as 2.5%. It is important to note that for organically modified clays, actual mineral content can vary depending on the amount of the organic modifier so although the results are reported for composites with 5 wt% clay, slight variations are expected. In addition, whenever different molecular weights of nylon6 are used (e.g., the study by Fornes [47]), a range is included for comparison. What is noticeable from the table is that the results of the current study compare favorably in terms of the absolute value of the Young's modulus observed. This is believed to be due to the success of the solution compounding process in dispersing and exfoliating the different types of clays in the current study. It is also believed that the improved matrix crystallinity contributes to the observed enhancements, since it is known that the final properties of the nanocomposites will depend not only on the distribution of layered silicates but also on the crystal type, the degree of polymer crystallization, and crystallite morphology [37].

Table 4. Comparing results of the current study to some reported Moduli of nylon6-montmorillonite clays.

Clay Type	$E_{polymer}$ (GPa)	$E_{composite}$ (GPa)	Preparation Method	Modulus Evaluation Technique	Reference
Cloisite 30 B	3.5	4.2	Solution compounding	nanoindentation	Current study
Cloisite 15A	3.5	4.15	Solution compounding	nanoindentation	Current study
Cloisite Na$^+$	3.5	3.9	Solution compounding	nanoindentation	Current study
Cloisite 30B	3	3.7 (+23%)	Melt compounding	nanoindentation	[7]
Cloisite 15A	3	3.25 (+8.3%)	Melt compounding	nanoindentation	[7]
Cloisite Na$^+$	3	3.2 (+6.7%)	Melt compounding	nanoindentation	[7]
organically modified clay (Nanomerw I.30TC)	1.06	2 (+88.7%)	Melt compounding	nanoindentation	[44]
Cloisite Na$^+$	2.66	3.01 (+13.2%)	Melt compounding	Tensile testing	[45]
Organoclay SCPX 2004	2.66	3.66 (+37.6%)	Melt compounding	Tensile testing	[45]
Cloisite 30B	1.2	1.3 (8.3%)	Melt compounding	Tensile testing	[46]
MMT	1.11	1.93 (+73.9%)	(in situ polymerization)		[47]
MMT	2.82	4.2–4.8 (+59.6%)	Melt intercalation		[47]
Orgomodified MMT	1.99	3.12 (+57%)	Melt compounding	Tensile testing	[48]
Natural MMT	1.99	2.04 (+2.5%)	Melt compounding	Tensile testing	[48]
Organoclay	0.73–1.2	1.05–1.6 (+43%)	Melt compounding	Compression testing	[49]
OrganoMMT	2.9	4.1 (+41.4%)	Melt compounding	Tensile testing	[50]

4. Conclusions

XRD and TEM investigations conducted in the current study confirmed that solution compounding facilitates the separation and dispersion of the silicate layers for cases of low/no polar interactions between the clays and the polymer. The results also revealed that solution compounding enhances the crystallinity for neat nylon6 from 36 to 58% and results in the more ordered and stable α type crystal structure. This enhanced the nanoindentation hardness and modulus. Additionally, solution compounding allowed a good dispersion of the 15A and 30B clays in the composites. The mechanical testing results compared favorably with published results in the literature for melt compounded samples and makes the process an alternative worth considering. It is also worth noting that solution compounded composites exhibited nanoindentation hardness and modulus values that are 60–110% higher than those produced by static melt annealing. Static melt annealing was found to be somewhat effective in the case of the N6-30B samples only, which further confirms that the observed exfoliated structures in the composites together with the associated enhancement of mechanical properties were brought about by the solution compounding process.

5. Patents

A US patent resulting from the work reported in the manuscript is listed below: Gawad, Ahmed Abdel M., Ramadan, Adham R., Esawi, Amal M. K., Solution blending process for the fabrication of NYLON6-montmorillonite nanocomposites, United States Patent 10100175, 2018.

Author Contributions: Conceptualization, A.M.K.E. and A.R.R.; methodology, A.M.A.-G., A.M.K.E. and A.R.R.; formal analysis, A.M.K.E., A.R.R. and A.F.; investigation, A.M.A.-G.; data curation, A.M.A.-G. and A.F.; writing—original draft preparation, A.M.A.-G.; writing—review and editing, A.M.K.E., A.R.R. and A.F.; visualization, A.M.A.-G.; supervision, A.M.K.E. and A.R.R. All authors have read and agreed to the published version of the manuscript.

Funding: This research was funded by the Yousef Jameel Science and Technology Research Center at the American University in Cairo, Egypt. AF's contribution was funded by the MICINN (Ministerio de Ciencia e Innovación), Spain under grant PID2020-117573GB-I00. The APC was funded by the American University in Cairo.

Institutional Review Board Statement: Not applicable.

Conflicts of Interest: The authors declare no conflict of interest.

References

1. Okada, A.; Kawasumi, M.; Usuki, A.; Kojima, Y.; Kurauchi, T.; Kamigaito, O. Synthesis and properties of nylon-6/clay hybrids. In *Polymer Based Molecular Composites, Proceedings of Materials Research Society Symposium, Pittsburgh, PA, USA*; Schaefer, D.W., Mark, J.E., Eds.; Cambridge University Press: New York, NY, USA, 1990; Volume 171, p. 45.
2. Ahmadi, S.; Huang, Y.; Li, W. Synthetic routes, properties and future applications of polymer-layered silicate nanocomposites. *J. Mater Sci.* **2004**, *39*, 1919. [CrossRef]
3. Chen, B. Polymer-clay nanocomposites: An overview with emphasis on interaction mechanisms. *Br. Ceram. Trans.* **2004**, *103*, 241. [CrossRef]
4. Giannelis, E.P. Polymer-layered silicate nanocomposites: Synthesis, properties and applications. *Appl. Organomet. Chem.* **1998**, *12*, 675. [CrossRef]
5. LeBaron, P.C.; Wang, Z.; Pinnavaia, T.J. Polymer-layered silicate nanocomposites: An overview. *Appl. Clay Sci.* **1999**, *15*, 11. [CrossRef]
6. Abdel Gawad, A.; Esawi, A.; Ramadan, A. Structure and properties of nylon 6–clay nanocomposites: Effect of temperature and reprocessing. *J. Mater Sci.* **2010**, *45*, 6677. [CrossRef]
7. Bhat, A.H.; Rangreez, T.A.; Inamuddin; Chisti, H.-T.-N. Wastewater Treatment and Biomedical Applications of Montmorillonite Based Nanocomposites: A Review. *Curr. Anal. Chem.* **2022**, *18*, 269–287. [CrossRef]
8. Das, P.; Manna, S.; Behera, A.K.; Shee, M.; Basak, P.; Sharma, A.K. Current synthesis and characterization techniques for clay-based polymer nano-composites and its biomedical applications: A review. *Environ. Res.* **2022**, *212*, 113534. [CrossRef]
9. Zhu, T.T.; Zhou, C.H.; Kabwe, F.B.; Wu, Q.Q.; Li, C.S.; Zhang, J.R. Exfoliation of montmorillonite and related properties of clay/polymer nanocomposites. *Appl. Clay Sci.* **2019**, *169*, 48–66. [CrossRef]
10. Martinez-Colunga, J.G.; Sanchez-Valdes, S.; Blanco-Cardenas, A.; Ramírez-Vargas, E.; Ramos-De Valle, L.F.; Benavides-Cantu, R.; Espinoza-Martinez, A.B.; Sanchez-Lopez, S.; Lafleur, P.G.; Karami, S.; et al. Dispersion and exfoliation of nanoclays in itaconic acid funcionalized LDPE by ultrasound treatment. *J. Appl. Polym. Sci.* **2018**, *135*, 46260. [CrossRef]
11. Asgari, M.; Abouelmagd, A.; Sundararaj, U. Silane functionalization of sodium montmorillonite nanoclay and its effect on rheological and mechanical properties of HDPE/clay nanocomposites. *Appl. Clay Sci.* **2017**, *146*, 439–448. [CrossRef]
12. Wu, H.; Krifa, M.; Koo, J.H. Flame retardant polyamide 6/nanoclay/intumescent nanocomposite fibers through electrospinning. *Text. Res. J.* **2014**, *84*, 1106–1118. [CrossRef]
13. Fereydoon, M.; Tabatabaei, S.H.; Ajji, A. Rheological, crystal structure, barrier, and mechanical properties of PA6 and MXD6 nanocomposite films. *Polym. Eng. Sci.* **2014**, *54*, 2617–2631.
14. Briesenick, D.; Bremser, W. Synthesis of polyamide-imide-montmorillonite-nanocomposites via new approach of in situ, polymerization and solvent casting. *Prog. Org. Coat.* **2015**, *82*, 26–32. [CrossRef]
15. Osman, A.F.; Kalo, H.; Hassan, M.S.; Hong, T.W.; Azmi, F. Pre–dispersing of montmorillonite nanofiller: Impact on morphology and performance of melt compounded ethyl vinyl acetate nanocomposites. *J. Appl. Polym. Sci.* **2016**, *133*, 43204. [CrossRef]
16. Moreira, J.F.M.; Alves, T.S.; Barbosa, R.; De Carvalho, É.M.; Carvalho, L.H. Effect of cis–13–docosenamide in the properties of compatibilized polypropylene/clay nanocomposites. *Macromol. Symp.* **2016**, *367*, 68–75. [CrossRef]
17. Zabihi, O.; Ahmadi, M.; Naebe, M. Self-assembly of quaternized chitosan nano-particles within nanoclay layers for enhancement of interfacial properties in toughened polymer nanocomposites. *Mater. Des.* **2017**, *119*, 277–289. [CrossRef]
18. Zhang, G.Z.; Wu, T.; Lin, W.Y.; Tan, Y.B.; Chen, R.Y.; Huang, Z.X.; Yin, X.C.; Qu, J.P. Preparation of polymer/clay nanocomposites via melt intercalation under continuous elongation flow. *Compos. Sci. Technol.* **2017**, *145*, 157–164. [CrossRef]
19. Ammar, A.; Elzatahry, A.; Al-Maadeed, M.; Alenizi, A.M.; Huq, A.F.; Karim, A. Nanoclay compatibilization of phase separated polysulfone/polyimide films for oxygen barrier. *Appl. Clay Sci.* **2017**, *137*, 123–134. [CrossRef]
20. Filippi, S.; Marazzato, C.; Magagnini, P.; Famulari, A.; Arosio, P.; Meille, S.V. Structure and morphology of HDPE-g-MA/organoclay nanocomposites: Effects of the preparation procedures. *Eur. Polym. J.* **2008**, *44*, 987–1002. [CrossRef]
21. Vaia, R.A.; Giannelis, E.P. Lattice model of polymer melt intercalation in organically-modified layered silicates. *Macromolecules* **1997**, *30*, 7990. [CrossRef]

22. Paci, M.; Filippi, S.; Magagnini, P. Nanostructure development in nylon 6-Cloisite®30B composites. Effects of the preparation conditions. *Eur. Polym. J.* **2010**, *46*, 838.
23. Dennis, H.; Hunter, D.; Chang, D.; Kim, S.; White, J.; Cho, J.; Paul, D. Effect of melt processing conditions on the extent of exfoliation in organoclay-based nanocomposites. *Polymer* **2001**, *42*, 9513. [CrossRef]
24. Shen, Z.; Simon, G.P.; Cheng, Y.-B. Comparison of solution intercalation and melt intercalation of polymer ± clay nanocomposites. *Polymer* **2002**, *43*, 4251–4260. [CrossRef]
25. Strawhecker, K.; Manias, E. Structure and properties of poly (vinyl alcohol)/Na+ montmorillonite nanocomposites. *Chem. Mater.* **2000**, *12*, 2943. [CrossRef]
26. Aranda, P.; Ruiz-Hitzky, E. Poly (ethylene oxide)-silicate intercalation materials. *Chem. Mater.* **1992**, *4*, 1395. [CrossRef]
27. Wu, J.; Lerner, M.M. Structural, thermal, and electrical characterization of layered nanocomposites derived from sodium-montmorillonite and polyethers. *Chem. Mater.* **1993**, *5*, 835. [CrossRef]
28. Ogata, N.; Jimenez, G.; Kawai, H.; Ogihara, T. Structure and thermal/mechanical properties of poly (L-lactide)–clay blend. *J. Polym. Sci. Part B Polym. Phys.* **1997**, *35*, 389. [CrossRef]
29. Yano, K.; Usuki, A.; Okada, A.; Kurauchi, T.; Kamigaito, O. Synthesis and properties of polyimide-clay hybrid. *J. Polym. Sci. Part A Polym. Chem.* **1993**, *31*, 2493. [CrossRef]
30. Magaraphan, R.; Lilayuthalert, W.; Sirivat, A.; Schwank, J.W. Preparation, structure, properties and thermal behavior of rigid-rod polyimide/montmorillonite nanocomposites. *Compos. Sci. Technol.* **2001**, *61*, 1253. [CrossRef]
31. Jeon, H.; Jung, H.T.; Lee, S.; Hudson, S. Morphology of polymer/silicate nanocomposites High density polyethylene and a nitrile copolymer. *Polym. Bull.* **1998**, *41*, 107. [CrossRef]
32. Burnside, S.D.; Giannelis, E.P. Synthesis and properties of new poly (dimethylsiloxane) nanocomposites. *Chem. Mater.* **1995**, *7*, 1597. [CrossRef]
33. Hammersley, A.P.; Svensson, S.O.; Thomson, A. Calibration and correction of spatial distortions in 2D detector systems. *Nucl. Instrum. Methods Phys. Res. Sect. A Accel. Spectrometers Detect. Assoc. Equip.* **1994**, *346*, 312–321. [CrossRef]
34. Samon, J.M.; Schultz, J.M.; Hsiao, B.S. Study of the cold drawing of nylon 6 fiber by in-situ simultaneous small- and wide-angle X-ray scattering techniques. *Polymer* **2000**, *41*, 2169–2182. [CrossRef]
35. Lee, D.; Char, K. Effect of acidity on the deintercalation of organically modified layered silicates. *Langmuir* **2002**, *18*, 6445. [CrossRef]
36. Filippi, S.; Paci, M.; Polacco, G.; Dintcheva, N.T.; Magagnini, P. On the interlayer spacing collapse of Cloisite 30B organoclay. *Polym. Degrad. Stab.* **2011**, *96*, 823–832. [CrossRef]
37. Kotal, M.; Bhowmick, A.K. Polymer nanocomposites from modified clays: Recent advances and challenges. *Prog. Polym. Sci.* **2015**, *51*, 127–187. [CrossRef]
38. Kojima, Y.; Matsuoka, T.; Takahashi, H.; Kurauchi, T. Crystallization of nylon 6–clay hybrid by annealing under elevated pressure. *J. Appl. Polym. Sci.* **1994**, *51*, 683. [CrossRef]
39. Aharoni, S.M. *n-Nylons: Their Synthesis, Structure, and Properties*; Wiley: New York, NY, USA, 1997.
40. Tidick, P.; Fakirov, S.; Avramova, N.; Zachmann, H. Effect of the melt annealing time on the crystallization of nylon-6 with various molecular weights. *Colloid Polym. Sci.* **1984**, *262*, 445. [CrossRef]
41. Ramadan, A.R.; Esawi, A.M.K.; Gawad, A.A. Effect of ball milling on the structure of Na+-montmorillonite and organo-montmorillonite (Cloisite 30B). *Appl. Clay Sci.* **2010**, *47*, 196. [CrossRef]
42. Mani, G. Size reduction of clay particles in nanometer dimensions. *Mater. Res. Soc. Symp. Proc.* **2003**, *740*, I3.23. [CrossRef]
43. Pavlidou, S.; Papaspyrides, C.D. A review on polymer–layered silicate nanocomposites. *Prog. Polym. Sci.* **2008**, *33*, 1119–1198. [CrossRef]
44. Shen, L.; Tjiu, W.C.; Liu, T. Nanoindentation and morphological studies on injection-molded nylon-6 nanocomposites. *Polymer* **2005**, *46*, 11969–11977. [CrossRef]
45. Cho, J.W.; Paul, D.R. Nylon 6 nanocomposites by melt compounding. *Polymer* **2001**, *42*, 1083–1094. [CrossRef]
46. Ranade, A.; D'Souza, N.A.; Gnade, B.; Dharia, A. Nylon-6 and Montmorillonite-Layered Silicate (MLS) Nanocomposites. *J. Plast. Film Sheeting* **2003**, *19*, 271–286. [CrossRef]
47. Fornes, T.D.; Yoon, P.J.; Keskkula, H.; Paul, D.R. Nylon 6 nanocomposites: The effect of matrix molecular weight. *Polymer* **2001**, *42*, 9929–9940. [CrossRef]
48. Gonzalez, T.V.; Salazar, C.G.; De la Rosa, J.R.; Gonzalez, V.G. Nylon 6/organoclay nanocomposites by extrusion. *J. Appl. Polym. Sci.* **2008**, *108*, 2923–2933. [CrossRef]
49. Tsai, J.-L.; Huang, J.-C. Strain Rate Effect on Mechanical Behaviors of Nylon 6–Clay Nanocomposites. *J. Compos. Mater.* **2006**, *40*, 925–938. [CrossRef]
50. Xie, S.; Zhang, S.; Zhao, B.; Qin, H.; Wang, F.; Yang, M. Tensile fracture morphologies of nylon-6/montmorillonite nanocomposites. *Polym. Int.* **2005**, *54*, 1673–1680. [CrossRef]
51. Okada, A.; Usuki, A. Polymer-layered silicate nanocomposites: An overview. *Macromol. Mater. Eng.* **2006**, *291*, 1449–1476. [CrossRef]
52. Yan, T.; Chen, D.; Zhao, B.; Jiang, X.; Wang, L.; Li, Y. Percolation Network Formation in Nylon 6/Montmorillonite Nanocomposites: A Critical Structural Insight and the Impact on Solidification Process and Mechanical Behavior. *Polymers* **2022**, *14*, 3672. [CrossRef]

53. Bazmara, M.; Silani, M.; Dayyani, I. Effect of functionally-graded interphase on the elasto-plastic behavior of nylon-6/clay nanocomposites; a numerical study. *Def. Technol.* **2021**, *17*, 177–184. [CrossRef]
54. Yao, W.H. The preparation of modified polyamide clay nanocomposite/recycled maleic anhydride polyamide 6 and blending with low density polyethylene for film blowing application. *Polym. Polym. Compos.* **2021**, *29* (Suppl. 9), S631–S643. [CrossRef]

Article

Electrical Properties of Polyetherimide-Based Nanocomposites Filled with Reduced Graphene Oxide and Graphene Oxide-Barium Titanate-Based Hybrid Nanoparticles

Quimberly Cuenca-Bracamonte [1], Mehrdad Yazdani-Pedram [1,*] and Héctor Aguilar-Bolados [2,*]

1 Facultad de Ciencias Químicas y Farmacéuticas, Universidad de Chile, Olivos 1007, Santiago 8380544, Chile
2 Departamento de Polímeros, Facultad de Ciencias Químicas, Universidad de Concepción, Concepción 3349001, Chile
* Correspondence: myazdani@ciq.uchile.cl (M.Y.-P.); haguilar@e-mail.com or haguilar@udec.cl (H.A.-B.)

Abstract: The electrical properties of nanocomposites based on polyetherimide (PEI) filled with reduced graphene oxide (rGO) and a graphene oxide hybrid material obtained from graphene oxide grafted with poly(monomethyl itaconate) (PMMI) modified with barium titanate nanoparticles (BTN) getting (GO-g-PMMI/BTN) were studied. The results indicated that the nanocomposite filled with GO-g-PMMI/BTN had almost the same electrical conductivity as PEI (1×10^{-11} S/cm). However, the nanocomposite containing 10 wt.% rGO and 10 wt.% GO-g-PMMI/BTN as fillers showed an electrical conductivity in the order of 1×10^{-7} S/cm. This electrical conductivity is higher than that obtained for nanocomposites filled with 10% rGO (1×10^{-8} S/cm). The combination of rGO and GO-g-PMMI/BTN as filler materials generates a synergistic effect within the polymeric matrix of the nanocomposite favoring the increase in the electrical conductivity of the system.

Keywords: polyetherimide nanocomposite; hybrid graphene materials; reduced graphene oxide

1. Introduction

Graphene materials, such as graphene oxide (GO) and reduced graphene oxide (rGO), have been widely studied due to their ease of obtainment and excellent properties. For instance, rGO presents excellent electrical properties, while GO has oxygen moieties that impart reactivity to be functionalized using different compounds [1–3]. Moreover, the use of graphene oxide and reduced graphene oxide as fillers in polymer nanocomposites to improve their thermal, electrical, and mechanical properties has increased since the advent of graphene materials. This has generated interest in scientific and technological fields, as well as an important advance in different areas [4,5].

The GO corresponds to oxidized graphene sheets, which present oxygenated functional groups such as hydroxyl, epoxide, carboxylic acid, ketones, and cyclic esters [6,7]. The presence of these functional groups favors the exfoliation of graphene layers by using thermal reduction as a top-down method. On the other hand, GO is a less conductive material than graphite since the carbon atoms bonded to oxygen functional groups present sp^3-hybridization, which disrupts the continuity of the long-range sp^2 conjugated system, affecting the charge transport lattice characteristics [8].

The GO can be reduced using different methods to obtain the rGO. One of these methods is thermal reduction, which consists of heating GO above 600 °C, which results in a fast reduction and rapid exfoliation of graphene oxide layers. The exfoliation is the result of the thermal decomposition of oxygenated functional groups present in GO, which yield gaseous molecules, such as CO_2, CO, and H_2O [6,7].

As is known, there are a wide variety of graphene materials, where among them polymer-grafted graphene oxide is highlighted because the polymer assists the compatibility of the graphene with the environment where it is dispersed [9]. Polymer-grafted graphene oxides can be obtained by the grafting-from, as well as the grafting-to methods [10]. In this regard, previously our group reported the obtaining of a grafted-from graphene oxide, where the grafting polymer was poly (monomethyl itaconate), and the route to obtain consisted of atom transfer radical polymerization [11]. It is important to notice that the monomer, monomethyl itaconate, corresponds to a derivative of the itaconic acid. Itaconic acid is obtained by a fermentation process of renewable sources [12].

On the other hand, graphene materials, such as graphene oxide and its derivatives, are used as fillers in polymers to improve the performance of properties such as electrical, mechanical, and barrier properties [13]. One of the challenges in this matter is the filler dispersion. In this respect, poly (monomethyl itaconate) grafted onto graphene oxide can be an alternative to assist the dispersion of the filler in a polymeric matrix.

On the other hand, barium titanate (BT) is a perovskite-type oxide considered a ferroelectric material with high relative permittivity [14]. Barium titanate nanoparticles (BTN) are generally used for capacitor design or as filler in dielectric elastomers to improve their dielectric permittivity [15,16]. Some studies have shown improvements in dielectric, stress-strain mechanical, and energy storage properties as results of the use of barium titanate nanoparticles for obtaining polymer-based nanocomposites [17]. In particular, an improvement in the dielectric breakdown strength was observed, which is attributed to the increase in the degree of charge dispersion and the mitigation of the local electric field concentration promoted by GO. The introduction of ceramic fillers with BTNs into a polymer matrix is an effective procedure to obtain dielectric nanocomposites with high energy storage density [18]. Other authors have reported that by using BTN and GO, composites with high dielectric strength can be obtained [19].

Although studies have been carried out regarding GO and BTN-filled nanocomposites, these have been prepared by physically mixing the components of nanocomposites. Hence, it is possible to think about the modification of GO with BTN, where BTN is covalently linked to GO.

To obtain hybrid materials based on functionalized graphene oxide and barium titanate nanoparticles, it is necessary to modify the structure of both materials. Therefore, a novel approach to synthesize these hybrid materials is to use polymer-grafted graphene oxide and functionalized barium titanate nanoparticles that, by a reaction yield, graphene oxide covalently bonds to the BTNs. In this regard, the poly (monomethyl itaconate)-grafted-graphene oxide can react with barium titanate functionalized with amine groups by nucleophilic substitution reaction. The amine-functionalized barium titanate is prepared by subsequent reactions of hydroxylation and silanization by using 3-aminopropyltriethoxysilane [20]. We hypothesize that the carboxylic acid and ester side groups of poly(monomethyl itaconate) are susceptible to the nucleophilic attack of amine groups, promoting the formation of covalent bonds that links graphene oxide sheets with barium titanate nanoparticles (Figure 1). In this work, the hybrid material prepared by this approach is designated as GO-g-PMMI/BTN and its use as filler for the preparation of nanocomposites based on polyether imide (PEI) was investigated. PEI corresponds to a high-performance thermoplastic polymer with excellent mechanical and thermal properties. The effect of the use of the hybrid graphene materials in combination with reduced graphene oxide was also evaluated. The aim of this work is to evaluate the electrical properties, such as electrical conductivity and dielectric permittivity, of nanocomposites based on polyetherimide and hybrid graphene and/or reduced graphene oxide prepared by a solvent casting method.

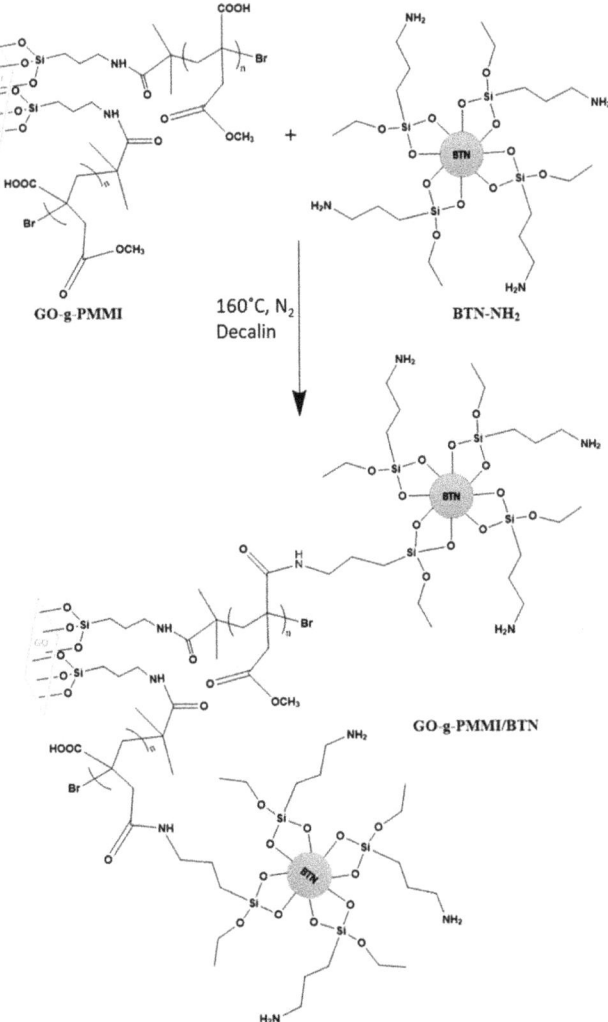

Figure 1. Proposed reaction scheme between silanized barium titanate nanoparticles (BTN-NH2) and GO-g-PMMI to yield hybrid GO-g-PMMI/BTN.

2. Materials and Methods

2.1. Materials

Graphite (+100 mesh (≥75% min)), polyetherimide (melt index 18 g/10 min (337 °C/6.6 kG)), and barium titanate (IV) nanopowder (cubic) 50 nm 99.9% were supplied by Aldrich. Nitric acid (≥99.0%), potassium chlorate (≥99.0%), (3-aminopropyl) triethoxysilane (APTES) (≥98.0%), α-bromoisobutyryl bromide (BIBB) (98.0%), triethylamine (≥98.0%), N,N,N′,N″,N″-pentamethyldiethylenetriamine (PMDETA, 99%), ascorbic acid (≥99.0%), decalin (decahydronaphthalene, mixture of cis + trans) (98%), dichloromethane, N,N-dimethylformamide (DMF), sulfuric acid (95–97%) and sodium nitrate were supplied by Sigma-Aldrich. Potassium permanganate and copper (II) bromide were obtained from Merck.

Monomethyl itaconate (MMI) [21], graphene oxide (GO) [16], reduced graphene oxide (rGO) [16], and poly (monomethyl Itaconate) grafted graphene oxide (GO-g-PMMI) [16] was synthesized using the previously reported method.

2.2. Synthesis of Hybrid Nanomaterial Based on Graphene Oxide Grafted with PMMI and Barium Titanate Nanoparticles (GO-g-PMMI/BTN)

2.2.1. Hydroxylation of Barium Titanate Nanoparticles

15.250 g of BTN in 80 mL of 30% H2O2 was added to a 2-neck flask. This mixture was sonicated for 30 min (25% amplitude). Next, it was left to react using a refluxing system at 105 °C for 4 h. The resulting suspension was centrifuged and then the obtained solid was washed with three portions of 50 mL distilled water. The solid was dried in a vacuum oven at 80 °C for 12 h. The weight of the white solid was of 14.000 g, and the resulting solid was designated as BTN-OH.

2.2.2. Silanization of Barium Titanate Nanoparticles

In a 250 mL three-necked flask, 2.000 g of BTN-OH and 120 mL of ethanol were added, and the mixing was sonicated for 30 min. Next, the flask was connected to a reflux condenser, a thermometer, and an addition funnel containing a 3 vol.% solution of APTES, which was slowly added drop by drop. Afterwards, the flask was heated at 80 °C, and left to react for 2 h under magnetic stirring. Once the reaction time was elapsed, the reaction mixture was centrifuged and the solid was washed with three portions of 50 mL of distilled water and then dried at 80 °C for 24 h. The weight of the resulting white-beige colored solid was 1.945 g and it was designated as BTN-NH$_2$.

2.2.3. Synthesis of the GO-g-PMMI/BTN

In a 250 mL two-necked flask, 0.052 g of GO-g-PMMI, 0.107 g of BTN-NH2, 150 mL of decahydronaphthalene (decalin) and 5 g of 4Å molecular sieve were added. The mixture was conditioned under an inert atmosphere and allowed to react at 160 °C for 4 h. Next, the reaction mixture was filtered under reduced pressure. The solid obtained was washed three times with 50 mL portions of petroleum ether and subsequently three times with 50 mL portions of acetone. Finally, the solid product was dried at 80 °C in a vacuum oven for 12 h, yielding 0.525 g of a gray solid designated as GO-g-PMMI/BTN.

2.3. Preparation of Nanocomposites

First, a 18 w/v% solution of PEI was prepared by dissolving 0.9 g of PEI in 5 mL of dichloromethane. The mixture was stirred for 30 min, achieving complete dissolution of the polymer. On the other hand, the filler material was weighed and added to a beaker according to the filler percentage stipulated for each nanocomposite and 5 mL of dichloromethane was also added. The masses of the fillers are shown in Table 1.

Table 1. Mass of PEI and fillers used for the preparation of the nanocomposites.

Compound	PEI (g)	GO-g-PMMI/BTN (g)	rGO (g)
PEI	0.9000	0	0
GO-g-PMMI/BTN	0.9200	0.0937	0
rGO	0.9300	0	0.0936
GO-g-PMMI/BTN + rGO	0.9300	0.0946	0.0937

This mixture was sonicated for 5 min at 25% amplitude to disperse the filler particles and was mixed with the polymer solution under magnetic stirring. The resulting mixture was sonicated for 5 min at 25% amplitude and poured into a Petri dish. The Petri dish was covered with a beaker to favor slow evaporation of the solvent. This procedure was used to obtain all the nanocomposites for evaluation.

2.4. Characterization

FTIR spectra were recorded using a Thermo Scientific Nicolet IS50 spectrophotometer (Waltham, MA, USA) with the attenuated total reflectance technique (ATR). The Raman spectra were registered using Raman WITec Alpha 300 RA spectrometer equipped with a 532 nm wavelength laser and 0.2 cm^{-1} resolution. X-ray diffraction analysis was recorded using a Bruker D8 Advance diffractometer (Billerica, MA, USA). The radiation frequency was the Kα line from Cu (1.5406 Å) with a power supply of 40 kV and 40 mA. The morphologies of the samples were obtained by scanning electron microscopy (SEM) and (STEM) using JSM-IT300LV microscope, Jeol (Tokyo, Japan). The samples were coated with an ultra-thin gold (Au) layer. The accelerating voltage was 10 kV. The DC electrical conductivity of samples were determined by using 8.0 cm diameter disc-shaped films using a Keithley High Resistivity Tester model 6517B (Cleveland, OH, USA) and a Resistivity Test Fixture 8009. Broadband dielectric spectroscopic data were obtained using a broadband dielectric spectrometer model BDS-40, Novocontrol Technologies GmbH (Hundsangen, Germany), over a frequency range window of 10^{-1} Hz to 10^6 Hz and at room temperature. The applied amplitude of the alternating current (A.C.) was 1 V.

3. Results and Discussion

Figure 2 presents the FTIR spectra of reduced graphene oxide (rGO), barium titanate nanoparticles (BTN) and hybrid material based on graphene oxide grafted with poly(monomethyl itaconate) modified with barium titanate (GO-g-PMMI/BTN).

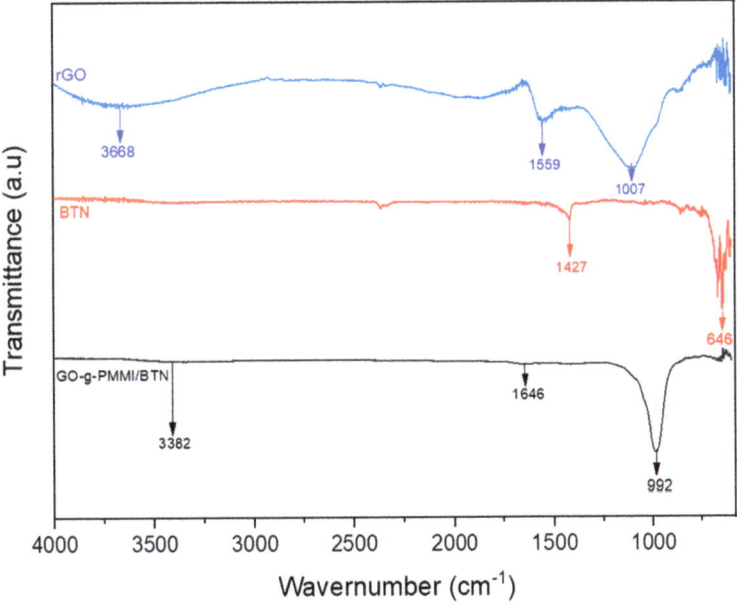

Figure 2. FTIR spectrum of rGO, BTN, and GO-g-PMMI/BTN.

A wide absorption band between 3200 and 3600 in the spectrum of rGO is observed, which is attributed to the presence of hydroxyl groups [22]. Likewise, the absorption band at 1559 cm^{-1} corresponds to the aromatic C=C, which indicates the presence of a conjugated sp^2 system [23]. Moreover, a band at 1007 cm^{-1} corresponds to hydroxyl and ether groups [24].

On the other hand, the BTN shows only two absorption bands at 1427 cm^{-1} and 646 cm^{-1}. The band at 1427 cm^{-1} corresponds to the stretching vibration of $-CO_3^{2-}$

from residual calcium carbonate from barium titanate synthesis and the band at 646 cm^{-1} represents the Ti-O bond vibration in BaTiO$_3$ [25].

In the case of the GO-g-PMMI/BTN spectrum, a band ca. 3382 cm^{-1} is observed, which is attributed to the stretching of the N-H bond associated with the secondary amide yielded by the reaction between BTN-NH$_2$ and GO-g-PMMI. The absorption band appearing at 1646 cm^{-1} corresponds to the C=O stretching of the amide groups. Additionally, the band at 992 cm^{-1} was attributed to the Si-O-C group.

The Raman spectra of rGO and GO-g-PMMI/BTN are depicted in Figure 3. As expected, these materials present characteristic bands, namely D-band and G-band. The D-band is due to the breathing mode of k-point photons of A$_{1g}$ symmetry, and the G-band arises from the first order scattering of E$_{2g}$ phonons with sp$_2$ carbon atoms [26]. The intensity of these bands is different, which is attributed to the differences between the graphenic material. Although GO-g-PMMI/BTN presents D and G bands with higher intensity than rGO, the D and G bands appear narrower, and their width differs (see Table 2). Considering the study reported by J. Liu et al. [27], these differences suggest that the charge transport of both graphenic materials will be different.

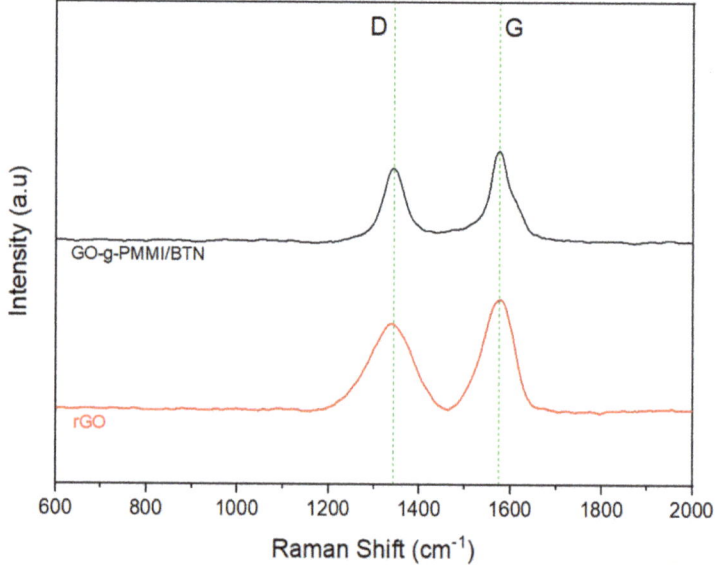

Figure 3. Raman spectrum of rGO and GO-g-PMMI/BTN.

Table 2. Values of I$_D$/I$_G$, FWHM and the Raman shifts of the D-band and G-band.

Sample	Raman Shift (cm^{-1})		FWHM		I$_D$/I$_G$
	D	G	D	G	
rGO	1335	1576	58.898	83.643	0.837
GO-g-PMMI/BTN	1343	1574	51.320	54.610	0.962

Figure 4 shows the XRD patterns of rGO and GO-g-PMMI/BTN. The characteristic diffraction peaks at 24.48° and 42.71° correspond to the (002) and (101) characteristic diffraction planes of the graphitic structure of rGO [28], which are randomly rearranged after the graphene oxide (GO) reduction process to rGO. For the GO-g-PMMI/BTN, the characteristic diffraction peaks of BTN appeared at 2θ = 22.05°, 31.42°, 38.86°, 45.18°, 50.80°, 56.15°, 65.64°, and 74.89°, which are associated with the typical structure of a perovskite, such as barium titanate, of (100), (110), (111), (200), (201), (211), (220) and (310) diffraction planes, respectively [29,30].

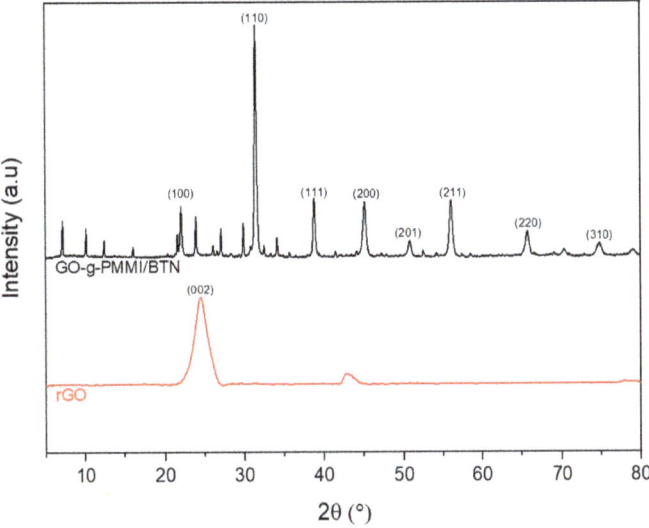

Figure 4. X-ray diffraction patterns of rGO and GO-g-PMMI/BTN.

These X-ray diffraction results suggest that in the process of incorporating barium titanate nanoparticles in GO-g-PMMI, the crystal structure of BTN was not modified.

The surface morphology of rGO and GO-g-PMMI/BTN was investigated by using SEM. As seen in Figure 5, there are notable morphological differences between the filler materials.

Figure 5a,b correspond to the rGO showing a compact structure of randomly stacked sheets because of the reduction process to which the graphene oxide was subjected. The rGO morphology appears as an interlinked layered structure [22]. Figure 5c,d correspond to the GO-g-PMMI/BTN. This material has a heterogenous shape, where flakes and particles appeared randomly dispersed.

STEM images of rGO and GO-g-PMMI/BTN are presented in Figure 6. Figure 6a shows the micrograph of rGO at a magnification of 100,000×, where rGO layers are stacked together. The micrograph of GO-g-PMMI/BTN is shown in Figure 5b at a magnification of 100,000×.

A dense structure formed by several stacked layers of GO is observed. Moreover, an amorphous structure corresponding to the presence of poly(monomethyl itaconate) on the surface of graphene oxide is observed. The presence of small particles randomly distributed onto the graphene oxide layer in Figure 6b are attributed to the barium titanate nanoparticles. As shown in Figure 7, the sizes of these particles are in the order of 50 to 76 nm. The nanoparticles tend to occupy the edges of the graphene oxide layers, probably due to the oxygen functional groups and, consequently, the brushes of poly(monomethyl itaconate) to which the nanoparticles would be linked [16].

The prepared nanocomposites using polyetherimide as the polymeric matrix and rGO and GO-g-PMMI/BTN as filler materials was characterized by FTIR. The results are shown in Figure 8.

The FTIR spectra of PEI show an absorption band at 1723 cm^{-1} typical of imide carbonyl C=O asymmetrical and symmetrical stretching. The absorption bands at 1603 cm^{-1} and 1475 cm^{-1} correspond to the stretching vibration of the C=C aromatic system. The other two absorption bands at 1358 cm^{-1} and 1239 cm^{-1} were assigned to C-N stretching and bending, and to the C-O-C aromatic ether group, respectively [31]. A band at

1075 cm^{-1} was attributed to the tertiary -O-C stretching and the band at 850 cm^{-1} to the of the cyano group (-CN).

Figure 5. SEM images of (**a**) rGO 800×, (**b**) rGO 12,000×, (**c**) GO-g-PMMI/BTN 800× and (**d**) GO-g-PMMI/BTN 12,000×.

The GO-g-PMMI/BTN + rGO nanocomposite shows the absorption bands corresponding to PEI. However, an additional wide band is observed at ca. 992 cm^{-1} attributed to the vibration of the Si-O-C and the alkoxy groups present in the GO-g-PMMI/BTN structure. These facts demonstrate the incorporation of the fillers in the polymer matrix.

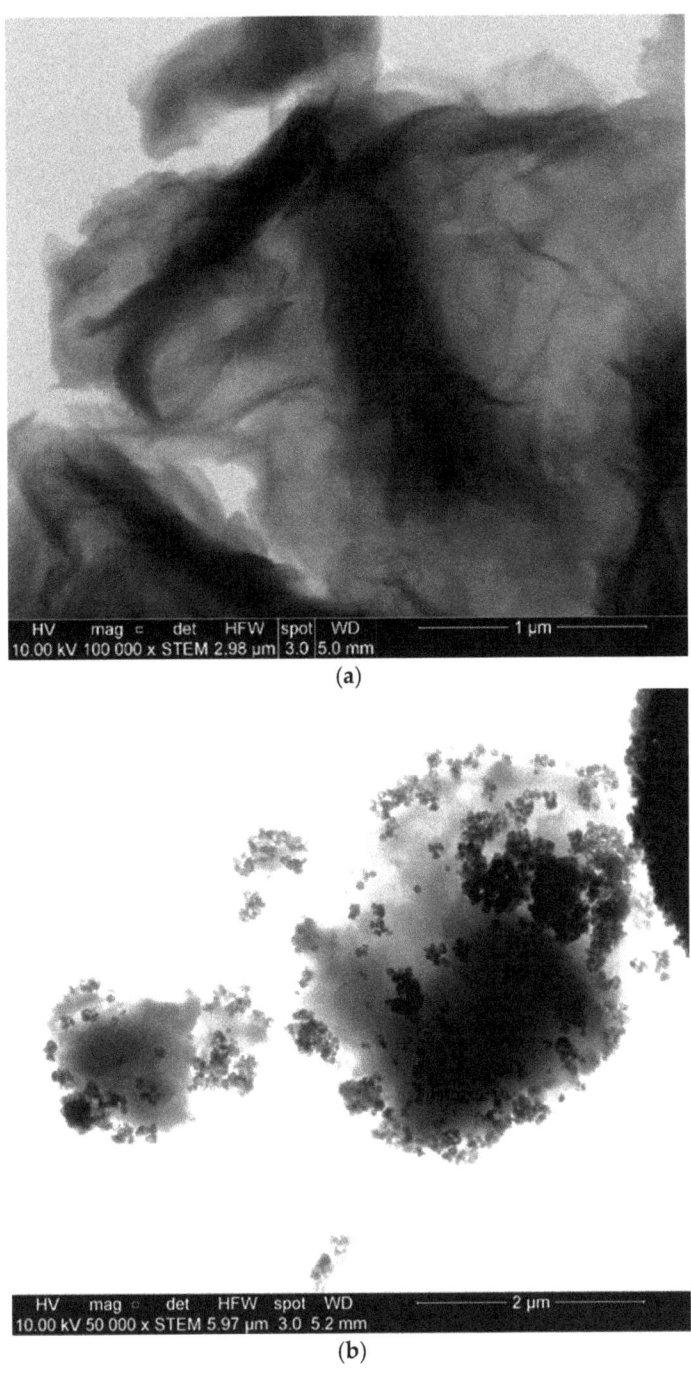

Figure 6. STEM images of rGO (**a**) and GO-g-PMMI/BTN (**b**), both at a magnification of 100,000×.

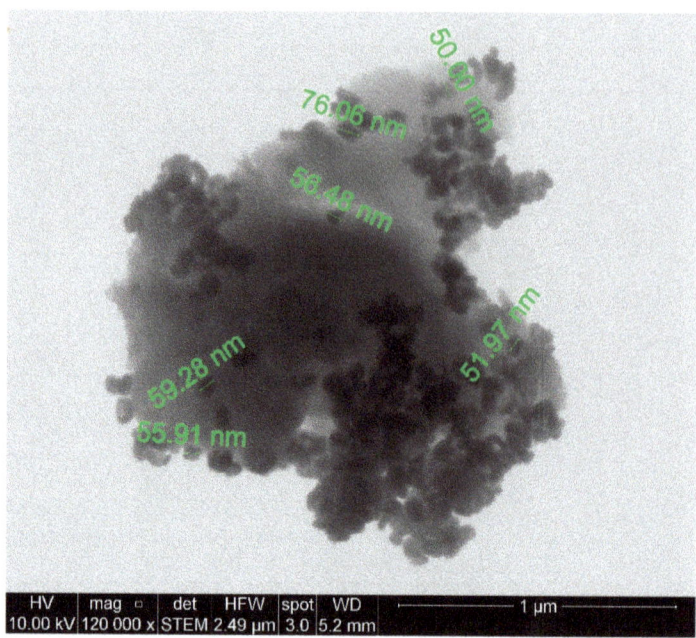

Figure 7. STEM image of GO-g-PMMI/BTN 120,000×, showing the presence of barium titanate nanoparticles on the surface of the GO-g-PMMI.

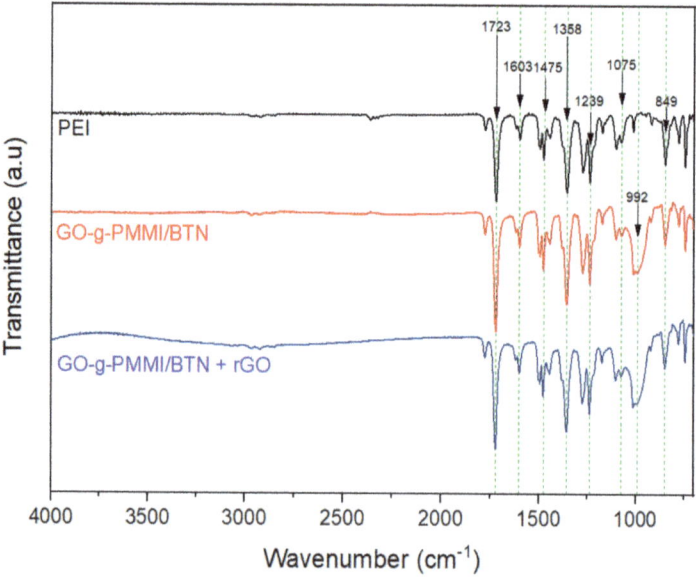

Figure 8. FTIR spectra of GO-g-PMMI/BTN nanocomposite and GO-g-PMMI/BTN + rGO nanocomposite.

The direct current (DC) electrical conductivity of the nanocomposites was evaluated and the results for the nanocomposites filled with 10 wt.% GO-g-PMMI/BTN, 10 wt.% rGO, and 10 wt.% GO-g-PMMI/BTN plus 10 wt.% rGO are shown in Figure 9.

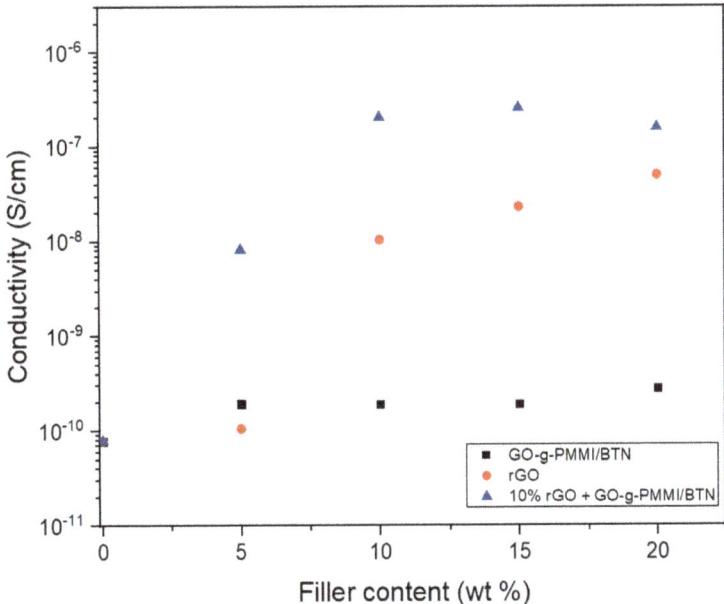

Figure 9. Electrical conductivities of rGO, GO-g-PMMI/BTN and GO-g-PMMI/BTN + 10 wt.% rGO nanocomposites.

As is known, electrical conductivity, σ, is a measure of the ability of a substance or medium to conduct electricity [32]. The conductivity of the PEI used as polymeric matrix is 1×10^{-11} S/cm, however, by incorporating different amounts of GO-g-PMMI/BTN and/or rGO as fillers, the electrical conductivity of the resulting nanocomposites showed variations in respect to that of PEI. The nanocomposite containing a fixed concentration of 10 wt.% rGO and various concentrations of GO-g-PMMI/BTN (5, 10, 15, and 20) wt.% showed the highest conductivity value, being in the order of 1×10^{-7} S/cm, followed by the nanocomposite filled with only 10 wt.% rGO, whose conductivity value was in the order of 1×10^{-8} S/cm. In the case of the nanocomposite containing 10 wt.% of GO-g-PMMI/BTN, no increase in conductivity was observed, remaining in the order of 1×10^{-11} S/cm. From these results, it can be highlighted that GO-g-PMMI/BTN does not improve the conductivity of the nanocomposite. However, GO-g-PMMI/BTN in combination with rGO improves the conductive properties of the nanocomposites above the values obtained for nanocomposites containing rGO as unique filling material (Figure 9). The increase of the electrical conductivity with the increase of the filler content is attributed to a percolation effect imparted by the graphene and hybrid graphene fillers. In fact, the presence of the filler leads to the formation of an electrical percolation network, and the charge transport becomes more efficient with the increase of the conductive filler [3].

Figure 10a shows the real part (σ') of the complex electrical conductivities of PEI and different nanocomposites as functions of the frequency of the electric field. It was observed that the electrical conductivity of PEI increased dramatically by adding 10 wt.% GO-g-PMMI/BTN+10 wt.% rGO ($\sigma' = 1 \times 10^{-5}$ S/cm, recorded at $\nu = 10^{-1}$ Hz) followed by the nanocomposite filled with 10 wt.% rGO ($\sigma' = 1 \times 10^{-7}$ S/cm, recorded at $\nu = 10^{-1}$ Hz), where in both cases the conductivity behavior was independent of the frequency. This indicates that the charge transport is favored by the presence of graphene material [33]. Contrary to the nanocomposite filled only with GO-g-PMMI/BTN, which presented an electrical conductivity close to the conductivity of the polymeric matrix ($\sigma' = 1 \times 10^{-15}$ S/cm, recorded at $\nu = 10^{-1}$ Hz) and the behavior of its conductivity increased with increasing frequency ($\sigma' = 1 \times 10^{-8}$ S/cm, recorded at $\nu = 10^{6}$ Hz). Besides, the discontinuity of the conductivity curve ca. 10x Hz is due to the logarithmic scale applied to the negative values registered. The negative value of the electrical conductivity could be attributed to the electrical current flows against the direction of the electric field [34].

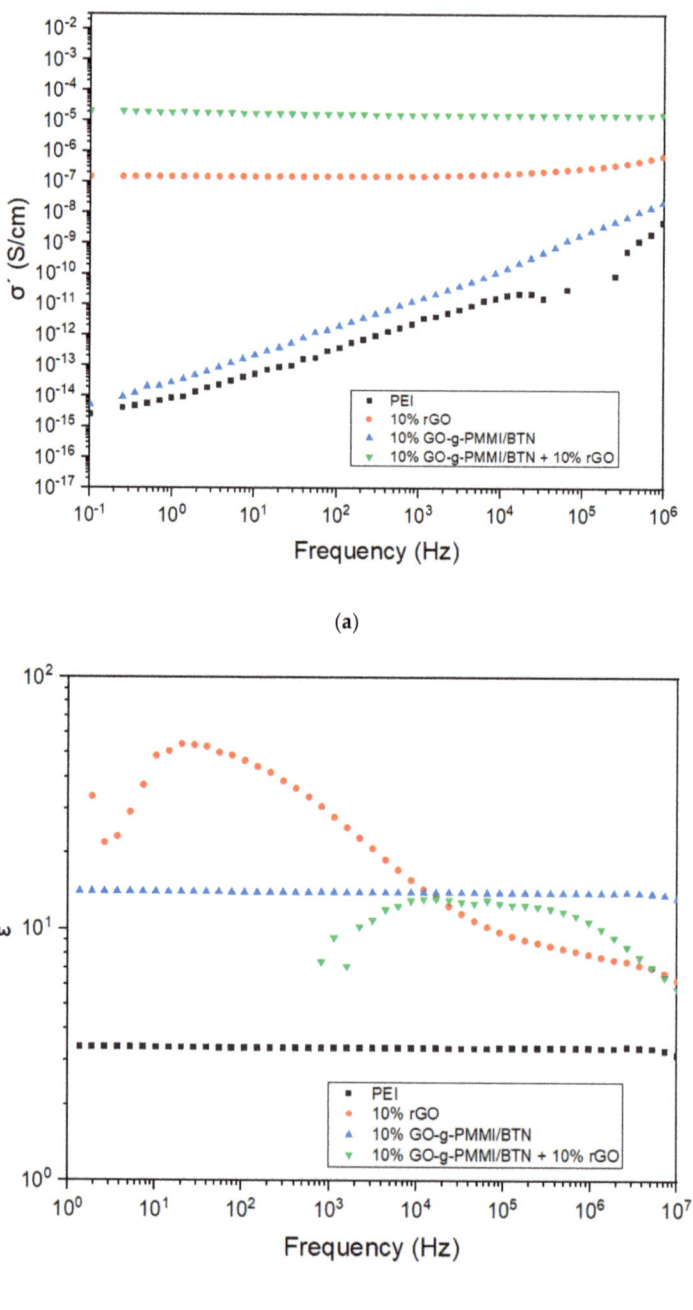

(a)

(b)

Figure 10. Conductivity (**a**) and dielectric permittivity (**b**) of PEI and nanocomposites filled with rGO, GO-g-PMMI/BTN and GO-g-PMMI/BTN+10% rGO as a function of the frequency.

Permittivity is a measure of the electrical polarizability of a material. The charges of a material exposed to an electric field undergo temporary displacement; this generates an induced field within the material [35]. Dielectric permittivity (ε'), which is also known as the relative dielectric constant, is the real part of the complex dielectric permittivity ($\varepsilon = \varepsilon' - j\varepsilon''$) [36]. Figure 10b presents ε' as function of the frequency of the applied electric field. It is possible to notice that the nanocomposites filled with 10 wt.% by weight of rGO and 10 wt.% of GO-g-PMMI/BTN + 10 wt.% rGO show a decrease in their ε' values with increasing frequency. The opposite was the case for the nanocomposite filled with 10 wt.% of GO-g-PMMI/BTN, whose value of ε' remained constant with the variation of the frequency and showed a dielectric permittivity one order higher than PEI. This indicates the effective contribution of barium titanate to the permittivity of the nanocomposite. It is important to notice that the dielectric curves of the nanocomposites containing hybrid filler and rGO do not present registers below 10^3 Hz. This is because the curves are represented in a logarithmic scale, and the permittivity values at these frequencies are negative. The negative values of the permittivity are being related with opposed polarization to the electric field. The materials that present negative dielectric constant are considered as revolutionary materials for electronic and photonic applications [37].

Figure 11 shows the results obtained of the dielectric loss (ε'') for PEI and the nanocomposites filled with rGO, GO-g-PMMI/BTN, or GO-g-PMMI/BTN+10 wt.% rGO as a function of the frequency of the applied electric field in the range between 10^{-1} and 10^6 Hz, observing a gradual decrease for GO-g-PMMI-filled nanocomposites ($\varepsilon'' = 3.818 \times 10^8$, recorded at $v = 10^{-1}$ to $\varepsilon'' = 3.053 \times 10^1$, recorded at $v = 10^{-6}$), as well as for those filled with rGO ($\varepsilon'' = 2.666 \times 10^6$ S/cm, recorded at $v = 10^{-1}$ to $\varepsilon'' = 1.35$ S/cm, recorded at $v = 10^{-6}$). The incorporation of 10 wt.% of rGO to the nanocomposite containing GO-g-PMMI/BTN improved the dielectric permittivity of the material since the nanocomposites with GO-g-PMMI/BTN as filler presented a lower dielectric permittivity ($\varepsilon'' = 0.097$, recorded at $v = 10^{-1}$ Hz) whose value did not vary with the increase in frequency.

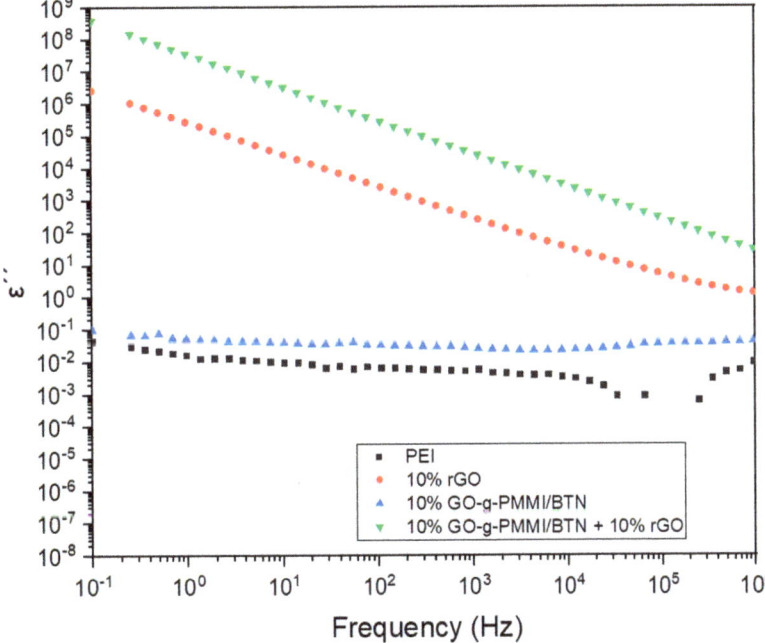

Figure 11. Loss factors of PEI and nanocomposites filled with rGO, GO-g-PMMI/BTN, and GO-g-PMMI/BTN+10 wt.% rGO as a function of frequency.

4. Conclusions

Nanocomposites based on polyetherimides filled with rGO, GO-g-PMMI/BTN or GO-g-PMMI/BTN+10 wt.% rGO, a graphene oxide hybrid material obtained from GO grafted with poly(monomethyl itaconate) modified with barium titanate nanoparticles, were prepared and their electrical properties were studied. It was observed that the electrical properties of the nanocomposites varied depending on the type of the filler material used. The nanocomposites filled with 10wt.% of GO-g-PMMI/BTN presented the same conductivity of 1×10^{-11} S/cm shown by the polymeric matrix. On the other hand, the nanocomposites filled with 10wt.% of rGO presented a conductivity in the order of 1×10^{-8} S/cm. The incorporation of 10 wt.% rGO with 10 wt.% GO-g-PMMI/BTN as fillers in PEI allowed us to obtain a new nanocomposite material with improved electrical properties compared with the nanocomposite containing only 10wt.% rGO. The conductivity of this new nanocomposite was in the order of 1×10^{-7} S/cm. The combination of rGO and GO-g-PMMI/BTN hybrid material improved the electrical conductivity of the resulting nanocomposite.

Author Contributions: Conceptualization, Q.C.-B., H.A.-B. and M.Y.-P.; methodology, H.A.-B. and M.Y.-P.; formal analysis, Q.C.-B. and H.A.-B.; investigation, Q.C.-B. and H.A.-B.; resources, M.Y.-P.; writing—original draft preparation, Q.C.-B.; writing—review and editing, Q.C.-B., H.A.-B. and M.Y.-P. All authors have read and agreed to the published version of the manuscript.

Funding: Q.C.-B. thanks the National Agency for Research and Development (ANID) for the Doctoral Scholarships 21191487. M.Y.-P. and H.A.-B. thank the ANID-FONDECYT 1191566 project. H.A.-B. also thanks the ANID-FONDECYT 11200437 project.

Institutional Review Board Statement: Not applicable.

Informed Consent Statement: Not applicable.

Data Availability Statement: Data sharing not applicable.

Conflicts of Interest: The authors declare no conflict of interest.

References

1. Aguilar-Bolados, H.; Vargas-Astudillo, D.; Yazdani-Pedram, M.; Acosta-Villavicencio, G.; Fuentealba, P.; Contreras-Cid, A.; Verdejo, R.; López-Manchado, M.A. Facile and Scalable One-Step Method for Amination of Graphene Using Leuckart Reaction. *Chem. Mater.* **2017**, *29*, 6698–6705. [CrossRef]
2. Aguilar-Bolados, H.; Contreras-Cid, A.; Yazdani-Pedram, M.; Acosta-Villavicencio, G.; Flores, M.; Fuentealba, P.; Neira-Carrillo, A.; Verdejo, R.; López-Manchado, M.A. Synthesis of fluorinated graphene oxide by using an easy one-pot deoxyfluorination reaction. *J. Colloid Interface Sci.* **2018**, *524*, 219–226. [CrossRef] [PubMed]
3. Aguilar-Bolados, H.; Yazdani-Pedram, M.; Verdejo, R. Thermal, electrical, and sensing properties of rubber nanocomposites. In *High-Performance Elastomeric Materials Reinforced by Nano-Carbons*; Elsevier: Amsterdam, The Netherlands, 2020; pp. 149–175. [CrossRef]
4. Yang, T.; Lin, H.; Loh, K.P.; Jia, B. Fundamental Transport Mechanisms and Advancements of Graphene Oxide Membranes for Molecular Separation. *Chem. Mater.* **2019**, *31*, 1829–1846. [CrossRef]
5. Faysal Hossain, M.D.; Akther, N.; Zhou, Y. Recent advancements in graphene adsorbents for wastewater treatment: Current status and challenges. *Chinese Chem. Lett.* **2020**, *31*, 2525–2538. [CrossRef]
6. Gao, W.; Alemany, L.B.; Ci, L.; Ajayan, P.M. New insights into the structure and reduction of graphite oxide. *Nat. Chem.* **2009**, *1*, 403–408. [CrossRef] [PubMed]
7. Potts, J.R.; Dreyer, D.R.; Bielawski, C.W.; Ruoff, R.S. Graphene-based polymer nanocomposites. *Polymer* **2011**, *52*, 5–25. [CrossRef]
8. Wang, S.; Dong, Y.; He, C.; Gao, Y.; Jia, N.; Chen, Z.; Song, W. The role of sp2/sp3 hybrid carbon regulation in the nonlinear optical properties of graphene oxide materials. *RSC Adv.* **2017**, *7*, 53643–53652. [CrossRef]
9. Georgakilas, V.; Otyepka, M.; Bourlinos, A.B.; Chandra, V.; Kim, N.; Kemp, K.C.; Hobza, P.; Zboril, R.; Kim, K.S. Functionalization of graphene: Covalent and non-covalent approach. *Chem. Rev.* **2012**, *112*, 6156–6214. [CrossRef] [PubMed]
10. Rubio, N.; Au, H.; Leese, H.S.; Hu, S.; Clancy, A.J.; Shaffer, M.S.P. Grafting from versus Grafting to Approaches for the Functionalization of Graphene Nanoplatelets with Poly(methyl methacrylate). *Macromolecules* **2017**, *50*, 7070–7079. [CrossRef]
11. Eskandari, P.; Abousalman-Rezvani, Z.; Roghani-Mamaqani, H.; Salami-Kalajahi, M.; Mardani, H. Polymer grafting on graphene layers by controlled radical polymerization. *Adv. Colloid Interface Sci.* **2019**, *273*, 102021. [CrossRef]
12. Zhao, M.; Lu, X.; Zong, H.; Li, J.; Zhuge, B. Itaconic acid production in microorganisms. *Biotechnol. Lett.* **2018**, *40*, 455–464. [CrossRef] [PubMed]

3. Itapu, B.M.; Jayatissa, A.H. A Review in Graphene/Polymer Composites. *Chem. Sci. Int. J.* **2018**, *23*, 1–16. [CrossRef]
4. Tewatia, K.; Sharma, A.; Sharma, M.; Kumar, A. Factors affecting morphological and electrical properties of Barium Titanate: A brief review. *Mater. Today Proc.* **2020**, *44*, 4548–4556. [CrossRef]
5. Romasanta, L.J.; Lopez-Manchado, M.A.; Verdejo, R. Increasing the performance of dielectric elastomer actuators: A review from the materials perspective. *Prog. Polym. Sci.* **2015**, *51*, 188–211. [CrossRef]
6. Aguilar-Bolados, H.; Yazdani-Pedram, M.; Quinteros-Jara, E.; Cuenca-Bracamonte, Q.; Quijada, R.; Carretero-González, J.; Avilés, F.; Lopez-Manchado, M.A.; Verdejo, R. Synthesis of sustainable, lightweight and electrically conductive polymer brushes grafted multi-layer graphene oxide. *Polym. Test.* **2021**, *93*, 106986. [CrossRef]
7. Zheng, M.S.; Zheng, Y.T.; Zha, J.W.; Yang, Y.; Han, P.; Wen, Y.Q.; Dang, Z.M. Improved dielectric, tensile and energy storage properties of surface rubberized BaTiO3/polypropylene nanocomposites. *Nano Energy* **2018**, *48*, 144–151. [CrossRef]
8. Zhang, J.; Ma, J.; Zhang, L.; Zong, C.; Xu, A.; Zhang, Y.; Geng, B.; Zhang, S. Enhanced breakdown strength and suppressed dielectric loss of polymer nanocomposites with BaTiO3 fillers modified by fluoropolymer. *RSC Adv.* **2020**, *10*, 7065–7072. [CrossRef]
9. Yaqoob, U.; Iftekhar Uddin, A.S.M.; Chung, G.S. The effect of reduced graphene oxide on the dielectric and ferroelectric properties of PVDF–BaTiO 3 nanocomposites. *RSC Adv.* **2016**, *6*, 30747–30754. [CrossRef]
20. Wang, S.; Chi, H.; Chen, L.; Li, W.; Li, Y.; Li, G.; Ge, X. Surface Functionalization of Graphene Oxide with Polymer Brushes for Improving Thermal Properties of the Polymer Matrix. *Adv. Polym. Technol.* **2021**, *2021*, 5591420. [CrossRef]
21. Molecular, L.D.F. Properties of polyelectrolytes: Poly(mono-methyl itaconate). conformational and viscometric behaviour in dilute soluton. *Eur. Polym. J.* **1989**, *25*, 1059–1063. [CrossRef]
22. Al-Mufti, S.M.S.; Almontasser, A.; Rizvi, S.J.A. Influence of temperature variations on the dielectric parameters of thermally reduced graphene oxide. *Mater. Today Proc.* **2022**, *57*, 1713–1718. [CrossRef]
23. Johra, F.T.; Jung, W.G. Hydrothermally reduced graphene oxide as a supercapacitor. *Appl. Surf. Sci.* **2015**, *357*, 1911–1914. [CrossRef]
24. Bychko, I.; Abakumov, A.; Didenko, O.; Chen, M.; Tang, J.; Strizhak, P. Differences in the structure and functionalities of graphene oxide and reduced graphene oxide obtained from graphite with various degrees of graphitization. *J. Phys. Chem. Solids* **2022**, *164*, 110614. [CrossRef]
25. Chang, S.J.; Liao, W.S.; Ciou, C.J.; Lee, J.T.; Li, C.C. An efficient approach to derive hydroxyl groups on the surface of barium titanate nanoparticles to improve its chemical modification ability. *J. Colloid Interface Sci.* **2009**, *329*, 300–305. [CrossRef] [PubMed]
26. Ran, J.; Guo, M.; Zhong, L.; Fu, H. In situ growth of BaTiO3 nanotube on the surface of reduced graphene oxide: A lightweight electromagnetic absorber. *J. Alloys Compd.* **2019**, *773*, 423–431. [CrossRef]
27. Liu, J.; Li, Q.; Zou, Y.; Qian, Q.; Jin, Y.; Li, G.; Jiang, K.; Fan, S. The dependence of graphene Raman D-band on carrier density. *Nano Lett.* **2013**, *13*, 6170–6175. [CrossRef] [PubMed]
28. Liu, Y.; Shi, J.; Kang, P.; Wu, P.; Zhou, Z.; Chen, G.X.; Li, Q. Improve the dielectric property and breakdown strength of composites by cladding a polymer/BaTiO3 composite layer around carbon nanotubes. *Polymer* **2020**, *188*, 122157. [CrossRef]
29. Li, L.; Zheng, S. Enhancement of dielectric constants of epoxy thermosets via a fine dispersion of barium titanate nanoparticles. *J. Appl. Polym. Sci.* **2016**, *133*, 1–10. [CrossRef]
30. Woudenberg, F.C.M.; Sager, W.F.C.; Ten Elshof, J.E.; Verweij, H. Nanostructured barium titanate thin films from nanoparticles obtained by an emulsion precipitation method. *Thin Solid Films* **2005**, *471*, 134–139. [CrossRef]
31. Chen, B.K.; Su, C.T.; Tseng, M.C.; Tsay, S.Y. Preparation of polyetherimide nanocomposites with improved thermal, mechanical and dielectric properties. *Polym. Bull.* **2006**, *57*, 671–681. [CrossRef]
32. Oliveira, R.; Georgieva, P.; Feyo de Azevedo, S. Plant and Equipment | Instrumentation and Process Control: Instrumentation. In *Encyclopedy of Diary Sciences*, 2nd ed.; Fuquay, J.W., Ed.; Academic Press: San Diego, CA, USA, 2002; pp. 234–241. ISBN 978-0-12-374407-4.
33. Cuenca-bracamonte, Q.; Yazdani-pedram, M.; Hernandez Santana, M.; Aguilar-Bolados, H. Electrical Properties of Poly(Monomethyl Itaconate)/Few-Layer Functionalized Graphene Oxide/Lithium Ion Nanocomposites. *Polymers* **2020**, *12*, 2673. [CrossRef] [PubMed]
34. Bormashenko, E. Negative Electrical Conductivity Metamaterials. *Preprints* **2022**, 2022030357. [CrossRef]
35. Schön, J.H. Chapter 8—Electrical Properties. In *Physical Properties of Rocks*; Schön, J.H., Ed.; Elsevier: Amsterdam, The Netherlands, 2015; Volume 65, pp. 301–367. ISBN 0376-7361.
36. Wang, D.; Zhang, X.; Zha, J.W.; Zhao, J.; Dang, Z.M.; Hu, G.H. Dielectric properties of reduced graphene oxide/polypropylene composites with ultralow percolation threshold. *Polymer* **2013**, *54*, 1916–1922. [CrossRef]
37. Yan, H.; Zhao, C.; Wang, K.; Deng, L.; Ma, M.; Xu, G. Negative dielectric constant manifested by static electricity. *Appl. Phys. Lett.* **2013**, *102*, 62904. [CrossRef]

Article

Magnetic Field Influence on the Microwave Characteristics of Composite Samples Based on Polycrystalline Y-Type Hexaferrite

Svetoslav Kolev [1,2,*], Borislava Georgieva [1], Tatyana Koutzarova [1], Kiril Krezhov [1], Chavdar Ghelev [1], Daniela Kovacheva [3], Benedicte Vertruyen [4], Raphael Closset [4], Lan Maria Tran [5], Michal Babij [5] and Andrzej J. Zaleski [5]

[1] Institute of Electronics, Bulgarian Academy of Sciences, 72 Tsarigradsko Chaussee, 1784 Sofia, Bulgaria
[2] Department of Physics, Faculty of Mathematics and Natural Science, Neofit Rilski South-Western University, 66 Ivan Mihailov Str., 2700 Blagoevgrad, Bulgaria
[3] Institute of General and Inorganic Chemistry, Bulgarian Academy of Sciences, Acad. Georgi Bonchev Str., bld. 11, 1113 Sofia, Bulgaria
[4] Greenmat, Chemistry Department, University of Liege, 11 Allée du 6 Août, 4000 Liège, Belgium
[5] Institute of Low Temperature and Structure Research, Polish Academy of Sciences, Ul. Okólna 2, 50-422 Wroclaw, Poland
* Correspondence: svet_kolev@yahoo.com or skolev@ie.bas.bg; Tel.: +359-2-979-5871

Citation: Kolev, S.; Georgieva, B.; Koutzarova, T.; Krezhov, K.; Ghelev, C.; Kovacheva, D.; Vertruyen, B.; Closset, R.; Tran, L.M.; Babij, M.; et al. Magnetic Field Influence on the Microwave Characteristics of Composite Samples Based on Polycrystalline Y-Type Hexaferrite. *Polymers* 2022, *14*, 4114. https://doi.org/10.3390/polym14194114

Academic Editors: Vineet Kumar and Xiaowu Tang

Received: 5 August 2022
Accepted: 28 September 2022
Published: 30 September 2022

Publisher's Note: MDPI stays neutral with regard to jurisdictional claims in published maps and institutional affiliations.

Copyright: © 2022 by the authors. Licensee MDPI, Basel, Switzerland. This article is an open access article distributed under the terms and conditions of the Creative Commons Attribution (CC BY) license (https://creativecommons.org/licenses/by/4.0/).

Abstract: Here, we report results on the magnetic and microwave properties of polycrystalline Y-type hexaferrite synthesized by sol-gel auto-combustion and acting as a filler in a composite microwave-absorbing material. The reflection losses in the 1–20 GHz range of the Y-type hexaferrite powder dispersed homogeneously in a polymer matrix of silicon rubber were investigated in the absence and in the presence of a magnetic field. A permanent magnet was used with a strength of 1.4 T, with the magnetic force lines oriented perpendicularly to the direction of the electromagnetic wave propagation. In the case of using an external magnetic field, an extraordinary result was observed. The microwave reflection losses reached a maximum value of 35.4 dB at 5.6 GHz in the Ku-band without a magnetic field and a maximum value of 21.4 dB at 8.2 GHz with the external magnetic field applied. The sensitivity of the microwave properties of the composite material to the external magnetic field was manifested by the decrease of the reflected wave attenuation. At a fixed thickness, t_m, of the composite, the attenuation peak frequency can be adjusted to a certain value either by changing the filling density or by applying an external magnetic field.

Keywords: Y-type hexaferrite; magnetic properties; microwave properties; reflection losses; external magnetic field

1. Introduction

The continuing development of microwave (MW) technologies and the extensive use of high-frequency electromagnetic radiation, not only in various industries but also in medicine and everyday life, such as designing "smart" cities, could lead to a massive increase in electromagnetic interference that could damage sensitive electronic devices and affect human health in ways that are yet to be fully understood. During the past decade, the necessity for the large-scale use of the gigahertz range has resulted in the development of novel functional materials with qualitatively novel properties. Serious efforts have been directed towards producing materials that attenuate (absorb and screen) microwave radiation with high efficiency in order to alleviate problems related to electromagnetic interference and to control its biological impact [1–3]. With the continuous innovations in information technologies, higher expectations have been put forward for portable, implantable and wearable electronic devices. The new trends in developing electromagnetic (EM) devices are focused on miniaturization, integration and multifunctionality. However, all of this has led to increased EM radiation pollution, which interferes with the regular operation of precision electronic components; these problems are becoming very serious [4,5].

On the other hand, EM radiation may also heat human cells or interfere with the intrinsic EM field of the human body; thus, harming human health [6–8].

These are the main reasons driving the development of various advanced materials for EM wave absorption. The parameters indicating the quality of EM wave absorbers are high reflection losses (RL), small thickness, wide bandwidth and low density [9–13].

Recently, the attention of researchers in this field has been attracted by a large variety of nanomaterials (e.g., magnetic materials and magnetic composites), mainly due to the possibility of controlling their EM wave absorption properties by varying the shape and size of nanoparticles or the structure and composition of nanocomposites [14–19].

In view of responding to the growing requirements of the materials intended for such applications, researchers have increasingly focused their attention on developing magnetic composites based on M-, Y-, Z- and U-hexaferrites [20,21]. Hexagonal ferrites are ferrimagnetic materials that are widely investigated for microwave absorption in the GHz frequency band. These materials exhibit strong magnetocrystalline anisotropy, very good high-frequency magnetic properties, large resistivity, good chemical stability and low cost [22–24]. In general, a pure hexagonal ferrite exhibits low microwave absorption efficiency and a rather narrow absorbing bandwidth. This is due to their relatively weak EM attenuation ability and a large inequality in magnetic and dielectric losses. To solve these problems and to prepare hexagonal ferrite materials with very good absorbing properties, efforts have been directed towards reducing the materials' dimension and particle size [25,26], doping with rare earth or transition metal ions [27,28] and combining hexagonal ferrites with other absorbing materials to form composites [29].

Thus, it has been clearly demonstrated that some multiferroic materials, such as the Y-type hexagonal ferrites, could have practical applications with very good reflection losses and absorption bandwidth [30].

This is why the last decade saw strong research efforts geared towards studying the structural and magnetic properties of, and particularly, the magnetic phase transitions in Y-type hexaferrites. This is explained by the magnetoelectric effect at room temperature observed in some of them and the consequent potential practical applications of these compounds as multiferroics.

In this paper, the effects of substituting magnetic Ni^{2+} cations with non-magnetic Zn^{2+} cations in $Ba_{0.5}Sr_{1.5}Zn_{2-x}Ni_xFe_{12}O_{22}$-based composites on their room temperature microwave behavior, together with the influence of applying an external magnetic field, were studied. The results are presented and discussed concerning the following two Y-type hexaferrite compositions: $Ba_{0.5}Sr_{1.5}Zn_{0.5}Ni_{1.5}Fe_{12}O_{22}$, $Ba_{0.5}Sr_{1.5}ZnNiFe_{12}O_{22}$, $Ba_{0.5}Sr_{1.5}Zn_{1.2}Ni_{0.8}Fe_{12}O_{22}$ and $Ba_{0.5}Sr_{1.5}Zn_2Fe_{12}O_{22}$.

2. Materials and Methods

Powder samples of $Ba_{0.5}Sr_{1.5}Zn_{0.5}Ni_{1.5}Fe_{12}O_{22}$, $Ba_{0.5}Sr_{1.5}ZnNiFe_{12}O_{22}$, $Ba_{0.5}Sr_{1.5}Zn_{1.2}Ni_{0.8}Fe_{12}O_{22}$ and $Ba_{0.5}Sr_{1.5}Zn_2Fe_{12}O_{22}$ were prepared by a modified citric acid sol-gel auto-combustion technique using stoichiometric amounts of the precursors. The synthesis procedure is described in detail in [31], and further information on the phase content analysis is given in [32]. In brief, the corresponding metal nitrates were used as starting materials; a citric acid solution was slowly added to the mixed nitrates as a chelator to form stable complexes with the metal cations. The solution was then slowly evaporated to form a gel; thus, it turned into a fluffy mass and burned in a self-propagating combustion manner. The obtained auto-combusted powders were homogenized and thermally treated at 800 °C for three hours. Finally, the Y-type hexaferrite powders were produced as follows: The powders were pressed into bulk pellets with a diameter of 16 mm. The pellets were annealed at 1170 °C in air for 10 h. After cooling down to room temperature, the pellets were ground to obtain the powder materials for the magnetic measurement. For the microwave measurements, the powders were homogenized and used as a filler for composite preparations.

The composite samples were prepared by dispersing the magnetic Y-type hexaferrite ($Ba_{0.5}Sr_{1.5}Zn_{0.5}Ni_{1.5}Fe_{12}O_{22}$, $Ba_{0.5}Sr_{1.5}ZnNiFe_{12}O_{22}$, $Ba_{0.5}Sr_{1.5}Zn_{1.2}Ni_{0.8}Fe_{12}O_{22}$ and $Ba_{0.5}Sr_{1.5}Zn_2Fe_{12}O_{22}$) powders (fillers) in commercial silicon rubber (Mastersil, ASP, Sofia, Bulgaria) as a polymer matrix. Two types of composite samples were prepared as follows: **A** is a series of samples with amounts of magnetic filler of 1.5 g per 1 cm^3 of silicon rubber ($Ba_{0.5}Sr_{1.5}Zn_{0.5}Ni_{1.5}Fe_{12}O_{22}$, P1A; $Ba_{0.5}Sr_{1.5}ZnNiFe_{12}O_{22}$, P2A; $Ba_{0.5}Sr_{1.5}Zn_{1.2}Ni_{0.8}Fe_{12}O_{22}$, P3A; $Ba_{0.5}Sr_{1.5}Zn_2Fe_{12}O_{22}$, P4A) and **B** is a series of samples with amounts of magnetic filler of 1.8 g per 1 cm^3 of silicon rubber ($Ba_{0.5}Sr_{1.5}Zn_{0.5}Ni_{1.5}Fe_{12}O_{22}$, P1B; $Ba_{0.5}Sr_{1.5}ZnNiFe_{12}O_{22}$, P2B; $Ba_{0.5}Sr_{1.5}Zn_{1.2}Ni_{0.8}Fe_{12}O_{22}$, P3B; $Ba_{0.5}Sr_{1.5}Zn_2Fe_{12}O_{22}$, P4B). The samples were molded into a toroidal shape with an outside diameter of 7 mm, an inner diameter of 3 mm and a thickness of 4 mm. A referent sample of identical dimensions (denoted as R) made of silicon rubber was only used to study the polymer matrix behavior in the respective microwave range.

The microwave (MW) measurements were conducted using a Hewlett-Packard 8756A microwave scalar network analyzer in a frequency range of 1–20 GHz. To determine the MW characteristics of the composites, a technique was employed whereby the electromagnetic wave (TEM) impinges normally on a single-layer absorber backed by a perfect conductor [33]. The toroidal samples were tightly fitted into a 50-Ω coaxial measurement cell (APC 7) backed by a perfect conductor (short-circuit measurement). During part of the measurements, an external magnetic field was applied using a permanent magnet, providing a flux density of 1.4 T. The magnetic force lines were perpendicular to the direction of the electromagnetic wave propagation. The magnetic flux density in the air gap in the coaxial line was 0.3 T, as measured by a Model 475 Gaussmeter with an HMNT-4E04-VR Hall sensor. The hysteresis measurements were conducted at 300 K by a physical property measurement system (PPMS, Quantum Design Inc., San Diego, CA, USA). Scanning electron microscopy (SEM TESCAN LYRA I, with an EDX detector, Bruker Quantax, TESCAN, Brno, Czech Republic) was used to determine the samples' morphology and the particles' distribution in the silicon rubber.

3. Results and Discussion

Figures 1 and 2 illustrate the microstructure of the cut P1A and P4B samples. They have different filler densities and different ferrite compositions, but they appear very similar. The SEM images show that the particles are of a hexagonal shape, typical for hexaferrites, and are well agglomerated to form clusters of different sizes and shapes. These clusters are randomly distributed in the volume of the toroidal samples. Our previous investigations have shown that the average particle size is around 600 nm, with the size being within the 200–1000-nm range [34]. Since the same synthesis technique was used for all of the samples, it was only natural to assume that all of them would have a broad particle size distribution. Therefore, they are in a mixed mono- and poly-domain state, since the critical size for a mono-domain state for these types of materials is around 500 nm. Thus, it seemed speculative to discuss the size effect in this case where the size distribution was so broad. This was the main reason why our attention was focused on the Zn:Ni substitution in the samples and its influence on the magnetic and microwave properties rather than on the particle size effect.

Figure 3 presents the hysteresis curves of the $Ba_{0.5}Sr_{1.5}Zn_{0.5}Ni_{1.5}Fe_{12}O_{22}$, $Ba_{0.5}Sr_{1.5}ZnNiFe_{12}O_{22}$, $Ba_{0.5}Sr_{1.5}Zn_{1.2}Ni_{0.8}Fe_{12}O_{22}$ and $Ba_{0.5}Sr_{1.5}Zn_2Fe_{12}O_{22}$ powders. The curves are very narrow, with a coercivity of about a few oersteds. The magnetization values at a magnetic field of 50 kOe (here referred to as saturation magnetization) as a function of the Zn:Ni cation ratio of the samples are presented in Table 1.

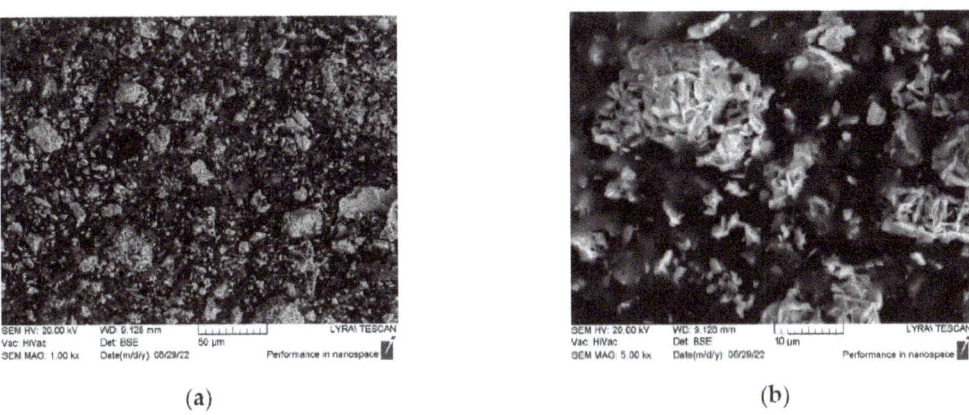

Figure 1. SEM images of (**a**) sample P1A and (**b**) zoomed area of a cluster.

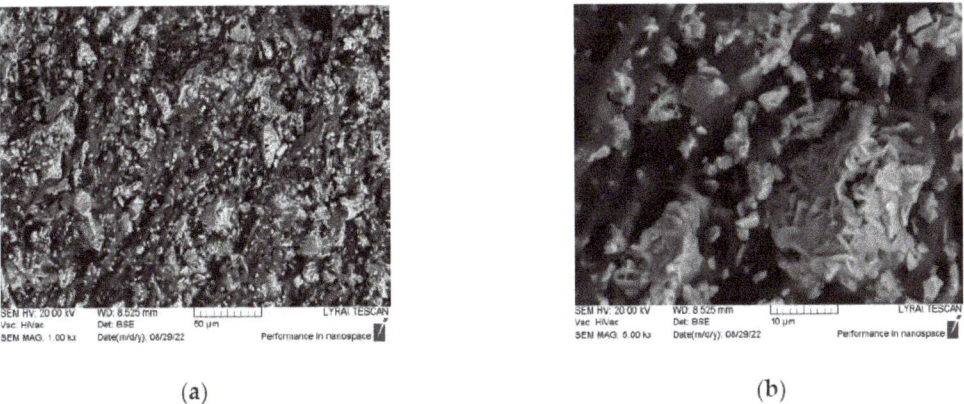

Figure 2. SEM images of (**a**) sample P4B and (**b**) zoomed area of a cluster.

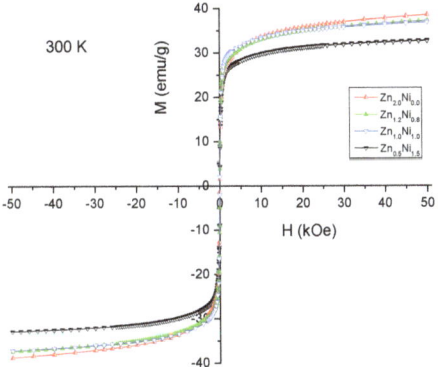

Figure 3. Hysteresis curves of $Ba_{0.5}Sr_{1.5}Zn_{2-x}Ni_xFe_{12}O_{22}$ (x = 0, 0.8, 1.2, 1.5) at room temperature.

Table 1. Magnetization values at a magnetic field of 50 kOe for the samples depending on the Zn:Ni cation ratio.

300 K	$Zn_{0.5}Ni_{1.5}$	ZnNi	$Zn_{1.2}Ni_{0.8}$	Zn_2Ni_0
saturation magnetization, emu/g	32.82	37.07	37.25	38.56

It is noteworthy that the sample with the highest content of nickel ions, which bear magnetic moments, has the lowest value of magnetization (50 kOe). This should be related to the known preference of the non-magnetic Zn^{2+} cations to occupy tetrahedral positions, whereas the magnetic Ni^{2+} cations prefer octahedral positions [35]. This entails the migration of iron cations between crystallographic sites with octahedral and tetrahedral oxygen configurations and is accompanied by a corresponding change in the magnetic structure. We also found a similar behavior at low temperatures in our earlier study [31]. One can also see that as the content of the non-magnetic Zn^{2+} cation is increased, the process of saturation is shifted to the higher values of the external magnetic field.

Our investigations were mainly aimed at measuring the microwave characteristics of the composite samples, in which the prepared powder material was dispersed in a polymer matrix (silicon rubber). Figure 4 presents the changes in the reflection losses (R_L) in the sweeping frequency range of 20 GHz for the composite **A** series samples with amounts of magnetic filler of 1.5 g per 1 cm^3 of silicon rubber (P1A, P2A, P3A and P4A) and the control polymer sample of the same thickness (4 mm) without a filler (R). Figure 5 presents the changes in the reflection losses (R_L) in the sweeping frequency range of 20 GHz for the composite **B** series samples with amounts of magnetic filler of 1.8 g per 1 cm^3 of silicon rubber (P1B, P2B, P3B and P4B) and the control polymer sample of the same thickness (4 mm) without a filler (R). Our previous experience has shown that the silicon rubber is transparent to electromagnetic waves in this microwave region [36,37], as it is confirmed in Figure 2 for the control (R). Consequently, in the case of a hexaferrite/silicon rubber composite, the microwave properties are due to the hexaferrite only. The curves representing R_L were obtained under the conditions of an electromagnetic wave incident perpendicularly to the surface of toroidal samples backed by a perfect conductor. According to the transmission line theory for such a case, the value of R_L as a function of the normalized input impedance is given by [38–40] and is as follows:

$$R_L(dB) = 20\log\left|\frac{Z_{in} - Z_0}{Z_{in} + Z_0}\right|, \quad (1)$$

where

$$Z_{in} = Z_0\sqrt{\frac{\mu_r}{\varepsilon_r}}\tanh\left[j\frac{2\pi}{c}\sqrt{\mu_r\varepsilon_r}fd\right] \quad (2)$$

where Z_{in} is the input impedance of the absorber; Z_0 is the impedance of free space; c is the speed of light in free space; f is the electromagnetic wave frequency; d is the absorber thickness; μ_r and ε_r are the complex permeability and permittivity, respectively.

An R_L value of -10 dB corresponds to a 90% attenuation of the electromagnetic wave. In general, materials with R_L values below -10 dB could be considered suitable microwave absorbers for practical applications [41].

At certain thicknesses and frequencies, minima are observed in the reflected wave. This takes place when the thickness of the absorber layer satisfies the quarter-wave thickness criterion, described by the quarter-wave theory [42,43] as follows:

$$t_m = \frac{nc}{4f_m\sqrt{|\varepsilon_r\mu_r|}}(n = 1, 3, 5, \ldots) \quad (3)$$

$$f_m = \frac{nc}{4t_m\sqrt{|\varepsilon_r\mu_r|}}(n = 1, 3, 5, \ldots) \quad (4)$$

where t_m and f_m are the matching thickness and the peak frequency [44].

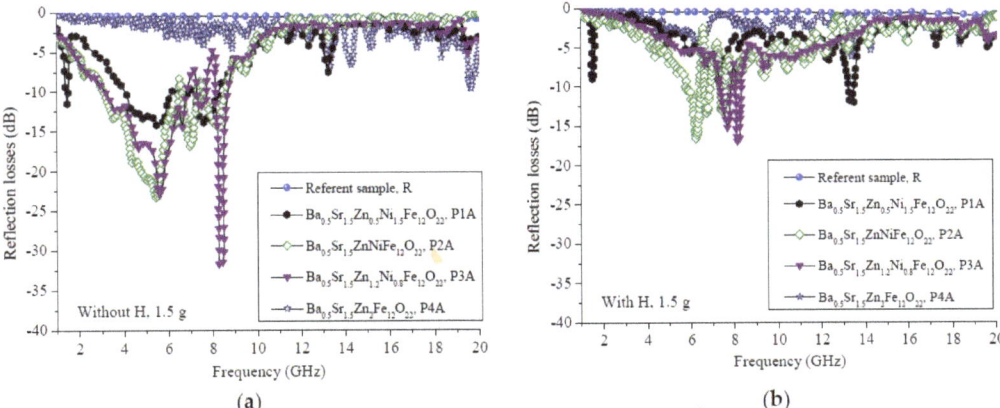

Figure 4. Reflection losses of the **A** series of samples with a filler ratio of 1.5 g per 1 cm³ (**a**) with a magnetic field and (**b**) without an external magnetic field applied.

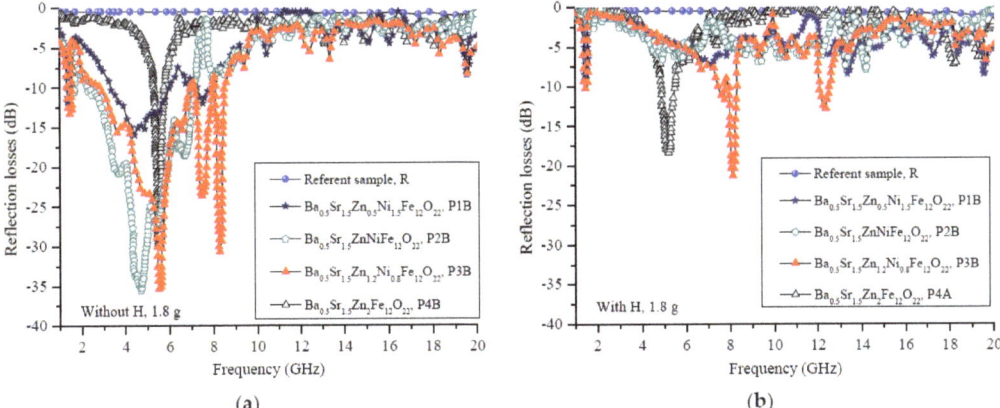

Figure 5. Reflection losses of the **B** series of samples with a filler ratio of 1.8 g per 1 cm³ (**a**) with a magnetic field and (**b**) without an external magnetic field applied.

To assess the EM wave attenuation, the intrinsic properties of the samples have to be accounted for. The incident microwave energy can generate heat in the material during the interaction of the electromagnetic field with the molecular and electronic structure of the material; the incident microwave energy can generate heat in the material. In this way, this process can convert the incident EM waves into thermal energy, and, finally, the energy will be dissipated.

In a sample, multiple scattering on inhomogeneities can take place [45]. Thus, by varying the material structure, in our case, the density of the hexaferrite particles, one could extend the EM wave propagation path within the sample and improve the absorption capacity of the EM wave absorber [46]. One can conclude that changing the electromagnetic parameters can also enhance the intrinsic EM absorption capabilities.

In the case reported, we measured the MW characteristics of two types of composite samples—with different powder compositions and with different weight ratios. We also added an external magnetic field perpendicular to the propagation of the electromagnetic wave to observe changes in the wave attenuation.

For both the **A** and **B** series of samples, two types of measurements were made—with and without an external magnetic field. The magnetic force lines of this field were perpendicular to the direction of the electromagnetic wave propagation; one would expect that the wave attenuation would be higher due to the orientation of the spins' magnetic moments. The results from the microwave measurements are opposite. In both cases, for the **A** and **B** series, the reflection losses decreased as a consequence of applying an external magnetic field. For the **A** series (Figure 4) without a magnetic field, the minimal reflection was observed at 8.2 GHz with an $R_L = -31.7$ dB. In the case with a magnetic field, the best value for the reflection losses was observed at 8.1 GHz with an $R_L = -17$ dB. An analogous behavior was identified for the **B** series (Figure 5), but the value of the reflection losses was higher and shifted to the lower frequencies due to the higher hexaferrite content in the composite samples. The minimal reflection without a magnetic field was observed at 4.6 GHz with an $R_L = -35.5$ dB, and it was observed with a magnetic field at 8.1 GHz with an $R_L = -21.4$ dB.

Following the main conclusion of the influence of the magnetic field on the reflection losses, the influence of the Zn:Ni ratio on them was discussed in Figure 6. For the **A** series samples, the main peak with the best value was observed for the Zn:Ni ratio of 1.2:0.8 ($Zn_{1.2}Ni_{0.8}$) at 8.3 GHz with an $R_L = -31.9$ dB without an external magnetic field. With the magnetic field applied, the sample with the best attenuation was of the same ratio of $Zn_{1.2}Ni_{0.8}$ at 8.1 GHz with an $R_L = -17.2$ dB.

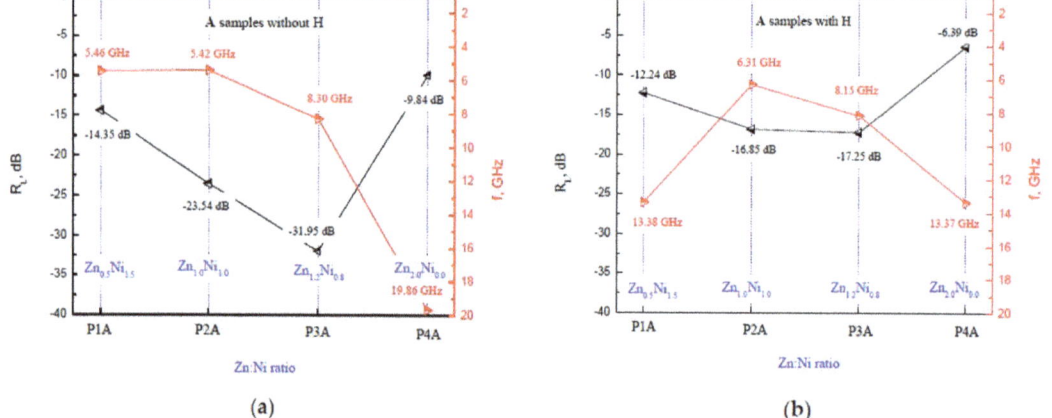

Figure 6. The main reflection loss peaks for the **A** series of samples depending on the Zn:Ni ratio (**a**) without a magnetic field and (**b**) with an external magnetic field applied.

A comparative view of the peaks for the **A** series of samples depending on the Zn:Ni ratio with and without a magnetic field is shown in Figure 7a. It is clear that applying the external magnetic field deteriorates the absorbing characteristics; namely, the attenuation decreases. Figure 7b shows the averaged results of ten measurements, with the standard deviation being 5%.

For the **B** series of samples, shown in Figure 8, two main peaks were observed with the same value of reflection losses as follows: for the ratio of Zn:Ni=1:1 at 4.7 GHz and for the ratio of $Zn_{1.2}Ni_{0.8}$ at 5.6 GHz with an $R_L = -35.4$ dB without an external magnetic field. When the magnetic field was applied, the sample with the best attenuation was that with the same ratio of $Zn_{1.2}Ni_{0.8}$ at 8.2 GHz with an $R_L = -21.4$ dB. Compared to the **A** series, the attenuation characteristics are improved, but, again, the external magnetic field plays a negative role and decreases the attenuation.

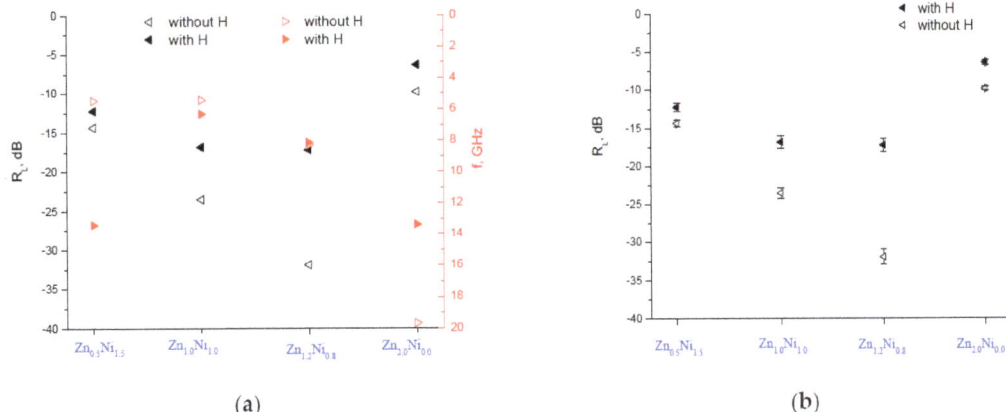

Figure 7. The main reflection loss peaks for the **A** series of samples depending on the Zn:Ni ratio. (**a**) Comparative view with and without a magnetic field and (**b**) an R_L with a standard deviation.

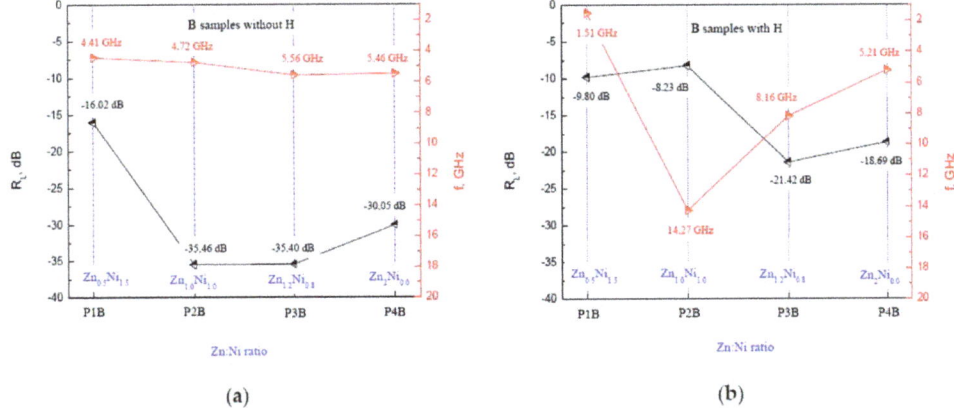

Figure 8. The main reflection loss peaks for the **B** series of samples depending on the Zn:Ni ratio (**a**) without a magnetic field and (**b**) with an external magnetic field applied.

A comparative view of the peaks is shown in Figure 9a for the **B** series of samples depending on the Zn:Ni ratio with and without a magnetic field. A similar behavior was observed when the external magnetic field was applied; the absorbing characteristics deteriorated. Figure 9b shows the results of ten repeated measurements. Here, again, the average value is obtained with a standard deviation of 5%. The most likely explanation for this error is the measurement setup since the samples were fixed mechanically without glue, so that minor differences might have arisen in their positioning.

One can, therefore, conclude that the material's saturation magnetization was raised by decreasing the degree of substitution with Ni, which, in turn, raised its reflection losses. Additionally, an optimal Zn:Ni ratio, namely, i.e., $Zn_{1.2}Ni_{0.8}$, existed there.

These results are opposite to the results observed by us for a Z-type hexaferrite ($Sr_3Co_2Fe_{24}O_{41}$) under the same experimental conditions [20]; applying a magnetic field resulted in an increased attenuation. In the reported case, for all of the investigated samples, the values of the reflection losses decreased when a magnetic field was applied. Thus, one might speculate that the external magnetic field changes the samples' permeability to negative or positive [47], depending on whether the material's resonance frequency is above or below the frequency range of the measurement. Additionally, for all of the samples, a magnetic phase transition from a helicoidal to a ferrimagnetic spin order occurs around room temperature, which was demonstrated in our earlier study [31,34]. This could play an important role in determining the microwave characteristics under an external magnetic field. Accordingly, applying a magnetic field most probably makes the sample more transparent, depending on the intrinsic properties of the Y-type hexaferrite.

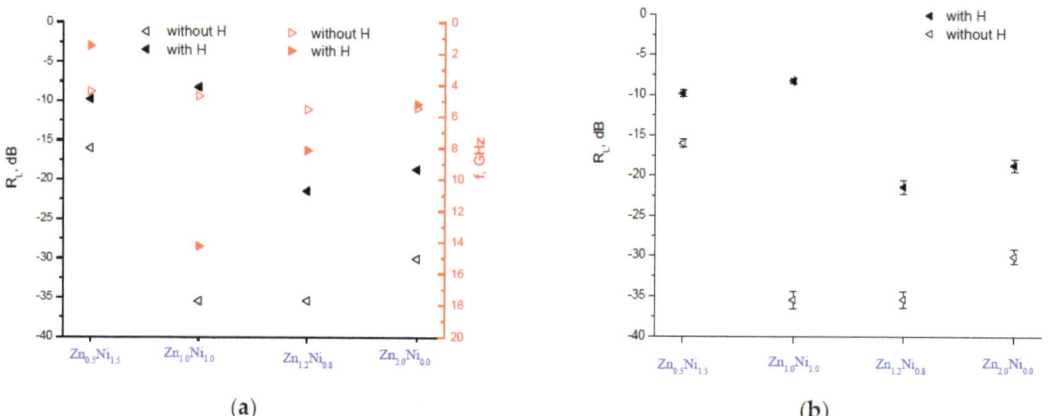

Figure 9. The main reflection loss peaks for the **A** series of samples depending on the Zn:Ni ratio (**a**) without a magnetic field and (**b**) with an external magnetic field applied.

4. Conclusions

Y-type hexaferrite materials were studied with different Ni substitutions, namely, $Ba_{0.5}Sr_{1.5}Zn_2Fe_{12}O_{22}$, $Ba_{0.5}Sr_{1.5}Zn_{1.2}Ni_{0.8}Fe_{12}O_{22}$, $Ba_{0.5}Sr_{1.5}ZnNiFe_{12}O_{22}$ and $Ba_{0.5}Sr_{1.5}Zn_{0.5}Ni_{1.5}Fe_{12}O_{22}$. The magnetic measurements at room temperature showed that the samples with an increased Ni substitution exhibited a decreased saturation magnetization. This effect impacted the results of the measurements of the composite samples' microwave characteristics when a Y-type hexaferrite was used as a filler material dispersed in a polymer matrix. It was estimated that composite samples with a higher hexaferrite content had better microwave properties, with the optimal Zn:Ni ratio for the best results being 1.2:0.8. The highest value of the reflection losses, $R_L = -35.4$ dB, was measured at 5.6 GHz without applying an external magnetic field. When a perpendicular external magnetic field was applied, the absorbing properties of both types of samples changed dramatically; namely, all main peaks were shifted to higher frequencies and the reflection losses decreased. The maximum value of 21.4 dB was found at 8.2 GHz with the external magnetic field applied. Thus, one might assume that the magnetic field makes the composite samples more transparent to the electromagnetic waves in this specific frequency range. This behavior should be related to the intrinsic properties of Y-type hexaferrite materials.

Author Contributions: Conceptualization, T.K., S.K. and K.K.; formal analysis, T.K., S.K., K.K., B.G., D.K., B.V., R.C., L.M.T., M.B. and A.J.Z.; funding acquisition, T.K., B.V. and A.J.Z.; investigation, T.K., S.K., K.K., D.K., B.V., R.C., L.M.T., M.B. and A.J.Z.; methodology, T.K. and B.G.; project administration, T.K., S.K. and K.K.; resources, T.K., S.K., B.G., D.K., C.G., B.V., R.C., L.M.T., M.B. and A.J.Z.; supervision, T.K. and S.K.; validation, T.K., S.K., K.K., B.G., D.K., B.V., R.C., L.M.T., M.B. and A.J.Z.; visualization, T.K. and S.K.; writing—original draft, T.K., S.K., K.K. and C.G.; writing—review and editing, C.G. All authors have read and agreed to the published version of the manuscript.

Funding: This research was funded by the Bulgarian National Science Fund (grant number KP-06-N48/5).

Institutional Review Board Statement: Not applicable.

Informed Consent Statement: Not applicable.

Data Availability Statement: Not applicable.

Acknowledgments: This work was supported by a joint research project between the Bulgarian Academy of Sciences and WBI, Belgium and by a joint research project between the Bulgarian Academy of Sciences and the Institute of Low Temperature and Structure Research, Polish Academy of Sciences.

Conflicts of Interest: The authors declare no conflict of interest.

References

1. Dankov, P. Material characterization in the microwave range, when the materials become composite, reinforced, 3d-printed, artificially mixed, nanomaterials and metamaterials, Nanomaterials and Metamaterials. *Forum Electromagn. Res. Methods Appl. Technol. (FERMAT J.)* **2020**, *41*, 1–37. Available online: https://www.e-fermat.org/articles.php (accessed on 20 December 2020).
2. Miszczyk, A. Protective and suppressing electromagnetic interference properties of epoxy coatings containing nano-sized NiZn ferrites. *Front. Mater.* **2020**, *7*, 183. [CrossRef]
3. Bera, S.C. *Microwave Active Devices and Circuits for Communication*, 1st ed.; Springer: Singapore, 2019.
4. Houbi, A.; Aldashevich, Z.A.; Atassi, Y.; Bagasharova Telmanovna, Z.; Saule, M.; Kubanych, K. Microwave absorbing properties of ferrites and their composites: A review. *J. Magn. Magn. Mater.* **2021**, *529*, 167839. [CrossRef]
5. Zeng, X.J.; Cheng, X.Y.; Yu, R.H.; Stucky, G.D. Electromagnetic microwave absorption theory and recent achievements in microwave absorbers. *Carbon* **2020**, *168*, 606–623. [CrossRef]
6. Cao, M.S.; Wang, X.X.; Zhang, M.; Cao, W.Q.; Fang, X.Y.; Yuan, J. Variable-temperature electron transport and dipole polarization turning flexible multifunctional microsensor beyond electrical and optical energy. *Adv. Mater.* **2020**, *32*, 1907156. [CrossRef]
7. Lee, S.; Jo, I.; Kang, S.; Jang, B.; Moon, J.; Park, J.B.; Lee, S.; Rho, S.; Kim, Y.; Hong, B.H. Smart contact lenses with graphene coating for electromagnetic interference shielding and dehydration protection. *ACS Nano* **2017**, *11*, 5318–5324. [CrossRef]
8. Wang, Z.; Mao, B.Y.; Wang, Q.L.; Yu, J.; Dai, J.X.; Song, R.G.; Pu, Z.H.; He, D.P.; Wu, Z.; Mu, S.C. Ultrahigh conductive copper/large flake size graphene heterostructure thin-film with remarkable electromagnetic interference shielding effectiveness. *Small* **2018**, *14*, 1704332. [CrossRef]
9. Meng, F.B.; Wang, H.G.; Huang, F.; Guo, Y.F.; Wang, Z.Y.; Hui, D.; Zhou, Z.W. Graphene-based microwave absorbing composites: A review and prospective. *Compos. Part B Eng.* **2018**, *137*, 260–277. [CrossRef]
10. Wang, F.Y.; Sun, Y.Q.; Li, D.R.; Zhong, B.; Wu, Z.G.; Zuo, S.Y.; Yan, D.; Zhuo, R.F.; Feng, J.J.; Yan, P.X. Core-shell FeCo@carbon nanoparticles encapsulated in polydopamine-derived carbon nanocages for efficient microwave absorption. *Carbon* **2018**, *134*, 264–273. [CrossRef]
11. Wang, L.N.; Jia, X.L.; Li, Y.F.; Yang, F.; Zhang, L.Q.; Liu, L.P.; Ren, X.; Yang, H.T. Synthesis and microwave absorption property of flexible magnetic film based on graphene oxide/carbon nanotubes and Fe_3O_4 nanoparticles. *J. Mater. Chem. A* **2014**, *2*, 14940–14946. [CrossRef]
12. Li, W.; Wei, J.; Wang, W.; Hu, D.; Li, Y.; Guan, J. Ferrite-based metamaterial microwave absorber with absorption frequency magnetically tunable in a wide range. *Mater. Des.* **2016**, *110*, 27–34. [CrossRef]
13. Pang, Y.; Li, Y.; Wang, J.; Yan, M.; Qu, S.; Xia, S.; Xu, Z. Electromagnetic reflection reduction of carbon composite materials mediated by collaborative mechanisms. *Carbon* **2019**, *147*, 112–119. [CrossRef]
14. Zhao, B.; Zhao, W.Y.; Shao, G.; Fan, B.B.; Zhang, R. Morphology-control synthesis of a core–shell structured NiCu alloy with tunable electromagnetic-wave absorption capabilities. *ACS Appl. Mater. Interfaces* **2015**, *7*, 12951–12960. [CrossRef]
15. Li, H.; Cao, Z.M.; Lin, J.Y.; Zhao, H.; Jiang, Q.R.; Jiang, Z.Y.; Liao, H.G.; Kuang, Q.; Xie, Z.X. Synthesis of u-channelled spherical $Fe_x(Co_yNi_{1-y})_{100-x}$ Janus colloidal particles with excellent electromagnetic wave absorption performance. *Nanoscale* **2018**, *10*, 1930–1938. [CrossRef]
16. Wang, C.H.; Ding, Y.J.; Yuan, Y.; He, X.D.; Wu, S.T.; Hu, S.; Zou, M.C.; Zhao, W.Q.; Yang, L.S.; Cao, A.Y.; et al. Graphene aerogel composites derived from recycled cigarette filters for electromagnetic wave absorption. *J. Mater. Chem. C* **2015**, *3*, 11893–11901. [CrossRef]
17. Zhou, X.F.; Jia, Z.R.; Feng, A.L.; Wang, X.X.; Liu, J.J.; Zhang, M.; Cao, H.J.; Wu, G.L. Synthesis of fish skin-derived 3D carbon foams with broadened bandwidth and excellent electromagnetic wave absorption performance. *Carbon* **2019**, *152*, 827–836. [CrossRef]
18. Xiang, J.; Li, J.L.; Zhang, X.H.; Ye, Q.; Xu, J.H.; Sheng, X.Q. Magnetic carbon nanofibers containing uniformly dispersed Fe/Co/Ni nanoparticles as stable and highperformance electromagnetic wave absorbers. *J. Mater. Chem. A* **2014**, *2*, 16905–16914. [CrossRef]
19. Li, X.A.; Du, D.X.; Wang, C.S.; Wang, H.Y.; Xu, Z.P. In situ synthesis of hierarchical rose-like porous Fe@C with enhanced electromagnetic wave absorption. *J. Mater. Chem. C* **2018**, *6*, 558–567. [CrossRef]
20. Kolev, S.; Peneva, P.; Krezhov, K.; Malakova, T.; Ghelev, C.; Koutzarova, T.; Kovacheva, D.; Vertruyen, B.; Closset, R.; Tran, L.-M.; et al. Structural, magnetic and microwave characterization of polycrystalline Z-type $Sr_3Co_2Fe_{24}O_{41}$ hexaferrite. *Materials* **2020**, *13*, 2355. [CrossRef] [PubMed]
21. Almessiere, M.; Genc, F.; Sozeri, H.; Baykal, A.; Trukhanov, S.V.; Trukhanov, A.V. Influence of charge disproportionation on microwave characteristics of Zn-Nd substituted Sr-hexaferrites. *J. Mater. Sci. Mater. Electron.* **2019**, *30*, 6776–6785. [CrossRef]
22. Kong, L.B.; Li, Z.W.; Liu, L.; Huang, R.; Abshinova, M.; Yang, Z.H.; Tang, C.B.; Tan, P.K.; Deng, C.R.; Matitsine, S. Recent progress in some composite materials and structures for specific electromagnetic applications. *Int. Mater. Rev.* **2013**, *58*, 203–259. [CrossRef]
23. Pullar, R.C. Hexagonal Ferrite Fibres and Nanofibres. *Solid State Phenom.* **2016**, *241*, 1–68. [CrossRef]

24. Kong, L.B.; Liu, L.; Yang, Z.; Li, S.; Zhang, T.; Wang, C. Ferrite-based composites for microwave absorbing applications. In *Metal Oxides, Magnetic, Ferroelectric, and Multiferroic Metal Oxides*; Stojanovic, B.D., Ed.; Elsevier: Amsterdam, The Netherlands, 2018; pp. 361–385. [CrossRef]
25. Bahadur, A.; Saeed, A.; Iqbal, S.; Shoaib, M.; Ahmad, I.; ur Rahman, M.S.; Bashir, M.I.; Yaseen, M.; Hussain, W. Morphological and magnetic properties of $BaFe_{12}O_{19}$ nanoferrite: A promising microwave absorbing material. *Ceram. Int.* **2017**, *43*, 7346–7350. [CrossRef]
26. Wei, S.C.; Liu, Y.; Tian, H.L.; Tong, H.; Liu, Y.X.; Xu, B.S. Microwave absorption property of plasma spray W-type hexagonal ferrite coating. *J. Magn. Magn. Mater.* **2015**, *377*, 419–423. [CrossRef]
27. Kumar, S.; Meena, R.S.; Chatterjee, R. Microwave absorption studies of Cr-doped Co–U type hexaferrites over 2–18 GHz frequency range. *J. Magn. Magn. Mater.* **2016**, *418*, 194–199. [CrossRef]
28. Singh, J.; Singh, C.; Kaur, D.; Narang, S.B.; Jotania, R.; Joshi, R. Investigation on structural and microwave absorption property of Co^{2+} and Y^{3+} substituted M-type Ba-Sr hexagonal ferrites prepared by a ceramic method. *J. Alloy. Comp.* **2017**, *695*, 792–798. [CrossRef]
29. Tang, X.T.; Wei, G.T.; Zhu, T.X.; Sheng, L.M.; An, K.; Yu, L.M.; Liu, Y.; Zhao, X.L. Microwave absorption performance enhanced by high-crystalline graphene and $BaFe_{12}O_{19}$ nanocomposites. *J. Appl. Phys.* **2016**, *119*, 204301. [CrossRef]
30. Stergiou, C.A.; Litsardakis, G. Y-type hexagonal ferrites for microwave absorber and antenna applications. *J. Magn. Magn. Mater.* **2016**, *405*, 54–61. [CrossRef]
31. Koutzarova, T.; Kolev, S.; Krezhov, K.; Georgieva, B.; Ghelev, C.; Kovacheva, D.; Vertruyen, B.; Closset, R.; Tran, L.-M.; Babij, M.; et al. Ni-substitution effect on the properties of $Ba_{0.5}Sr_{1.5}Zn_{2-x}Ni_xFe_{12}O_{22}$ powders. *J. Magn. Magn. Mater.* **2020**, *505*, 166725. [CrossRef]
32. Koutzarova, T.; Kolev, S.; Krezhov, K.; Georgieva, B.; Ghelev, C.; Kovacheva, D.; Vertruyen, B.; Closset, R.; Tran, L.-M.; Babij, M.; et al. Data supporting the results of the characterization of the phases and structures appearing during the synthesis process of $Ba_{0.5}Sr_{1.5}Zn_{2-x}Ni_xFe_{12}O_{22}$ by auto-combustion. *Data Brief* **2020**, *31*, 105803. [CrossRef] [PubMed]
33. Kim, S.; Jo, S.; Gueon, K.; Choi, K.; Kim, J.; Churn, K. Complex permeability and permittivity and microwave absorption of ferrite-rubber composite at X-band frequencies. *IEEE Trans. Magn.* **1991**, *27*, 5462–5464. [CrossRef]
34. Georgieva, B.; Kolev, S.; Krezhov, K.; Ghelev, C.; Kovacheva, D.; Vertruyen, B.; Closset, R.; Tran, L.-M.; Babij, M.; Zaleski, A.J.; et al. Structural and magnetic characterization of Y-type hexaferrite powders prepared by sol-gel auto-combustion and sonochemistry. *J. Magn. Magn. Mater.* **2019**, *477*, 131–135. [CrossRef]
35. El Hiti, M.; Abo El Ata, A. Semiconductivity in $Ba_2Ni_{2-x}Zn_xFe_{12}O_{22}$ Y-type hexaferrites. *J. Magn. Magn. Mater.* **1999**, *195*, 667–678. [CrossRef]
36. Kolev, S.; Yanev, A.; Nedkov, I. Microwave absorption of ferrite powders in a polymer matrix. *Phys. Stat. Sol.* **2006**, *3*, 1308–1315. [CrossRef]
37. Kolev, S.; Koutzarova, T. Nanosized ferrite materials for absorption and protection from microwave radiation. In *Advanced Nanotechnologies for Detection and Defence against CBRN Agents NATO Science for Peace and Security Series B: Physics and Biophysics*; Petkov, P., Tsiulyanu, D., Popov, C., Kulisch, W., Eds.; Springer: Dordrecht, The Netherlands, 2018; pp. 273–283. [CrossRef]
38. Naito, Y.; Suetake, K. Application of ferrite to electromagnetic wave absorber and its characteristics. *IEEE Trans. Microw. Theory Tech* **1971**, *19*, 65–72. [CrossRef]
39. Liu, W.; Liu, L.; Yang, Z.; Xu, J.; Hou, Y.; Ji, G. A versatile route toward the electromagnetic functionalization of metal–organic framework-derived three-dimensional nanoporous carbon composites. *ACS Appl. Mater. Interfaces* **2018**, *10*, 8965–8975. [CrossRef] [PubMed]
40. Quan, B.; Shi, W.H.; Ong, S.J.H.; Lu, X.C.; Wang, P.L.Y.; Ji, G.B.; Guo, Y.F.; Zheng, L.R.; Xu, Z.C.J. Defect engineering in two common types of dielectric materials for electromagnetic absorption applications. *Adv. Funct. Mater.* **2019**, *29*, 1901236. [CrossRef]
41. Wang, G.Z.; Gao, Z.; Tang, S.W.; Chen, C.Q.; Duan, F.F.; Zhao, S.C.; Lin, S.W.; Feng, Y.H.; Zhou, L.; Qin, Y. Microwave absorption properties of carbon nanocoils coated with highly controlled magnetic materials by atomic layer deposition. *ACS Nano* **2012**, *6*, 11009–11017. [CrossRef]
42. Qiao, L.; Wang, T.; Mei, Z.L.; Li, X.L.; Sui, W.B.; Tang, L.Y.; Li, F.S. Analyzing bandwidth on the microwave absorber by the interface reflection model. *Chin. Phys. Lett.* **2016**, *33*, 027502. [CrossRef]
43. Zhao, H.; Xu, S.Y.; Tang, D.M.; Yang, Y.; Zhang, B.S. Thin magnetic coating for low-frequency broadband microwave absorption. *J. Appl. Phys.* **2014**, *116*, 243911. [CrossRef]
44. Kong, I.; Ahmad, S.H.; Abdullah, M.H.; Hui, D.; Yusoff, A.N.; Puryanti, D. Magnetic and microwave absorbing properties of magnetite thermoplastic natural rubber nanocomposites. *J. Magn. Magn. Mater.* **2010**, *322*, 3401–3409. [CrossRef]
45. Cao, M.S.; Song, W.L.; Hou, Z.L.; Wen, B.; Yuan, J. The effects of temperature and frequency on the dielectric properties, electromagnetic interference shielding and microwave-absorption of short carbon fiber/silica composites. *Carbon* **2010**, *48*, 788–796. [CrossRef]
46. Zhu, L.Y.; Zeng, X.J.; Li, X.P.; Yang, B.; Yu, R.H. Hydrothermal synthesis of magnetic Fe_3O_4/graphene composites with good electromagnetic microwave absorbing performances. *J. Magn. Magn. Mater.* **2017**, *426*, 114–120. [CrossRef]
47. Smit, J.; Wijn, H.P.J. *Ferrites: Physical Properties of Ferrimagnetic Oxides in Relation to Their Technical Applications*; Wiley: New York, NY, USA, 1959.

Article

Advancement of the Power-Law Model and Its Percolation Exponent for the Electrical Conductivity of a Graphene-Containing System as a Component in the Biosensing of Breast Cancer

Yasser Zare [1,*], Kyong Yop Rhee [2,*] and Soo Jin Park [3,*]

1. Biomaterials and Tissue Engineering Research Group, Department of Interdisciplinary Technologies, Breast Cancer Research Center, Motamed Cancer Institute, ACECR, Tehran 1125342432, Iran
2. Department of Mechanical Engineering (BK21 Four), College of Engineering, Kyung Hee University, Yongin 17104, Korea
3. Department of Chemistry, Inha University, Incheon 22212, Korea
* Correspondence: y.zare@aut.ac.ir (Y.Z.); rheeky@khu.ac.kr (K.Y.R.); sjpark@inha.ac.kr (S.J.P.)

Citation: Zare, Y.; Rhee, K.Y.; Park, S.J. Advancement of the Power-Law Model and Its Percolation Exponent for the Electrical Conductivity of a Graphene-Containing System as a Component in the Biosensing of Breast Cancer. *Polymers* 2022, *14*, 3057. https://doi.org/10.3390/polym14153057

Academic Editors: Vineet Kumar and Xiaowu Tang

Received: 8 July 2022
Accepted: 23 July 2022
Published: 28 July 2022

Publisher's Note: MDPI stays neutral with regard to jurisdictional claims in published maps and institutional affiliations.

Copyright: © 2022 by the authors. Licensee MDPI, Basel, Switzerland. This article is an open access article distributed under the terms and conditions of the Creative Commons Attribution (CC BY) license (https://creativecommons.org/licenses/by/4.0/).

Abstract: The power-law model for composite conductivity is expanded for graphene-based samples using the effects of interphase, tunnels and net on the effective filler fraction, percolation start and "b" exponent. In fact, filler dimensions, interphase thickness, tunneling distance and net dimension/density express the effective filler fraction, percolation start and "b" exponent. The developed equations are assessed by experimented values from previous works. Additionally, the effects of all parameters on "b" exponent and conductivity are analyzed. The experimented quantities of percolation start and conductivity confirm the predictability of the expressed equations. Thick interphase, large tunneling distance, high aspect ratio and big nets as well as skinny and large graphene nano-sheets produce a low "b" and a high conductivity, because they improve the conduction efficiency of graphene nets in the system. Graphene-filled nanocomposites can be applied in the biosensing of breast cancer cells and thus the developed model can help optimize the performance of biosensors.

Keywords: graphene; polymer nanocomposite; percolation theory; conductivity; interphase; tunneling distance

1. Introduction

Carbon nanotubes (CNT) have a tubular structure of carbon atoms [1–9]. However, graphene 2D nano-sheets in the form of sp^2 carbon show wonderful electronic, unique mechanical, significant thermal and good chemical properties [10–18]. Thus, polymer nanocomposites containing graphene can be applied in different technologies such as transparent electronics, electromagnetic interference shielding, energy devices, light emitting diodes and lightning protection [19–23]. These applications mainly need to the electrical conductivity justifying the wide research on the conductivity graphene-filled products. The conductivity in nanocomposites is achieved when the filler percentage reaches an essential level as percolation start [24–26]. Actually, the significant effect of graphene on the conductivity is obtained after percolation start and the formation of conductive graphene nets. The polymer nanocomposites containing graphene nano-sheets present lower percolation start and more conductivity compared to CNT systems [27], because the graphene has big aspect ratio and very giant specific superficial zone. However, some undesirable phenomena such as aggregation, crimping and difficult networking may weaken the efficiency of graphene for conductivity [28].

The conductivity of graphene-based polymer systems has been extensively studied by experimental studies [29–31]. They focused on the physical and processing factors to obtain the little percolation start and promote the conductivity by low volume fraction of graphene. However, the influence of main terms on the percolation start and conductivity of graphene-filled samples was not studied. This matter can be evaluated by the modeling methods, but the theoretical studies of graphene nanocomposites mainly include the application of the simple power-law model to calculate the percolation start and conductivity [29,32,33]. So, the roles of main parameters such as nano-dimensions and interphase regions in the conductivity graphene systems were not evaluated.

The interphase regions are commonly formed in nanocomposites, due to the large interfacial zone amongst polymer medium and nanofiller [34–39]. The characters of interphase dimensions and toughness in the rigidity of nanocomposites have been conferred in previous articles [40–46]. Furthermore, it was indicated that the interphase regions can facilitate the production of conductive nets in the samples in advance the real attachment of particles [47–51]. Thus, the interphase regions can positively decrease the percolation level to low filler fractions. However, the conventional model such as the power-law model cannot consider the important issues in nanocomposites such as nanoscale, interphase and tunnels.

The power-law model shows respectable arrangement with the experimented conductivity of graphene products [32,52]. Nonetheless, this model disrespects the main attributes of graphene nanocomposites. Additionally, there is no accurate equation for the "b" exponent in this model. In this work, this conventional model is advanced for graphene-filled systems assuming the impacts of interphase, tunnels and the dimensions/density of filler nets on the effective filler fraction, percolation start and "b" exponent. The established equations are assessed using experimental results from previous papers. Likewise, the impact of all factors on the "b" exponent and conductivity is analyzed to confirm the advanced technique.

2. Theoretical Views

The simple power equation for calculating the conductivity of composites was suggested [29] as:

$$\sigma = \sigma_f (\phi_f - \phi_p)^b \qquad (1)$$

where "σ_f" is the filler conduction, "ϕ_f" is the filler volume portion, "ϕ_p" is the volume share at percolation start and "b" is the exponent. Additionally, "b" was reported to be 1.6-2 and 1-1.3 for 3D and 2D systems, respectively [52], although more "b" value was calculated for polymer graphene nanocomposites.

This equation only reflects the effects of conduction, amount and percolation start of particles on the conductivity, but it neglects the interphase and tunneling zones, as mentioned. Undoubtedly, these terms affect the effective volume fraction and percolation start of nanofiller, which change the conductivity of whole system.

The interphase zones of the nanoparticles increase the efficiency of nanofiller, because they can decrease the distance between nano-sheets and contribute to the net. The interphase volume portion in graphene-based nanocomposites [53] is predicted by:

$$\phi_i = \phi_f \left(\frac{2t_i}{t} \right) \qquad (2)$$

where "t" and "t_i" are the thicknesses of the nano-sheets and interphase, respectively.

The effective graphene volume portion in the samples can be calculated by the total sum of the interphase and filler as:

$$\phi_{eff} = \phi_f + \phi_i = \phi_f \left(1 + \frac{2t_i}{t} \right) \qquad (3)$$

which highlights that the interphase thickness and graphene thickness control the effectiveness of nanoparticles in the nanocomposite.

The percolation start of 3D unsystematically organized graphite sheets in the nanocomposite was also recommended [31] as:

$$\phi_p = \frac{27\pi D^2 t}{4(D+d)^3} \quad (4)$$

"D" is the diameter of the nano-sheets and "d" is the tunneling distance. Nonetheless, D >> d eliminates the role of tunnels in the latter equation as:

$$\phi_p = \frac{27\pi t}{4D} \quad (5)$$

The interphase regions are formed on both sides of the graphene nano-sheets. In addition, the tunneling spaces comprise the distance between the adjacent nano-sheets. The impacts of interphase and tunnels on the percolation start can be suggested by the development of the above equation as:

$$\phi_{pi} = \frac{27\pi t}{4D + 2(Dt_i + Dd)} \quad (6)$$

The predictability of this equation for the percolation start of polymer graphene nanocomposites is examined by the tentative facts in the next section.

Assuming the graphene aspect ratio ($\alpha = D/t$), "ϕ_{pi}" is given by:

$$\phi_{pi} = \frac{13.5\pi}{\alpha(2 + t_i + d)} \quad (7)$$

expressing an opposite relation amid percolation start and aspect ratio.

The "b" exponent was insufficiently defined in the previous articles for polymer composites and nanocomposites. Some authors have correlated the "b" to particle diameter and distribution [54,55]. Shao et al. [56] also defined the "b" as a function of universal critical exponent, a structure factor and the number fractions of hanging ends and backbone framework. More recently, Mutlay and Tudoran [30] have developed the Shao approach and suggested that the "b" exponent depends on the dimensional particle distribution, structure factor and aspect ratio of nanoparticles. They yielded good agreement between the predictions and experimental data in graphene and graphite nanocomposites [30]. However, their equation does not assume the interphase and tunnels as well as net dimensions, which undoubtedly affect the "b" exponent.

The "b" exponent can be defined for graphene samples by the mentioned terms by mathematical operations as:

$$b = 4 + \frac{10}{t_i + 1} + \frac{10}{d+1} + \frac{500}{\alpha} - \frac{N}{5} \quad (8)$$

where "N" shows the dimensionality, dimension and density of filler nets in the nanocomposite. The correctness of this equation is also examined in the next section by the tested data of conductivity in dissimilar examples.

Supposing the impacts of interphase and tunnels on the effective graphene amount (Equation (3)), percolation start (Equation (7)) and "b" exponent (Equation (8)), the power-law model in Equation (1) is:

$$\sigma = \sigma_f (\phi_{eff} - \phi_{pi})^{4 + \frac{10}{t_i+1} + \frac{10}{d+1} + \frac{500}{\alpha} - \frac{N}{5}} \quad (9)$$

Figure 1 depicts the effects of various factors on the forecasts of this model. In Figure 1a, the best conductivity is obtained as 0.14 S/m at $\phi_{eff} = 0.07$ and $\phi_{pi} = 0.001$, while low "ϕ_{eff}" significantly decreases the conductivity. Figure 1b also reveals that the greatest conductivity of 12 S/m is found by $\sigma_f = 3 \times 10^5$ S/m and b = 3, whereas the conductivity mainly falls at b > 4. As a result, the highest nanocomposite conductivity is gained by the uppermost grades of effective filler fraction and graphene conduction as well as by the smallest ranges of percolation start and "b" exponent. Moreover, it is observed that both filler conduction and "b" affect the conductivity more compared to other parameters.

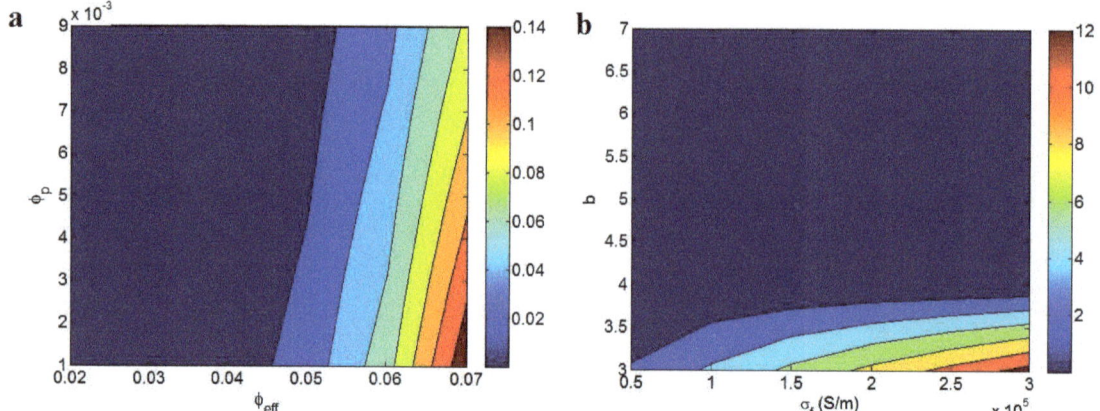

Figure 1. Two-dimensional plots showing the estimates of conductivity using Equation (9) at (**a**) various ranges of "ϕ_{eff}" and "ϕ_p" ($\sigma_f = 10^5$ S/m and b = 5) and (**b**) different levels of "σ_f" and "b" ($\phi_{eff} = 0.04$ and $\phi_p = 0.005$).

The developed model assumes the influence of graphene agglomeration on the conductivity when the average dimensions of agglomerations are considered. The agglomerates of graphene have different sizes and aspect ratios from a graphene layer, which affect the effective graphene concentration (Equation (3)), percolation start (Equation (7)), "b" (Equation (8)) and conductivity (Equation (9)). Therefore, it is possible to take into account the agglomeration of graphene in the conductivity of nanocomposites using the developed model.

3. Results and Discussion

3.1. Assessment of Equations by Experimented Records

The obtained equations for percolation start, "b" and electrical conductivity are assessed using the experimental facts of graphene systems from literature.

Table 1 shows the reported specimens and the levels of "t", "D" as well as "ϕ_p" from the measurements of electrical conductivity at different filler concentrations. By comparing the experimental "ϕ_p" to Equation (6), the values of "t_i" and "d" are calculated and observed in Table 1. The dissimilar values of "t_i" and "d" show the existence of unlike interphase and tunnels in the examples. The densest interphase (8 nm) and the largest tunnels (10 nm) are witnessed in samples No. 4 and 3, respectively. It should also be indicated that disregarding these parameters results in the incorrect estimation of percolation start. In other words, only the geometries of graphene nano-sheets cannot yield the very small percolation start in nanocomposites, but the interphase around the nanoparticles and the tunneling spaces between neighboring nano-sheets play a role in the percolating of nanoparticles. Accordingly, Equation (6) finely predicts the percolation start in graphene-filled nanocomposites, considering the impacts of the interphase and tunneling zones.

Table 1. Selected examples and the outputs of numerous terms by the equations.

No.	Samples [Ref.]	t (nm)	D (μm)	ϕ_p	t_i (nm)	d (nm)	N	b
1	PI[1]/graphene [57]	3	5	0.0015	7	9	13.0	4.0
2	PET[2]/graphene [58]	2	2	0.0050	3	4	22.0	4.6
3	PS[3]/graphene [33]	1	4	0.0005	7	10	7.00	4.9
4	PS/graphene [59]	1	2	0.0010	8	8	4.50	5.6
5	PVA[4]/graphene [52]	2	2	0.0035	5	5	10.5	5.7
6	epoxy/graphene [60]	2	2	0.0050	2	4	14.5	7.0
7	PVDF[5]/graphene [29]	1	2	0.0030	2	3	15.5	7.0
8	SAN[6]/graphene [61]	1	2	0.0017	5	5	1.50	7.3
9	ABS[7]/graphene [61]	1	4	0.0013	3	3	8.00	7.5

[1]: polyimide; [2]: poly (ethylene terephthalate); [3]: polystyrene; [4]: poly (vinyl alcohol); [5]: poly (vinylidene fluoride); [6]: acrylonitrile butadiene styrene; [7]: styrene acrylonitrile.

The tested conductivity of the examples is applied to the innovative model and the values of the "b" exponent are calculated. Figure 2 shows the tested conductivity and the model's calculations for the examples. The model's estimates acceptably agree with the tested results. Thus, it is logical to apply the developed form of the power-law model (Equation (9)), supposing the interphase and tunnels for the approximation of conductivity in the graphene systems. The calculated values of "b" for the reported samples are shown in Table 1. The smallest and the highest levels of "b" are obtained as 4 and 7.5 for samples No. 1 and 9, respectively. As a result, "b" changes from 4 to 7.5 for the examples. This range is greater than the values of "b" calculated for graphene nanocomposites using the conventional power-law model (Equation (1)), disregarding the interphase and tunneling parts.

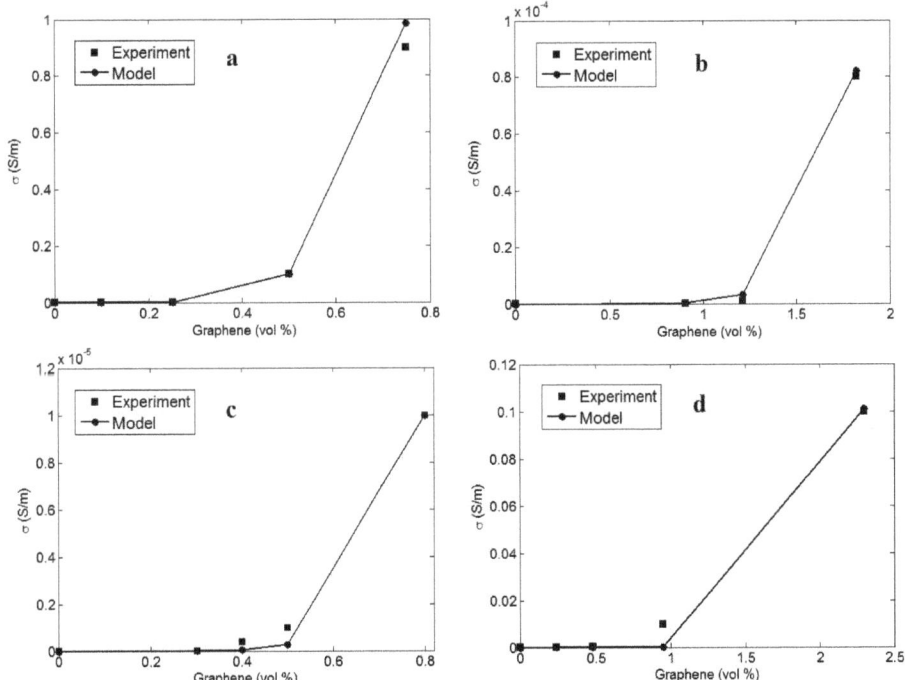

Figure 2. The experimented and calculated conductivity using the novel model for (a) PS [59], (b) epoxy [60], (c) PVDF [29] and (d) ABS [61] graphene systems.

The values of the "b" exponent can be applied using Equation (8) to approximate the level of "N". The calculated "N" for all samples is reported in Table 1. "N" ranges from 1.5 to 22 for the reported examples. As recommended, "N" is a representative of the dimension and density of filler net. So, a higher "N" shows the creation of bigger and thicker nets in the sample. It can be suggested that sample No. 2 contains the biggest and the densest nets among the reported samples. Additionally, an inverse relation between "N" and "b" is extracted from the reported calculations in Table 1. The samples with high "N" illustrate small "b", while a low "N" results in a high "b". This evidence is logical, because the big and dense nets of graphene produce a strong conductivity in the nanocomposite as predicted by low "b" (see Figure 1b). Conclusively, Equation (8) successfully states the possessions of interphase depth, tunneling distance and net dimensions on the "b" exponent. In other words, the suggested equation for "b" considers the influence of all main factors, which may govern the percolation start and the nets of graphene nano-sheets in the nanocomposite. In the absence of accurate experimental techniques for the characterization of interphase, tunneling and net dimensions/density, the developed equations for percolation start and "b" exponent in this study can help approximate these parameters in polymer graphene nanocomposites.

3.2. Parameters' Effects on the "b"

The stimuli of parameters on the "b" exponent are discussed using Equation (8).

Figure 3 exemplifies the characters of "t_i" and "d" in the "b" at t = 2 nm, D = 1 μm and N = 10. The highest value of "b" as 9.5 is observed at t_i = d = 2 nm, while "b" decreases to about 4.15 at t_i > 10 nm and d > 8 nm. Consequently, the high values of both "t_i" and "d" decrease "b". In other words, thick interphase and long tunneling distance can produce a low "b", whereas thin interphase and short tunneling distance undesirably enhance it.

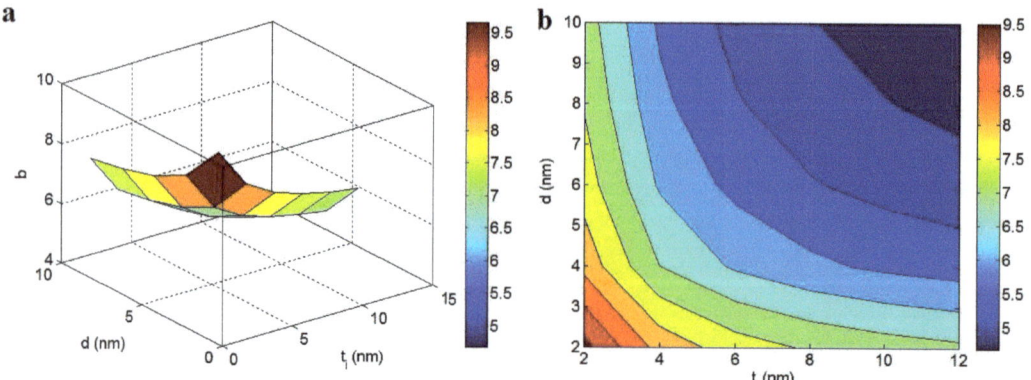

Figure 3. Effects of "t_i" and "d" on "b" at normal t = 2 nm, D = 1 μm and N = 10 by (a) 3D and (b) 2D pictures.

It was mentioned that the "b" exponent inversely depends on the properties of graphene nets in the nanocomposite. Both "t_i" and "d" significantly affect the mentioned terms. The interphase adjoining the nanoparticles can participate in the filler nets; thus, they facilitate the percolation of nanoparticles and enhance the size and compactness of the nets. Likewise, the tunneling spaces between adjacent nanoparticles can contribute to the networking of graphene nano-sheets, because the nanoparticles can form the nets in the presence of tunneling regions [62,63]. As a result, thick interphase and large tunneling distance can raise the scale and density of conductive nets in the nanocomposites, which diminishes the "b".

"b" exponent at different values of "α" and "N" and average t_i = 4 nm and d = 5 nm are also depicted in Figure 4. The high levels of both "α" and "N" decrease the "b", but the highest "b" is projected by the minimum values of these factors. As shown, α = 900 and N = 20 produce b = 4.2, while b = 8 is obtained by α = 300 and N = 5. So, the high levels of these parameters can positively reduce the "b", highlighting that the large aspect ratio and high net properties can produce a desirable "b".

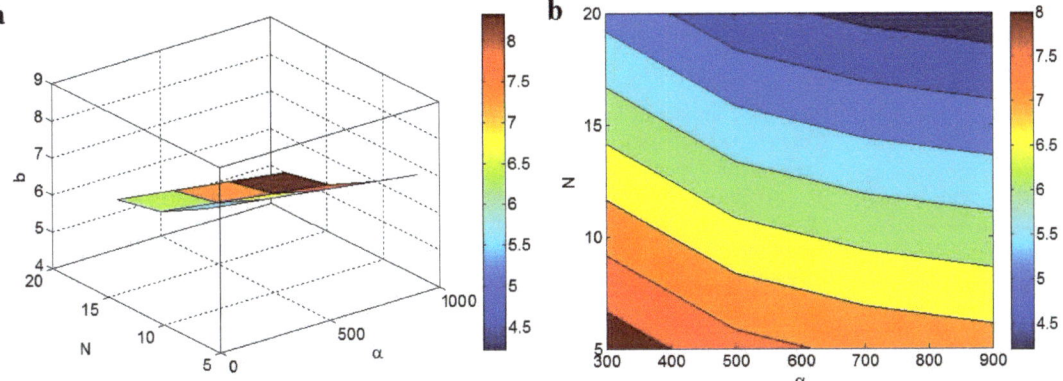

Figure 4. Variations in the "b" exponent at dissimilar grades of "α" and "N" and t_i = 4 nm and d = 5 nm by (**a**) 3D and (**b**) 2D schemes.

The associations of "b" to these factors are expected, due to the direct influence of "α" and "N" on the performance of graphene nets in the nanocomposite. As predicted, a high aspect ratio of the nano-sheets can produce a slight percolation start, which desirably affects the magnitudes of the nets [60,64]. In fact, the large aspect ratio of the nanoparticles improves the scale and density of the conductive nets. Conversely, a high rank of "N" obviously increases the net properties, because "N" reveals the dimensions/density of nets. Thus, a small "b" is observed due to the big aspect ratio and "N".

The calculations of the "b" exponent at unlike arrays of "t" and "D" are also seen in Figure 5. A small "t" and large "l" decrease the "b" exponent. As shown, t = 5 nm and D = 1 µm result in b = 8, while b = 5.8 is achieved by t < 1.5 nm and D > 2.5 µm. It can be suggested that the thin and large graphene nano-sheets positively influence the "b", while thick and small nano-sheets detrimentally affect it. So, it is necessary to control the dimensions of graphene nano-sheets in the nanocomposite to obtain a good "b".

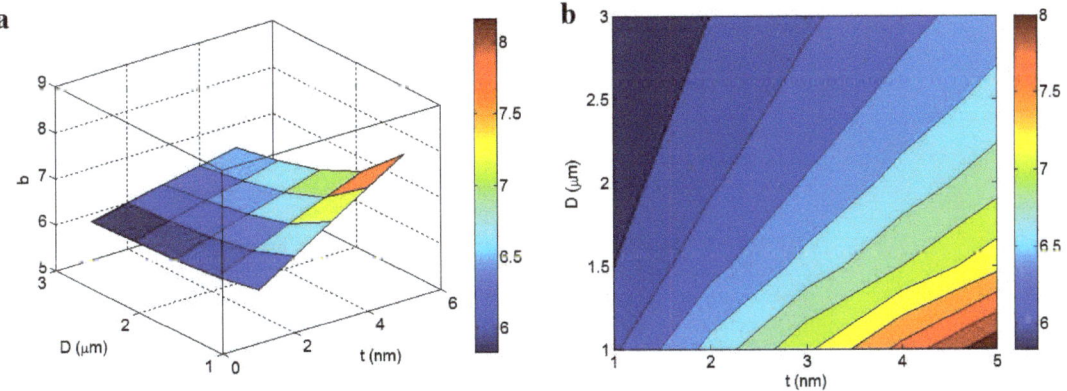

Figure 5. Dependencies of "b" exponent on "t" and "D" at t_i = 4 nm, d = 5 nm and N = 10 by (**a**) 3D and (**b**) 2D charts.

Skinny and large nano-sheets beneficially manage the size of nets in the products, because they cause a poor percolation start and also produce dense nets. At a constant volume of nanoparticles, thin nano-sheets show a high number. Therefore, the nanocomposites containing thinner nano-sheets contain a larger number of nanoparticles. Additionally, the contacts between larger nano-sheets are more than those of short ones. Thus, skinny and big nano-sheets can develop the attributes of filler nets in the nanocomposite; therefore, the developed equation accurately forecasts the "b".

3.3. Parameters' Effects on the Conductivity

The impact of the parameters on the conductivity of graphene is evaluated by the developed equation (Equation (9)) based on interphase and tunneling regions. In all calculations, $\sigma_f = 10^5$ S/m is considered.

Figure 6 shows the conductivity of the nanocomposite correlating to "t_i" and "d" at $t = 2$ nm, $\phi_f = 0.01$, $D = 1$ μm and $N = 10$. The top conductivity as 6 S/m is witnessed at the extreme levels of $t_i = 12$ nm and $d = 10$ nm. However, $t_i < 8.5$ nm and $d < 5$ nm induce very little conductivity adjacent to 0. So, profuse interphase and high tunneling distance can harvest a high conductivity. Instead, thin interphase and short tunnels cannot significantly improve the conductivity.

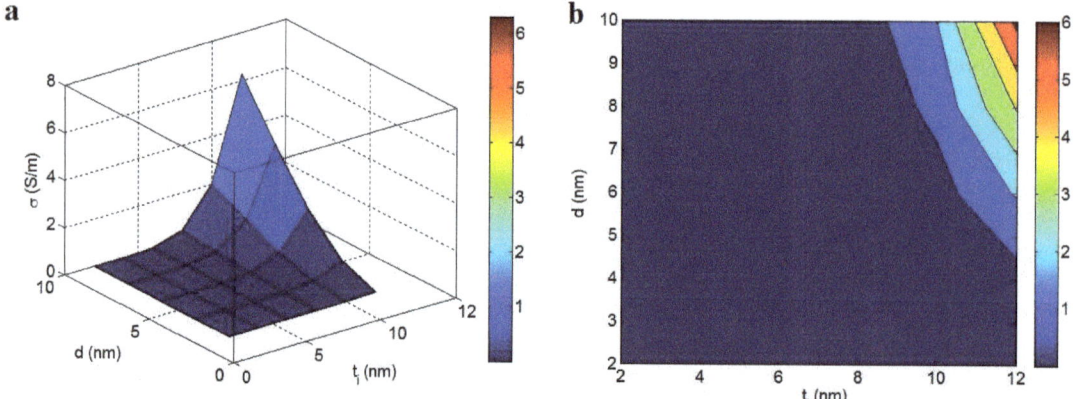

Figure 6. Impact of "t_i" and "d" on the nanocomposite's conductivity using (**a**) 3D and (**b**) 2D designs.

As mentioned, the interphase and tunnels take part in the nets because they surround the nanoparticles. Therefore, profuse interphase and large tunnels significantly increase the net size/density, which positively improves the conductivity. It can be said that a thick interphase and a large tunneling space can involve more nanoparticles in the conductive nets. In contrast, thin interphase and short tunneling distance negligibly manipulate the net size/density. So, the advanced model shows reasonable impacts of interphase deepness and tunnel size on the nanocomposite's conductivity. However, it should be said that a very large tunneling distance between adjacent nano-sheets weakens the tunneling effect, producing insulation.

The impacts of "α" and "N" on the conductivity of the system at $t = 2$ nm, $t_i = 4$ nm, $\phi_f = 0.01$, and $d = 5$ nm are also illustrated in Figure 7. The finest results are gained by the peak values of "α" and "N", though a pitiable conductivity is witnessed at low levels of these factors. The upper conductivity of 0.22 S/m is calculated at $\alpha = 900$ and $N = 20$, while $N < 15$ only decreases the conductivity to about 0. As a result, only a low level of "N" can decrease the conductivity, but the highest ranges of both the aspect ratio and "N" produce the highest conductivity.

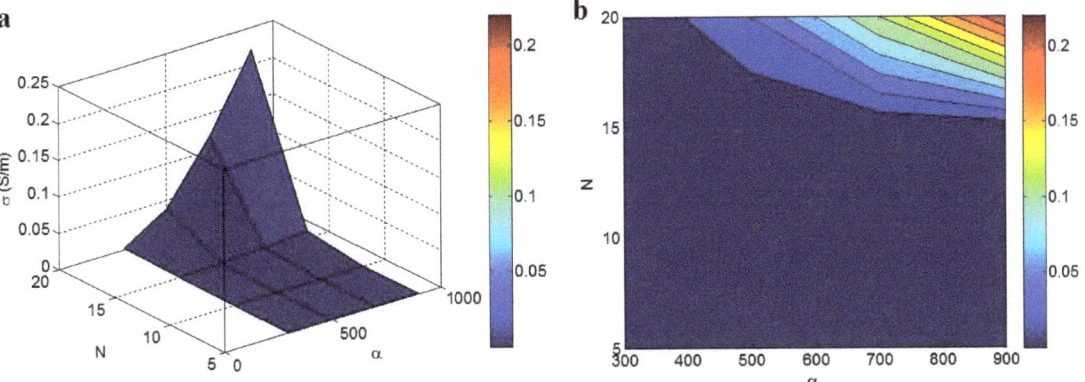

Figure 7. Impact of "α" and "N" on the conductivity by (**a**) 3D and (**b**) 2D schemes.

The optimistic role of the aspect ratio in the sample's conductivity is attributed to its impacts on the percolation start and net dimensions. A high aspect ratio results in a low percolation start in the graphene nanocomposites, as mentioned. Additionally, a high aspect ratio causes a big surface zone, which creates large nets. A low percolation start boosts the conductivity of the system, as seen in Figure 1a. Moreover, the efficiency of electron transferring throughout the nanocomposite is improved by the formation of large nets. Consequently, the correlation of conductivity to the aspect ratio is logical. In addition, the "N" shows the magnitude of filler nets in the nanocomposite. A small "N" shows the foundation of short and weak nets in the specimens, whereas a high "N" depicts the large and dense nets. Therefore, "N" rightly manages the conductivity of the graphene nanocomposite, as recommended by the new methodology.

Figure 8 also reveals the influence of graphene dimensions on the conductivity of nanocomposites ($t_i = 4$ nm, $\phi_f = 0.01$, d = 5 nm and N = 10). t > 2 nm reduces the conductivity to about 0, but the smallest "t" (t = 1 nm) and the highest "D" (D = 3 μm) harvest the uppermost conductivity as 0.07 S/m. Accordingly, the best conductivity is obtained by very thin and large graphene nano-sheets. On the other hand, thick nano-sheets cannot increase the conductivity of the nanocomposite.

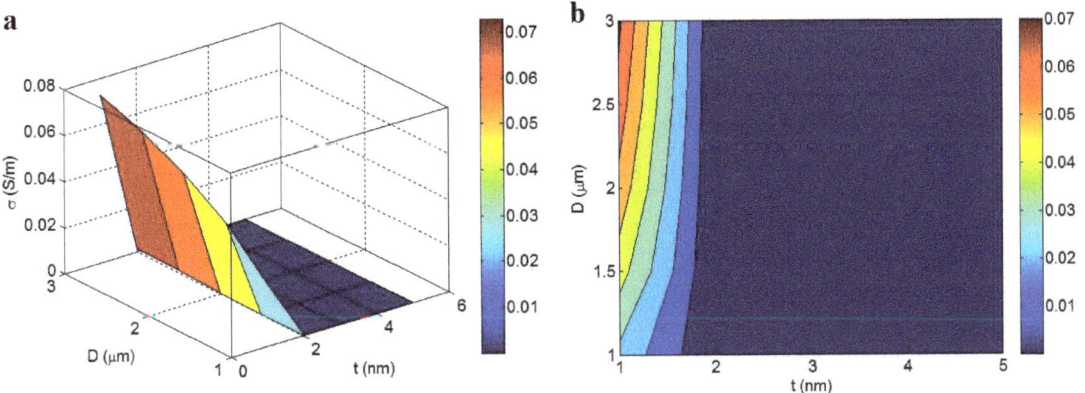

Figure 8. (**a**) Three-dimensional and (**b**) two-dimensional charts for the characters of "t" and "D" in the conductivity.

The thin and large nano-sheets control the percolation start and net scale in the nanocomposite based on Equations (6) and (8). Additionally, polymers are frequently insulated and the conductive nanofillers handle the whole conductivity. Since the conductivity of the nanocomposite depends on the percolation level and the net dimensions, skinny and large nano-sheets can play an optimistic role in the conductivity. In fact, the thin and large nano-sheets can cover a high fraction of the nanocomposite, which produces big nets and high conductivity. Alternatively, the efficiency of dense and small nano-sheets is insignificant, because they produce small nets, which cannot effectively transfer the electrons. As a result, the optimistic effects of skinny and large nano-sheets on the conductivity of samples are meaningful.

4. Conclusions

The power equation for composite conductivity was developed for graphene-filled samples, determining the effects of interphase, tunnels and net dimension/density on the effective filler fraction, percolation start and "b" exponent. Additionally, the measured records of percolation start and conductivity were applied to confirm the predictability of the established equations. A high conductivity is found using a large filler amount, slight percolation start, significant filler conduction and small "b"; however, the impact of the filler conduction and the "b" exponent is more significant compared to the other parameters. The experimental data of percolation start have good arrangement with the predictions. So, the interphase depth and tunneling size play a main role in the percolation value of graphene in the system. Moreover, the innovative model adequately predicts the conductivity of the examples. Generally, thick interphase, large tunneling distance, high aspect ratio and dense nets as well as thin and big graphene nano-sheets produce a low "b" exponent. In addition, these factors cause high conductivity, suggesting that they considerably increase electron transfer in the system. This model was only developed for polymer graphene nanocomposites. Since graphene-filled nanocomposites can be used in the biosensing of breast cancer cells, the developed model can help enhance the performance of biosensors.

Author Contributions: Conceptualization, Y.Z. and K.Y.R.; methodology, Y.Z.; software, Y.Z.; validation, Y.Z., K.Y.R. and S.J.P.; formal analysis, Y.Z.; investigation, Y.Z.; resources, Y.Z., K.Y.R. and S.J.P.; data curation, Y.Z. and K.Y.R.; writing—original draft preparation, Y.Z.; writing—review and editing, Y.Z., K.Y.R. and S.J.P.; visualization, Y.Z.; supervision, K.Y.R. and S.J.P.; project administration, K.Y.R. and S.J.P.; funding acquisition, K.Y.R. and S.J.P. All authors have read and agreed to the published version of the manuscript.

Funding: This work was supported by the Basic Science Research Program through the National Research Foundation of Korea (NRF) funded by the Ministry of Education, Science and Technology (2022R1A2C1004437). It was also supported by the Korea government (MSIT) (2022M3J7A1062940).

Institutional Review Board Statement: Not applicable.

Informed Consent Statement: Not applicable.

Data Availability Statement: All data are available in the paper.

Conflicts of Interest: The authors declare no conflict of interest.

References

1. Farzaneh, A.; Rostami, A.; Nazockdast, H. Thermoplastic polyurethane/multiwalled carbon nanotubes nanocomposites: Effect of nanoparticle content, shear, and thermal processing. *Polym. Compos.* **2021**, *42*, 4804–4813. [CrossRef]
2. Farzaneh, A.; Rostami, A.; Nazockdast, H. Mono-filler and bi-filler composites based on thermoplastic polyurethane, carbon fibers and carbon nanotubes with improved physicomechanical and engineering properties. *Polym. Int.* **2022**, *71*, 232–242. [CrossRef]
3. Zare, Y.; Rhee, K.Y. Effects of interphase regions and filler networks on the viscosity of PLA/PEO/carbon nanotubes biosensor. *Polym. Compos.* **2019**, *40*, 4135–4141. [CrossRef]
4. Zare, Y.; Garmabi, H.; Rhee, K.Y. Structural and phase separation characterization of poly (lactic acid)/poly (ethylene oxide)/carbon nanotube nanocomposites by rheological examinations. *Compos. Part B Eng.* **2018**, *144*, 1–10. [CrossRef]
5. Zare, Y.; Rhee, K.Y. Modeling of viscosity and complex modulus for poly (lactic acid)/poly (ethylene oxide)/carbon nanotubes nanocomposites assuming yield stress and network breaking time. *Compos. Part B Eng.* **2019**, *156*, 100–107. [CrossRef]
6. Zare, Y. Modeling of tensile modulus in polymer/carbon nanotubes (CNT) nanocomposites. *Synth. Met.* **2015**, *202*, 68–72. [CrossRef]

7. Kazemi, F.; Mohammadpour, Z.; Naghib, S.M.; Zare, Y.; Rhee, K.Y. Percolation onset and electrical conductivity for a multiphase system containing carbon nanotubes and nanoclay. *J. Mater. Res. Technol.* **2021**, *15*, 1777–1788. [CrossRef]
8. Dassan, E.G.B.; Ab Rahman, A.A.; Abidin, M.S.Z.; Akil, H.M. Carbon nanotube–reinforced polymer composite for electromagnetic interference application: A review. *Nanotechnol. Rev.* **2020**, *9*, 768–788. [CrossRef]
9. Alkhedher, M. Hygrothermal environment effect on the critical buckling load of FGP microbeams with initial curvature integrated by CNT-reinforced skins considering the influence of thickness stretching. *Nanotechnol. Rev.* **2021**, *10*, 1140–1156. [CrossRef]
10. Yan, Y.; Nashath, F.Z.; Chen, S.; Manickam, S.; Lim, S.S.; Zhao, H.; Lester, E.; Wu, T.; Pang, C.H. Synthesis of graphene: Potential carbon precursors and approaches. *Nanotechnol. Rev.* **2020**, *9*, 1284–1314.
11. Krishnan, S.K.; Singh, E.; Singh, P.; Meyyappan, M.; Nalwa, H.S. A review on graphene-based nanocomposites for electrochemical and fluorescent biosensors. *RSC Adv.* **2019**, *9*, 8778–8881. [CrossRef] [PubMed]
12. Ezenkwa, O.E.; Hassan, A.; Samsudin, S.A. Comparison of mechanical properties and thermal stability of graphene-based materials and halloysite nanotubes reinforced maleated polymer compatibilized polypropylene nanocomposites. *Polym. Compos.* **2022**, *43*, 1852–1863. [CrossRef]
13. Keshvardoostchokami, M.; Piri, F.; Jafarian, V.; Zamani, A. Fabrication and antibacterial properties of silver/graphite oxide/chitosan and silver/reduced graphene oxide/chitosan nanocomposites. *JOM* **2020**, *72*, 4477–4485. [CrossRef]
14. Safamanesh, A.; Mousavi, S.M.; Khosravi, H.; Tohidlou, E. On the low-velocity and high-velocity impact behaviors of aramid fiber/epoxy composites containing modified-graphene oxide. *Polym. Compos.* **2021**, *42*, 608–617. [CrossRef]
15. Kim, S.-H.; Zhang, Y.; Lee, J.-H.; Lee, S.-Y.; Kim, Y.-H.; Rhee, K.Y.; Park, S.-J. A study on interfacial behaviors of epoxy/graphene oxide derived from pitch-based graphite fibers. *Nanotechnol. Rev.* **2021**, *10*, 1827–1837. [CrossRef]
16. Bhat, A.; Budholiya, S.; Raj, S.A.; Sultan, M.T.H.; Hui, D.; Shah, A.U.M.; Safri, S.N.A. Review on nanocomposites based on aerospace applications. *Nanotechnol. Rev.* **2021**, *10*, 237–253. [CrossRef]
17. Sagadevan, S.; Shahid, M.M.; Yiqiang, Z.; Oh, W.-C.; Soga, T.; Lett, J.A.; Alshahateet, S.F.; Fatimah, I.; Waqar, A.; Paiman, S. Functionalized graphene-based nanocomposites for smart optoelectronic applications. *Nanotechnol. Rev.* **2021**, *10*, 605–635. [CrossRef]
18. Pagnola, M.; Morales, F.; Tancredi, P.; Socolovsky, L. Radial Distribution Function Analysis and Molecular Simulation of Graphene Nanoplatelets Obtained by Mechanical Ball Milling. *JOM* **2021**, *73*, 2471–2478. [CrossRef]
19. Céspedes-Valenzuela, D.N.; Sánchez-Rentería, S.; Cifuentes, J.; Gantiva-Diaz, M.; Serna, J.A.; Reyes, L.H.; Ostos, C.; Cifuentes-De la Portilla, C.; Muñoz-Camargo, C.; Cruz, J.C. Preparation and Characterization of an Injectable and Photo-Responsive Chitosan Methacrylate/Graphene Oxide Hydrogel: Potential Applications in Bone Tissue Adhesion and Repair. *Polymers* **2021**, *14*, 126. [CrossRef] [PubMed]
20. Abdullah, A.; Al-Qahatani, A.; Alquraish, M.; Baily, C.; El-Mofty, S.; El-Shazly, A. Experimental Investigation of Fabricated Graphene Nanoplates/Polystyrene Nanofibrous Membrane for DCMD. *Polymers* **2021**, *13*, 3499. [CrossRef] [PubMed]
21. Ikram, R.; Mohamed Jan, B.; Abdul Qadir, M.; Sidek, A.; Stylianakis, M.M.; Kenanakis, G. Recent Advances in Chitin and Chitosan/Graphene-Based Bio-Nanocomposites for Energetic Applications. *Polymers* **2021**, *13*, 3266. [CrossRef] [PubMed]
22. Joynal Abedin, F.N.; Hamid, H.A.; Alkarkhi, A.F.; Amr, S.S.A.; Khalil, N.A.; Ahmad Yahaya, A.N.; Hossain, M.S.; Hassan, A.; Zulkifli, M. The effect of graphene oxide and SEBS-g-MAH compatibilizer on mechanical and thermal properties of acrylonitrile-butadiene-styrene/talc composite. *Polymers* **2021**, *13*, 3180. [CrossRef]
23. Díez-Pascual, A.M. Development of Graphene-Based Polymeric Nanocomposites: A Brief Overview. *Polymers* **2021**, *13*, 2978. [CrossRef]
24. Zare, Y.; Rhee, K. Evaluation and development of expanded equations based on Takayanagi model for tensile modulus of polymer nanocomposites assuming the formation of percolating networks. *Phys. Mesomech.* **2018**, *21*, 351–357. [CrossRef]
25. Zare, Y.; Rhee, K.Y. Simplification and development of McLachlan model for electrical conductivity of polymer carbon nanotubes nanocomposites assuming the networking of interphase regions. *Compos. Part B Eng.* **2019**, *156*, 64–71. [CrossRef]
26. Zare, Y.; Rhee, K.Y. Significances of interphase conductivity and tunneling resistance on the conductivity of carbon nanotubes nanocomposites. *Polym. Compos.* **2020**, *41*, 748–756. [CrossRef]
27. Xie, S.; Liu, Y.; Li, J. Comparison of the effective conductivity between composites reinforced by graphene nanosheets and carbon nanotubes. *Appl. Phys. Lett.* **2008**, *92*, 243121. [CrossRef]
28. Du, J.; Zhao, L.; Zeng, Y.; Zhang, L.; Li, F.; Liu, P.; Liu, C. Comparison of electrical properties between multi-walled carbon nanotube and graphene nanosheet/high density polyethylene composites with a segregated network structure. *Carbon* **2011**, *49*, 1094–1100. [CrossRef]
29. He, L.; Tjong, S.C. Low percolation threshold of graphene/polymer composites prepared by solvothermal reduction of graphene oxide in the polymer solution. *Nanoscale Res. Lett.* **2013**, *8*, 132. [CrossRef]
30. Mutlay, İ.; Tudoran, L.B. Percolation behavior of electrically conductive graphene nanoplatelets/polymer nanocomposites: Theory and experiment. *Fuller. Nanotub. Carbon Nanostructures* **2014**, *22*, 413–433. [CrossRef]
31. Li, J.; Kim, J.-K. Percolation threshold of conducting polymer composites containing 3D randomly distributed graphite nanoplatelets. *Compos. Sci. Technol.* **2007**, *67*, 2114–2120. [CrossRef]
32. Lan, Y.; Liu, H.; Cao, X.; Zhao, S.; Dai, K.; Yan, X.; Zheng, G.; Liu, C.; Shen, C.; Guo, Z. Electrically conductive thermoplastic polyurethane/polypropylene nanocomposites with selectively distributed graphene. *Polymer* **2016**, *97*, 11–19. [CrossRef]

33. Tu, Z.; Wang, J.; Yu, C.; Xiao, H.; Jiang, T.; Yang, Y.; Shi, D.; Mai, Y.-W.; Li, R.K. A facile approach for preparation of polystyrene/graphene nanocomposites with ultra-low percolation threshold through an electrostatic assembly process. *Compos. Sci. Technol.* **2016**, *134*, 49–56. [CrossRef]
34. Zare, Y.; Rhee, K.Y. Model progress for tensile power of polymer nanocomposites reinforced with carbon nanotubes by percolating interphase zone and network aspects. *Polymers* **2020**, *12*, 1047. [CrossRef]
35. Zare, Y.; Rhee, K.Y. Analysis of the connecting effectiveness of the interphase zone on the tensile properties of carbon nanotubes (CNT) reinforced nanocomposite. *Polymers* **2020**, *12*, 896. [CrossRef] [PubMed]
36. Zare, Y.; Rhee, K.Y. Development of expanded takayanagi model for tensile modulus of carbon nanotubes reinforced nanocomposites assuming interphase regions surrounding the dispersed and networked nanoparticles. *Polymers* **2020**, *12*, 233. [CrossRef] [PubMed]
37. Zare, Y.; Daraei, A.; Vatani, M.; Aghasafari, P. An analysis of interfacial adhesion in nanocomposites from recycled polymers. *Comput. Mater. Sci.* **2014**, *81*, 612–616. [CrossRef]
38. Zare, Y.; Garmabi, H. Modeling of interfacial bonding between two nanofillers (montmorillonite and CaCO3) and a polymer matrix (PP) in a ternary polymer nanocomposite. *Appl. Surf. Sci.* **2014**, *321*, 219–225. [CrossRef]
39. Power, A.J.; Remediakis, I.N.; Harmandaris, V. Interface and interphase in polymer nanocomposites with bare and core-shell gold nanoparticles. *Polymers* **2021**, *13*, 541. [CrossRef]
40. Baek, K.; Shin, H.; Cho, M. Multiscale modeling of mechanical behaviors of Nano-SiC/epoxy nanocomposites with modified interphase model: Effect of nanoparticle clustering. *Compos. Sci. Technol.* **2021**, *203*, 108572. [CrossRef]
41. Taheri, S.S.; Fakhrabadi, M.M.S. Interphase effects on elastic properties of polymer nanocomposites reinforced by carbon nanocones. *Comput. Mater. Sci.* **2022**, *201*, 110910. [CrossRef]
42. Zare, Y.; Rhee, K.Y. Multistep modeling of Young's modulus in polymer/clay nanocomposites assuming the intercalation/exfoliation of clay layers and the interphase between polymer matrix and nanoparticles. *Compos. Part A Appl. Sci. Manuf.* **2017**, *102*, 137–144. [CrossRef]
43. Zare, Y.; Rhee, K.Y. Development of Hashin-Shtrikman model to determine the roles and properties of interphases in clay/CaCO3/PP ternary nanocomposite. *Appl. Clay Sci.* **2017**, *137*, 176–182. [CrossRef]
44. Zare, Y.; Garmabi, H. A developed model to assume the interphase properties in a ternary polymer nanocomposite reinforced with two nanofillers. *Compos. Part B Eng.* **2015**, *75*, 29–35. [CrossRef]
45. Zare, Y. Modeling approach for tensile strength of interphase layers in polymer nanocomposites. *J. Colloid Interface Sci.* **2016**, *471*, 89–93. [CrossRef]
46. Zare, Y. Study on interfacial properties in polymer blend ternary nanocomposites: Role of nanofiller content. *Comput. Mater. Sci.* **2016**, *111*, 334–338. [CrossRef]
47. Shin, H.; Yang, S.; Choi, J.; Chang, S.; Cho, M. Effect of interphase percolation on mechanical behavior of nanoparticle-reinforced polymer nanocomposite with filler agglomeration: A multiscale approach. *Chem. Phys. Lett.* **2015**, *635*, 80–85. [CrossRef]
48. Xu, W.; Lan, P.; Jiang, Y.; Lei, D.; Yang, H. Insights into excluded volume and percolation of soft interphase and conductivity of carbon fibrous composites with core-shell networks. *Carbon* **2020**, *161*, 392–402. [CrossRef]
49. Zare, Y.; Rhee, K.Y. A simple model for electrical conductivity of polymer carbon nanotubes nanocomposites assuming the filler properties, interphase dimension, network level, interfacial tension and tunneling distance. *Compos. Sci. Technol.* **2018**, *155*, 252–260. [CrossRef]
50. Kim, S.; Zare, Y.; Garmabi, H.; Rhee, K.Y. Variations of tunneling properties in poly (lactic acid)(PLA)/poly (ethylene oxide)(PEO)/carbon nanotubes (CNT) nanocomposites during hydrolytic degradation. *Sens. Actuators A Phys.* **2018**, *274*, 28–36. [CrossRef]
51. Zare, Y.; Rhee, K.Y. Evaluation of the tensile strength in carbon nanotube-reinforced nanocomposites using the expanded Takayanagi model. *JOM* **2019**, *71*, 3980–3988. [CrossRef]
52. Goumri, M.; Lucas, B.; Ratier, B.; Baitoul, M. Electrical and optical properties of reduced graphene oxide and multi-walled carbon nanotubes based nanocomposites: A comparative study. *Opt. Mater.* **2016**, *60*, 105–113. [CrossRef]
53. Yanovsky, Y.G.; Kozlov, G.; Karnet, Y.N. Fractal description of significant nano-effects in polymer composites with nanosized fillers. Aggregation, phase interaction, and reinforcement. *Phys. Mesomech.* **2013**, *16*, 9–22. [CrossRef]
54. Balberg, I.; Azulay, D.; Toker, D.; Millo, O. Percolation and tunneling in composite materials. *Int. J. Mod. Phys. B* **2004**, *18*, 2091–2121. [CrossRef]
55. Harris, A.B.; Lubensky, T.C.; Holcomb, W.K.; Dasgupta, C. Renormalization-group approach to percolation problems. *Phys. Rev. Lett.* **1975**, *35*, 327. [CrossRef]
56. Shao, W.; Xie, N.; Zhen, L.; Feng, L. Conductivity critical exponents lower than the universal value in continuum percolation systems. *J. Phys. Condens. Matter* **2008**, *20*, 395235. [CrossRef]
57. Xu, L.; Chen, G.; Wang, W.; Li, L.; Fang, X. A facile assembly of polyimide/graphene core–shell structured nanocomposites with both high electrical and thermal conductivities. *Compos. Part Appl. Sci. Manuf.* **2016**, *84*, 472–481. [CrossRef]
58. Zhang, H.-B.; Zheng, W.-G.; Yan, Q.; Yang, Y.; Wang, J.-W.; Lu, Z.-H.; Ji, G.-Y.; Yu, Z.-Z. Electrically conductive polyethylene terephthalate/graphene nanocomposites prepared by melt compounding. *Polymer* **2010**, *51*, 1191–1196. [CrossRef]
59. Stankovich, S.; Dikin, D.A.; Dommett, G.H.; Kohlhaas, K.M.; Zimney, E.J.; Stach, E.A.; Piner, R.D.; Nguyen, S.T.; Ruoff, R.S. Graphene-based composite materials. *Nature* **2006**, *442*, 282–286. [CrossRef]

40. Li, Y.; Zhang, H.; Porwal, H.; Huang, Z.; Bilotti, E.; Peijs, T. Mechanical, electrical and thermal properties of in-situ exfoliated graphene/epoxy nanocomposites. *Compos. Part A Appl. Sci. Manuf.* **2017**, *95*, 229–236. [CrossRef]
41. Gao, C.; Zhang, S.; Wang, F.; Wen, B.; Han, C.; Ding, Y.; Yang, M. Graphene networks with low percolation threshold in ABS nanocomposites: Selective localization and electrical and rheological properties. *ACS Appl. Mater. Interfaces* **2014**, *6*, 12252–12260. [CrossRef] [PubMed]
42. Seidel, G.; Puydupin-Jamin, A.-S. Analysis of clustering, interphase region, and orientation effects on the electrical conductivity of carbon nanotube–polymer nanocomposites via computational micromechanics. *Mech. Mater.* **2011**, *43*, 755–774. [CrossRef]
43. Mortazavi, B.; Bardon, J.; Ahzi, S. Interphase effect on the elastic and thermal conductivity response of polymer nanocomposite materials: 3D finite element study. *Comput. Mater. Sci.* **2013**, *69*, 100–106. [CrossRef]
44. Kim, H.; Macosko, C.W. Morphology and properties of polyester/exfoliated graphite nanocomposites. *Macromolecules* **2008**, *41*, 3317–3327. [CrossRef]

Article

Soft Composites Filled with Iron Oxide and Graphite Nanoplatelets under Static and Cyclic Strain for Different Industrial Applications

Vineet Kumar, Md Najib Alam and Sang Shin Park *

School of Mechanical Engineering, Yeungnam University, 280, Daehak-ro, Gyeongsan 38541, Korea; vineetfri@gmail.com (V.K.); mdnajib.alam3@gmail.com (M.N.A.)
* Correspondence: pss@ynu.ac.kr

Abstract: Simultaneously exhibiting both a magnetic response and piezoelectric energy harvesting in magneto-rheological elastomers (MREs) is a win–win situation in a soft (hardness below 65) composite-based device. In the present work, composites based on iron oxide (Fe_2O_3) were prepared and exhibited a magnetic response; other composites based on the electrically conductive reinforcing nanofiller, graphite nanoplatelets (GNP), were also prepared and exhibited energy generation. A piezoelectric energy-harvesting device based on composites exhibited an impressive voltage of ~10 V and demonstrated a high durability of 0.5 million cycles. These nanofillers were added in room temperature vulcanized silicone rubber (RTV-SR) and their magnetic response and piezoelectric energy generation were studied both in single and hybrid form. The hybrid composite consisted of 10 per hundred parts of rubber (phr) of Fe_2O_3 and 10 phr of GNP. The experimental data show that the compressive modulus of the composites was 1.71 MPa (virgin), 2.73 (GNP), 2.65 MPa (Fe_2O_3), and 3.54 MPa (hybrid). Similarly, the fracture strain of the composites was 89% (virgin), 109% (GNP), 105% (Fe_2O_3), 133% (hybrid). Moreover, cyclic multi-hysteresis tests show that the hybrid composites exhibiting higher mechanical properties had the shortcoming of showing higher dissipation losses. In the end, this work demonstrates a rubber composite that provides an energy-harvesting device with an impressive voltage, high durability, and MREs with high magnetic sensitivity.

Keywords: graphite nanoplatelets; iron oxide; silicone rubber; composites; energy harvesting; magnetic sensitivity

Citation: Kumar, V.; Alam, M.N.; Park, S.S. Soft Composites Filled with Iron Oxide and Graphite Nanoplatelets under Static and Cyclic Strain for Different Industrial Applications. *Polymers* **2022**, *14*, 2393. https://doi.org/10.3390/polym14122393

Academic Editor: Luca Valentini

Received: 23 May 2022
Accepted: 10 June 2022
Published: 13 June 2022

Publisher's Note: MDPI stays neutral with regard to jurisdictional claims in published maps and institutional affiliations.

Copyright: © 2022 by the authors. Licensee MDPI, Basel, Switzerland. This article is an open access article distributed under the terms and conditions of the Creative Commons Attribution (CC BY) license (https://creativecommons.org/licenses/by/4.0/).

1. Introduction

A polymer composite containing an electrical-conducting reinforcing filler showing piezoelectric behavior, in a hybrid with iron particles, and which exhibits a magnetic response, is an interesting topic in research. Various new routes are under investigation to meet an increase in energy demands. The use of polymer composites in piezoelectric energy-harvesting-based devices tends to produce few volts via mechanical motion [1,2]. On the other hand, iron particles in a polymer composite are categorized as magneto-rheological elastomers (MREs) and their mechanical properties are sensitive to a magnetic field. Polymer composites with hybrid fillers containing iron particles and electrically conductive reinforcing fillers received great attention [3,4]. Research and review studies by Zaghloul et al. reveals fatigue and wear properties of polymer composites for various useful applications [5–7].

MREs are well known and have been investigated for many decades [8]. MREs contain magnetic particles that tend to orient in a magnetic field, thereby influencing mechanical properties such as the modulus. Various types of polymer matrix were used in designing MREs. These polymer matrixes are elastomers, such as natural rubber [9], butadiene rubber [10], or silicone rubber [11]. Among them, silicone rubber is frequently used, as it is easily processed and cured and has low hardness, all of which make it a candidate for soft composites [12]. It is well known that silicone rubber can be categorized on the basis of vulcanization type. These are high temperature vulcanized (HTV) and room temperature vulcanized (RTV) silicone rubber [13]. Room temperature vulcanized silicone rubber can be used in various applications, such as insulating coatings, strain sensors or actuators, and piezoelectric energy harvesting [14].

Recently, carbon-based secondary fillers, such as MWCNTs were added to MREs containing iron particles to improve the mechanical properties of MREs [15]. These secondary reinforcing fillers form synergy with the iron particles and produce high performance MREs [16]. The type of carbon-based secondary nanofiller used in MREs also influences their damping properties. MWCNTs are one-dimensional in nature and have a tube-shaped morphology and high aspect ratio that significantly improves mechanical and electrical properties; they act as a catalyst to properties of MREs in terms of energy harvesting [17]. Another way to produce high performance MREs is to produce them with iron particles characterized by small particle size, high surface area, high aspect ratio, and favorable morphology so that they orient easily in a magnetic field [18]. In addition to the type of iron particle, and the type of secondary reinforcing filler, the type of polymer matrix has a significant role in affecting properties of MREs as detailed above [19].

In this work, soft silicone rubber (hardness below 65) composites are used to produce high performance MREs which exhibit an improved magnetic response. The type of iron particle is Fe_2O_3 and type of secondary reinforcing nanofiller is GNP. MREs with different features were demonstrated as (a) MREs with iron particles only, which only show a magnetic response; and (b) Specimens with GNP only, which only exhibit piezoelectric energy harvesting. To the best of the authors' knowledge, the MREs exhibiting both a magnetic response and piezoelectric energy harvesting has not been fully understood and is, most recently, a hot topic of research and development in MREs. In previous work, Mannikkavel et al. show a robust piezoelectric energy-harvesting device based on a composite made from HTV–RTV silicone rubber with an MWCNT as the electrode. The voltage generation was, however, lower and was approximately 1 V [20]. In this work, with the addition of a hybrid filler in the substrate, the voltage increases to as high as 10 V and durability was enhanced to 0.5 million cycles. This work is, thus, advantageous in terms of voltage and durability.

2. Materials and Methods

2.1. Materials

The MREs were fabricated with room temperature vulcanized silicone rubber (RTV-SR, obtained from Shin-Etsu, Tokyo, Japan) as the elastomeric matrix (commercial name KE-441-KT, obtained from Shin-Etsu, Tokyo, Japan) which was transparent in appearance. The iron oxide (Fe_2O_3) and graphite nanoplatelets (GNP) were used as nanofillers. The Fe_2O_3 had a surface area of around 50 m^2/g, and particle size below 50 nm and was used as iron particles in MREs fabricated in this work; it was obtained from Alfa-Aesar, Ward Hill, Massachusetts, USA. The GNP, on the other hand, had a surface area of around 125 m^2/g, thickness of 4–5 nm and lateral size of ~2 μm. Both nanofillers (Fe_2O_3 and GNP) were used in single and hybrid forms in the MREs and their improved properties are demonstrated. The vulcanizing agent was "CAT-RM"—a whitish-blue liquid—used for curing and obtained from Shin-Etsu, Tokyo, Japan. The mold-releasing agent spray was white liquid powder, obtained from Nabakem, Gyeonggi-do, Korea.

2.2. Preparation of Composites

The MREs and piezoelectric energy-harvesting-based composites were prepared by the solution-mixing technique and the procedure is reported elsewhere [21]. The preparation of composites was initiated by spraying the molds with mold-releasing agent. The molds were sprayed and left for 1–2 h before use for composite sample preparation. An amount of 100 phr of RTV-SR solution (liquid state of rubber without any solvent) was then mixed with different concentrations of nanofillers, both in single and hybrid forms, as detailed in

Table 1. The filler–rubber mixing lasted for 10 min. Next, 2 phr of vulcanizing agent was added and mixed for 1 min. The composite was poured into the molds and kept at room temperature for vulcanization for 24 h before removal for testing.

Table 1. Formulation table of the RTV-SR based composites.

Formulation *	RTV-SR (Phr)	Fe_2O_3 (Phr)	GNP (Phr)	Vulcanizing Solution (Phr)
Virgin	100	-	-	2
Fe_2O_3 only	100	10	-	2
GNP only	100	-	10	2
Hybrid	100	10	10	2

* 10 phr of GNP and 10 phr of Fe_2O_3 were selected because these concentrations are filler percolation of these fillers and dominant effect of filler in reinforcing composites can be witnessed.

2.3. Characterization Technique

The morphology of Fe_2O_3 and GNP were studied by SEM microscopy (S-4800, Hitachi, Tokyo, Japan). All samples were coated with platinum before the SEM micrographs were taken. The coating was performed to facilitate the surface conduction for the SEM measurements. The SEM was also used to investigate filler dispersion. For these tests, the cylindrical sample (20 × 10 mm) was sectioned by surgical blade into thin slices of 0.5 mm-thick samples. The applied voltage for SEM measurements was 10 kV and the working distance was ~11 mm. The compressive cyclic and static mechanical measurements were performed by a universal testing machine (UTS, Lloyd Instruments, West Sussex, UK). The cylindrical samples were used in measurements of compressive tests at a strain rate of 2 mm/min.

The compressive strain applied for static and cyclic tests was 0–35% max and 30% max, respectively, for 100 cycles. The tensile mechanical measurements were taken at 100 mm/min on dumbbell shaped samples with a gauge length of 25 mm and thickness of 2 mm using the UTS machine (Lloyd instruments, West Sussex, UK). The mechanical properties were tested according to DIN 53 504 standards. The optical images of the set-up for compressive and tensile measurements are reported elsewhere [21]. Hardness tests were taken on cylindrical samples using a Westop durometer, according to ASTM D 2583 standards. The magnetic response measurements and stress–relaxation tests were performed on cylindrical tests using the UTS machine and the optical image of the set-up is displayed elsewhere [22]. The number of samples tested for mechanical properties were 3 for each tensile/compressive/hardness test. For the compressive and hardness tests, the sample was manually gripped, while a pneumatic grip was used for tensile measurements. The magnetic response of the MREs was performed at 90 mT and the stress–relaxation was performed under on–off of the magnetic field. The piezoelectric energy harvesting was performed using a mechanical testing machine (Samick-THK, Daegu, Korea) under cyclic strain and the optical images of the set-up, optical image of the sample, dimension of electrode and thickness of sample under compressive strain are detailed in Figure 1 above.

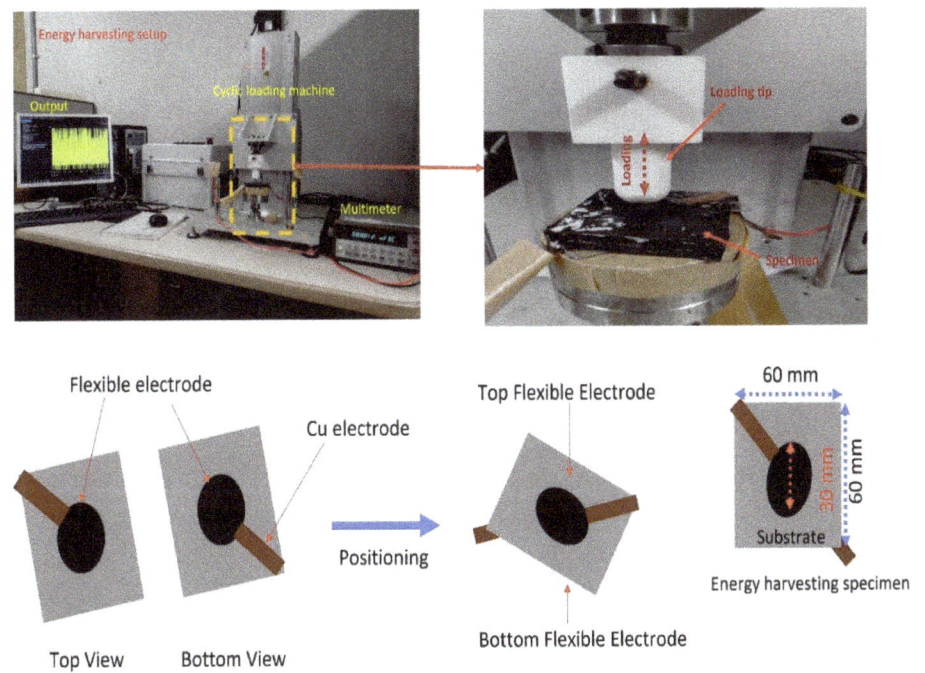

Figure 1. Piezoelectric energy-harvesting set-up with details on device dimensions and their assembly.

3. Results

3.1. Morphology of Fe_2O_3 and GNPs as Filler Particles

The morphology of fillers is well known to affect the properties of the composites. Fillers with favorable morphology are easily dispersed in the polymer matrix and their uniform dispersion simulates the properties [23]. Here, two types of nanofillers with different morphology were used (Figure 2) and their use as MREs and energy harvesters was studied. Fe_2O_3 with oval 0-Dimensional (0D) morphology was used as iron particles to make the composite magnetically active both in single and hybrid forms and their improved properties were studied. GNP with sheet-like morphology with 3D structure forms, 3D filler networks, and their synergy with Fe_2O_3, was investigated. The single and hybrid forms of GNPs create electrically conductive networks which are useful for energy harvesting. Briefly, MREs were prepared using Fe_2O_3 as iron particles and GNPs as secondary reinforcing and conductive fillers to make the MREs useful for energy harvesting.

Figure 2. SEM showing GNPs and Fe_2O_3.

3.2. Filler Dispersion of Composites

3.2.1. Through SEM Microscope

Filler dispersion plays an important role in influencing the properties of composites. A composite in which the filler dispersion is uniform tends to show better properties than those with poor filler dispersion [23]. Here, we employed the SEM technique to determine filler dispersion in composites. A number of SEM images were studied and their representative images per sample are shown in Figure 3. The filled composite SEM images show that the filler particles form long-range networks throughout the silicone rubber matrix. Moreover, uniform filler dispersion was noticed in the SEM images. The SEM images of composites based on both Fe_2O_3 and GNP, as the only filler, shows improved dispersion, and evidence of filler aggregation was absent. Moreover, the composites based on hybrid fillers tend to show synergy between the Fe_2O_3 and GNP particles and, in a few cases, evidence of exfoliation of GNP particles was noticed. In some cases, Fe_2O_3 particles were found in the vicinity of the GNP particles and in a few cases GNP platelets were found in the vicinity of bunches of Fe_2O_3 particles. The GNP platelets were held together by van der Waals forces which are weak and easily exfoliated against mechanical strain, such as during mixing and under mechanical compressive or tensile strain, and leads to distribution of the exfoliated GNP particles [24]. On the other hand, the Fe_2O_3 nanoparticles with a particle size below 50 nm were also distributed uniformly due to nano-dimensions and their favorable oval morphology that allows easy dispersal in the composite, both in single and hybrid states. Moreover, the exfoliated GNP particles help to prevent aggregation of Fe_2O_3 particles, thereby leading to their uniform dispersion as observed in the SEM images.

3.2.2. Through Elemental Mapping

The filler dispersion can be further assessed with the help of elemental mapping [25]. In the present hybrid composite, four types of elements were noticed, specifically, Si from silicone rubber, C from GNP, and Fe and O from Fe_2O_3 (Figure 4). All the maps show that the elements are densely distributed, whether Si from silicone rubber or C, Fe, or O from different nanofillers present in the hybrid composite. The element maps further justify the absence of aggregation of nanofillers in the composite. The SEMs in Figure 2 and element maps in Figure 3 agree with each other.

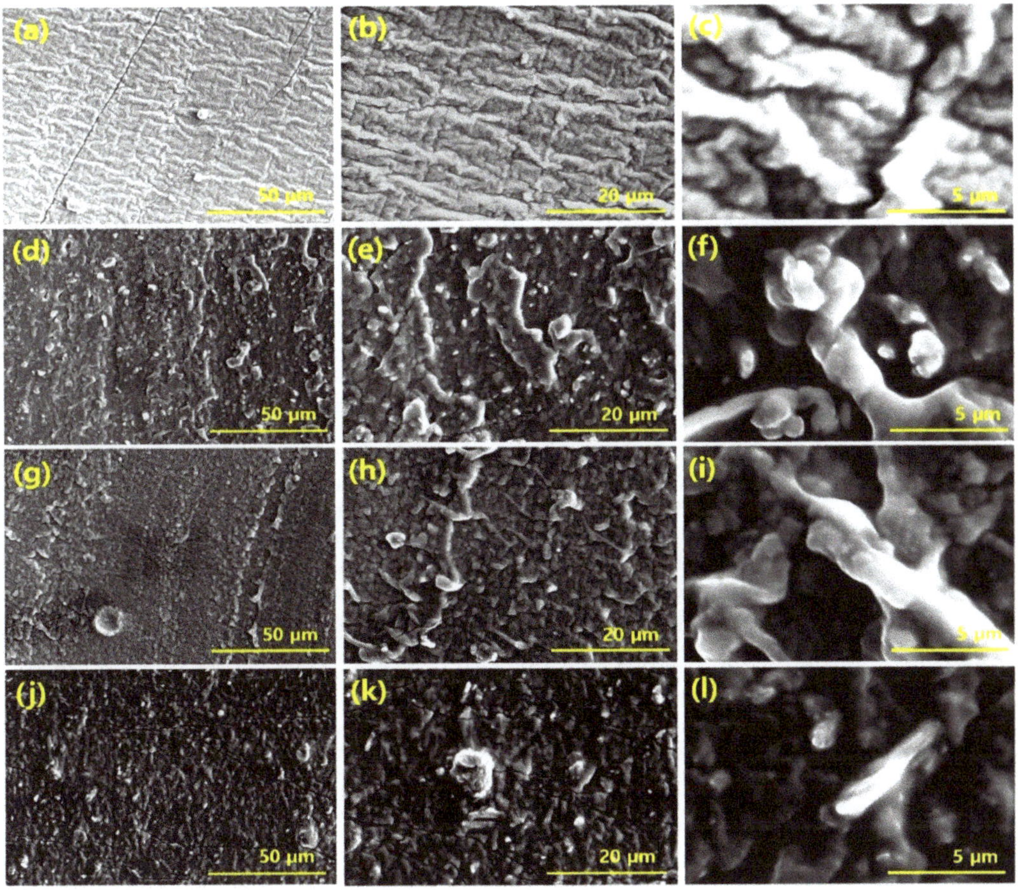

Figure 3. SEM micrographs of different composites; (**a–c**) virgin, (**d–f**) GNP composites, (**g–i**) Fe_2O_3 composites, (**j–l**) hybrid composite.

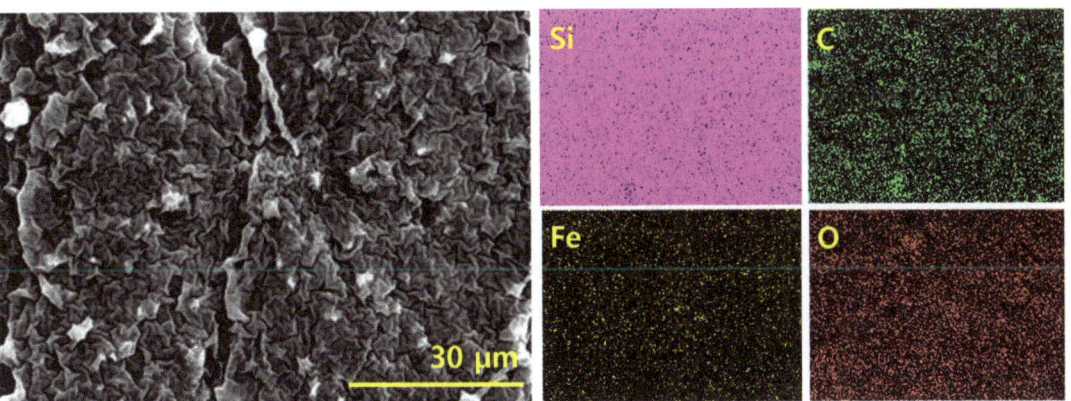

Figure 4. SEM and element maps of different elements in hybrid composite.

3.3. Mechanical Properties

There are three stages of filler network formation in rubber composites. Firstly, at the low filler volume fraction of the filler, filler network formation is initiated and the modulus increases linearly up to a certain amount of filler. The modulus then increases exponentially at a certain amount of filler volume fraction. This is the stage of the filler percolation threshold in which long-range and continuous filler networks are formed [26,27]. The third stage is the filler aggregation stage in which the modulus falls and results in the presence of excess filler particles in the composite. In this work, an amount of filler was added at the second stage; i.e., at the filler percolation threshold at which exponentially high properties are witnessed. Beside the amount of filler, the type of strain also affects the mechanical properties of the composites. In this work, compressive static and cyclic strains and tensile strain are applied, and the behavior of mechanical properties are studied and presented below.

3.3.1. Under Static Compressive Strain

The mechanical behavior under static compressive strain was studied and is presented in Figure 5. Figure 5a shows the compressive stress–strain for different composites prepared in this work. It was found that the stress increases with increasing compressive strain. It is also interesting to note that the compressive stress was linear up to 15% of compressive strain and then increases exponentially. Such a behavior is based on filler networks and the packing fraction of polymer chains and filler particles [28]. It is also interesting to note that as the volume fraction of the filler particles increased to 20 phr (hybrid composite), the mechanical properties increased at all strains. The mechanical properties were higher for the hybrid composite than for GNP and Fe_2O_3 as the sole filler in composites. In addition, the GNP as the sole filler shows higher reinforcement than Fe_2O_3 as the sole filler; this is due to the higher reinforcement exhibited by GNP in rubber composite [29]. On the other hand, Fe_2O_3 tends to exhibit lower reinforcing properties due to its poor reinforcing effect in composites.

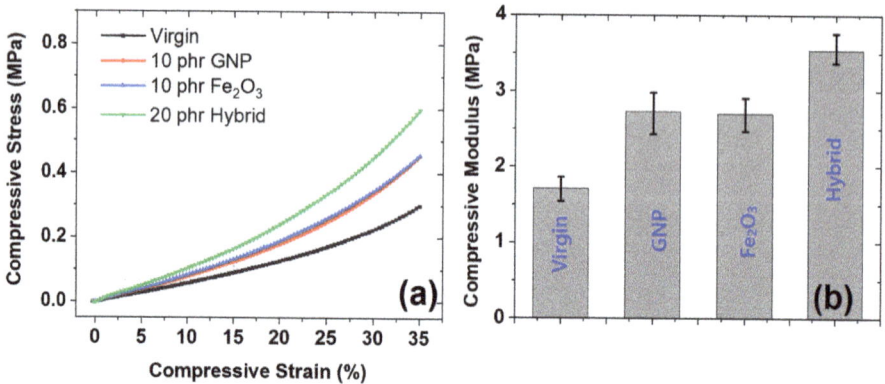

Figure 5. Compressive mechanical properties; (a) compressive stress–strain of the composites, (b) compressive modulus of different composites.

The behavior of compressive modulus was studied and is presented in Figure 5b. It was found that the reinforcing effect was higher for GNP-filled composites and highest for hybrid-filled composites. The higher mechanical compressive modulus for GNP is due to the higher reinforcing effect of GNP particles due to their favorable sheet-like morphology that allows easy dispersal and because they exhibit a higher compressive modulus [30]. Beside this, the hybrid-filled composites exhibit the highest compressive modulus; this is due to presence of the higher amount of filler (20 phr) in the hybrid composite than when

GNP and Fe$_2$O$_3$ are used as single fillers (10 phr each). The higher filler content leads to higher reinforcements and, thus, higher mechanical properties.

3.3.2. Under Cyclic Compressive Strain

Figure 6 shows multi-hysteresis measurements for different types of composites. The measurements are useful for determining the heat dissipation under cyclic strain. It was found from the tests that the dissipation losses were lower for virgin and Fe$_2$O$_3$ composite samples, higher for the GNP composite, and highest for the hybrid composite. The lower dissipation losses for virgin samples are due to the absence of filler particles and, therefore, lower hysteresis losses [31]. In the case of Fe$_2$O$_3$, the dissipation was also lower. This was attributed to the lower reinforcing effect of the Fe$_2$O$_3$ filler and, therefore, lower hysteresis losses. The higher hysteresis losses for the GNP and hybrid composites are due to the higher reinforcing effect of GNP in both GNP as a single filler and a hybrid filler [32].

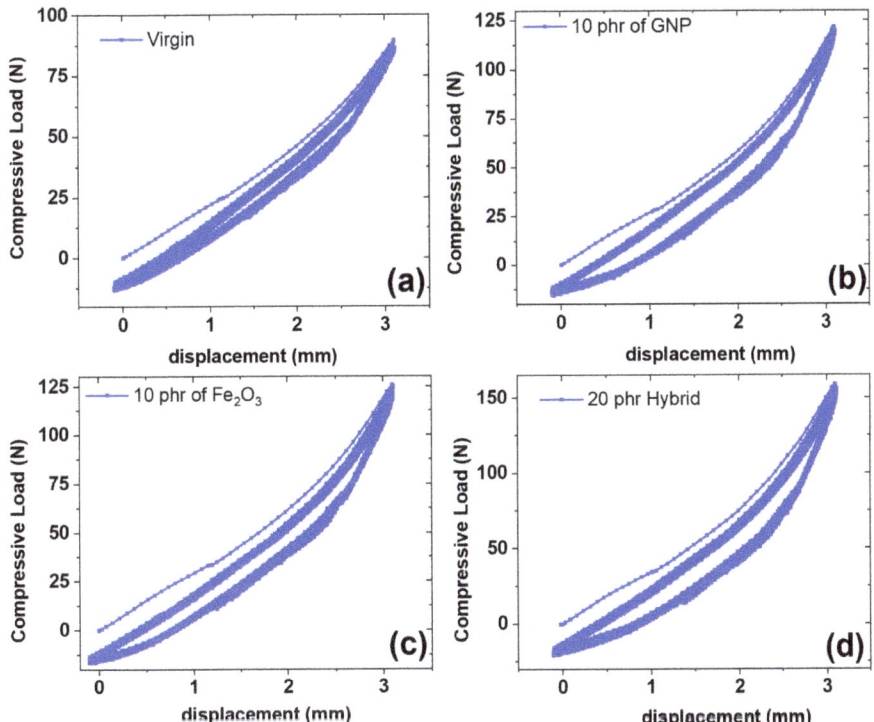

Figure 6. Compressive cyclic mechanical properties; (**a**) virgin composite, (**b**) GNP composite, (**c**) Fe$_2$O$_3$ composite, (**d**) hybrid composite.

The highest dissipation losses in the hybrid composite, which exhibits higher reinforcement, are due to the presence of higher filler content, as described in the mechanical properties' sections. Thus, although the hybrid exhibits higher mechanical properties, it shows the limitation of exhibiting higher dissipation losses, which is a drawback. It is also interesting to note that the first hysteresis cycle exhibits higher dissipation losses in all composites and stabilizes after subsequent cycles [33]. This can be attributed to the break-down of new bonds under strain for the first cycle and the formation of new bonds once the stress is removed. The stabilization of hysteresis losses in subsequent cycles is due to the attainment of equilibrium between breakdown and formation of bonds in the filler networks, as justified in Figure 6. Our future work will involve the functionalization of the filler to enhance polymer-filler interactions, thereby improving stress transfer from the polymer matrix to filler particles to suppress dissipation losses, which will be an advantage for the industrial application of composites.

3.3.3. Under Static Tensile Strain

The behavior of the mechanical properties under tensile strain until fracture was investigated. Properties, specifically, the tensile modulus, tensile strength, and fracture strain of the composites, were determined (Figure 7). It was found that the tensile stress increases linearly with increasing tensile strain until fracture strain (Figure 7a). It was also interesting to note that the tensile mechanical properties were higher for the GNP and hybrid composites and lower for the Fe_2O_3 and virgin composites. A similar trend was found in the compressive mechanical properties. Thus, our experimental data are consistent with each other. The higher tensile properties for GNP and hybrid composites are due to the higher reinforcing ability of GNP in both single and hybrid forms [34]. The highest mechanical properties of the hybrid are due to its higher filler content (20 phr) compared with GNP and Fe_2O_3 as single fillers (10 phr).

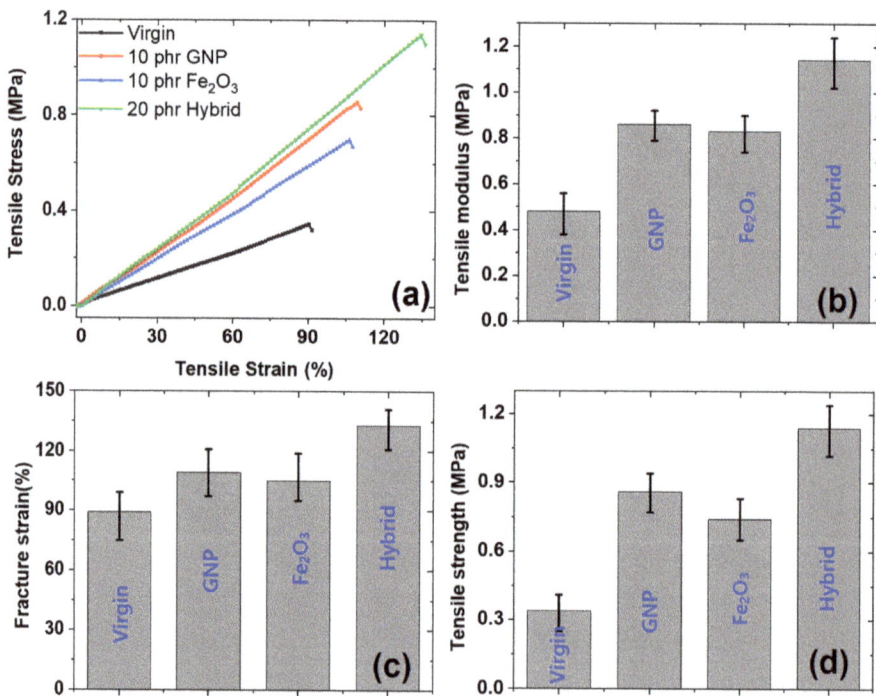

Figure 7. Tensile mechanical properties; (**a**) profiles of stress–strain of different composites, (**b**) tensile modulus for different composites, (**c**) fracture strain of different composites, (**d**) tensile strength of the composites.

The tensile modulus (Figure 7b), fracture strain (Figure 7c) and tensile strength (Figure 7d) were studied for different composites. The properties were lowest for the virgin and Fe_2O_3-filled composites, while they were higher for the GNP and hybrid composites. The higher tensile modulus for the GNP and hybrid-filled composites is due to the induced stiffness of GNP for a rubber matrix and the high reinforcing property of GNP, as detailed in the literature [35]. The higher fracture strain and tensile strength of the GNP and hybrid composites are due to the lubricating nature of GNP, in which the graphene planes stack together via van der Waals forces that tend to repel each other, leading to higher fracture strain and tensile strength [36]. It is also interesting to note that hybrid composites exhibit the highest mechanical properties, and a sort of synergism was noticed for composites, especially in fracture strain and tensile strength. The higher properties of the hybrid composites are also due to the higher filler content (20 phr) compared with other-filled composites, which cause higher reinforcements for rubber matrix.

3.4. Hardness of Rubber Composites

The hardness of composites is useful for determining whether a composite is soft or not. Generally, a composite with a hardness below 65 is termed a soft composite [25]. In this work, hardness was determined. It was found that the hardness was well below 65 (Figure 8). This means that the composites prepared in this work are soft and useful for various applications, such as soft robotics etc. It was also interesting to note that the behavior of hardness is in agreement with the behavior of mechanical properties, for example, the modulus of the composites and hardness are in agreement with each other. Moreover, the hardness of GNP was higher than that of Fe_2O_3 as the sole filler and the hardness was highest for the hybrid composite. The higher hardness of GNP is due to the higher reinforcing capacity of GNP and the highest hardness of the hybrid composite is due to the high filler content (20 phr) compared with other fillers.

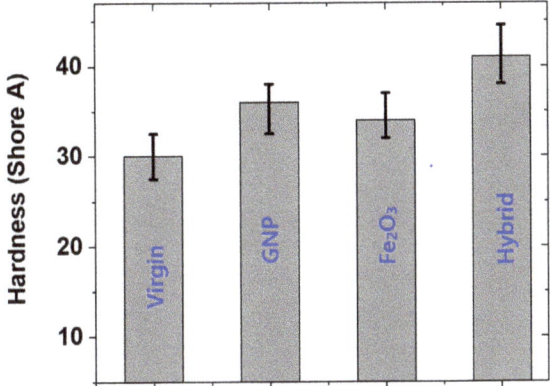

Figure 8. Hardness of different composites.

3.5. Industrial Applications

3.5.1. Piezoelectric Energy-Harvesting Tests

In this work the substrate was made of reinforcing GNP+Fe_2O_3 hybrid nanofillers with RTV-SR. The electrode was made of 2 phr of MWCNT+ 2 phr of MoS_2 reinforced with RTV-SR. The flexible energy harvester substrate thickness was 8 mm and a 0.2 mm-thick electrode was painted on both the sides. The load was applied for a displacement of 4 mm, with the hemispherical loader having a diameter of 21 mm. A constant amplitude of displacement was provided for the entire cyclic loading. During the starting cycles, the obtained voltage value was higher than 1.5 V. This value was constant for up to 40,000 cycles (Figure 9). This was due to the energy of charge carriers present in the dielectric material [37]. After 40,000 cycles, the voltage value started to enhance at the higher rate. This was because of the geometry of the specimen. The specimen construction made it a capacitor. The charging effect produced in the electrode was due to the continuous voltage production because of repeated loading. Once sufficient charging occurs, on further loading, the capacitive geometry specimen is able to multiply the voltage production. When the applied load is continuous, the saturation in the charging of the electrode occurs and a regular voltage output is obtained from the specimen [38]. Uneven voltage output results in the variation in the activation of charge carriers during loading of the specimen [39]. During 100,000 cycles of loading, the voltage output was above 5 V. This value steadily increased with a further increase in the number of cycles. There was a sharp increase during 100,000 to 200,000 cycles. The value of voltage output during 200,000 cycles was around 7 V. In the 200,000 to 300,000 phase, the voltage value steadily increased. This is due to the energy of charge carriers occurring as the rate increases. In the course of 300,000 cycles, the maximum voltage value of above 8 V obtained. From 300,000 to 400,000 cycles, the voltage value increased again. The voltage value of 9 V was obtained by the specimen during the repeated loading of 400,000 cycles. During 400,000 to 500,000 cycles, the voltage value increased slightly. The maximum obtained voltage value of around 10 V was achieved during 500,000 cycles. In the curve, the top portion is the voltage output due to loading of the specimen. The bottom portion of the voltage formation is due to the free movement of the specimen. The forced loading was able to achieve a higher voltage output than the free movement of the specimen. The MWCNT present in the electrode has the property of achieving higher conductivity for the electrode, which can collect the charge formation during loading. MoS$_2$ has a lubricating nature and is able to achieve higher fracture toughness when combined with the silicone rubber. This property reduces crack formation on the electrode during repeated loading. The reinforcement present in the substrate has an effect on the dielectric property. The nano reinforcements in the electrode and substrate collectively enhance the voltage output.

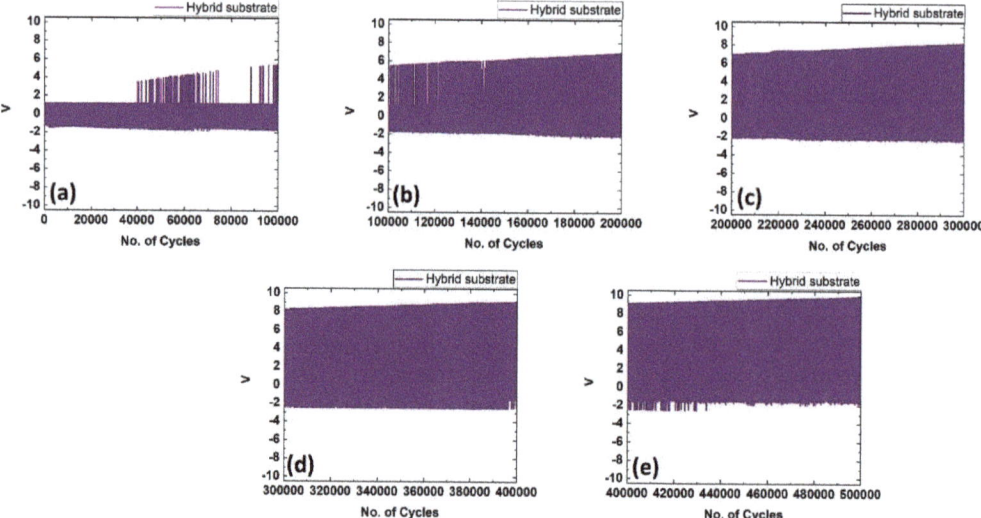

Figure 9. Piezoelectric energy-harvesting device cycles (a–e) output-voltages for number of cycles from 0 to 0.5 million.

3.5.2. Magnetic Response Tests of MREs

The above composites with lower Fe$_2$O$_3$ can sense the external magnetic field but it was not significant in the change of stiffness of the rubber composites. To study the magneto-mechanical response properties at a lower external magnetic field (90 mT), a higher amount of 40 phr Fe$_2$O$_3$ was utilized. Figure 10a indicates the compressive mechanical profile under one cycle of external magnetic field described elsewhere [38,39]. The lowest slope in the compressive profile is due to isotropic filler distribution [38,39], whereas the highest slope is due to a fast change in the force value because of the attraction of induced magnetic poles [38,39] and is regarded as a response force. The medium slope is regarded as the true value for the anisotropic filler orientation in the presence of an externally induced magnetic field. The response force without a sample was 7.5 N in our experimental system. A slight decrease in the response force was observed because some amount of force is utilized to orient the filler particles in the direction of the external magnetic field. After the complete orientation of filler particles, the composite shows anisotropic behavior with a higher mechanical modulus. Figure 10b shows the changes in the mechanical modulus from isotropic to anisotropic filler distribution. A greater change in the modulus indicates that the composites can be utilized as magnetically derived flexible actuators in many smart applications [40,41].

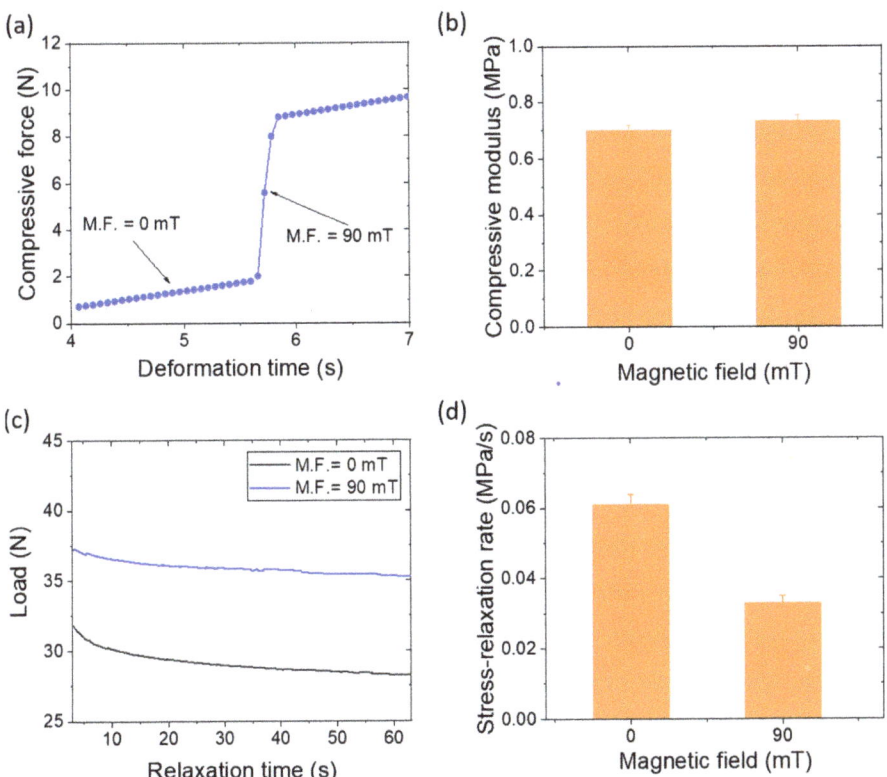

Figure 10. Magneto-mechanical properties of 40 phr Fe_2O_3-filled silicone rubber composite; (**a**) compressive force under a magnetic cycle, (**b**) magnetic effect on modulus, (**c**) load relaxation curves within 3–63 s with 10% deformation of cylindrical sample, and (**d**) magnetic effect on the stress–relaxation rate.

A stress–relaxation experiment was also undertaken to observe the filler–filler interactions in the presence of an externally applied magnetic field [42,43]. Viscoelastic materials such as rubber undergo stress relaxation due to the strain-generated structures in the materials. Depending upon the direction of the applied magnetic field, the stress–relaxation rate can be increased or decreased. In our experimental condition, we found a decrease in the stress–relaxation rate because the flow of the matrix and the applied magnetic field acted in the reverse direction. The load relaxation curves in the absence and presence of 90 mT external magnetic fields are provided in Figure 10c. A significant reduction in the stress–relaxation rate was observed in the presence of the external magnetic field (Figure 10d), which was due to induced filler–filler interactions [42,43] which can restrict the movement of polymer chains and improve the stiffness of the composite.

4. Conclusions

The composites prepared in this work exhibit magnetic sensitivity and act as an energy-harvesting device. The piezoelectric energy-harvesting device generates an impressive voltage of around 10 V, high stability, and durability of more than 0.5 million cycles. From the experimental measurements, the virgin and Fe_2O_3-filled composites show poor mechanical properties, while the GNP and hybrid-filled composites show higher mechanical properties. However, multi-hysteresis measurements show that GNP and hybrid composites exhibit higher dissipation losses, which is a disadvantage of using these composites for industrial applications. The stress-transfer from the polymer matrix to filler particles can, however, be enhanced by improving interfacial interaction. Our future work will address the improvement of interfacial interaction, thereby suppressing the dissipation losses to make the composites more promising.

In conclusion, this work provides the readers with a method of obtaining MREs which show magnetic sensitivity and an impressive energy-generating device with high durability which is an interesting current research. This study also highlights that RTV-SR is suitable for making MREs and stretchable flexible devices. The work demonstrates that with the addition of the hybrid filler in a substrate, the voltage is enhanced as high as 10 V and durability greatly improved to as high as 0.5 million cycles. Moreover, the RTV-SR and composites prepared from this rubber were soft, with a hardness below 65, and suitable for soft applications, such as flexible electronics. In conclusion, this work recommends that hybrid filler must be used to obtain higher mechanical properties even though obtaining higher dissipation losses is a demerit. Nevertheless, the substrate with the hybrid filler exhibits higher energy generation and shows a prolonged durability, as demonstrated in Figure 9.

Author Contributions: Conceptualization, V.K. and M.N.A.; methodology, V.K. and M.N.A.; software, V.K.; validation, V.K. and M.N.A.; formal analysis, V.K.; investigation, V.K. and M.N.A.; resources, S.S.P.; data curation, V.K.; writing—original draft preparation, V.K. and M.N.A.; writing—review and editing, V.K.; visualization, V.K.; supervision, S.S.P.; project administration, S.S.P. All authors have read and agreed to the published version of the manuscript.

Funding: This research received no external funding.

Institutional Review Board Statement: Not applicable.

Informed Consent Statement: Not applicable.

Data Availability Statement: The data presented in this study are available on request from the corresponding author.

Conflicts of Interest: The authors declare no conflict of interest.

References

1. Mishra, S.; Unnikrishnan, L.; Nayak, S.K.; Mohanty, S. Advances in piezoelectric polymer composites for energy harvesting applications: A systematic review. *Macromol. Mater. Eng.* **2019**, *304*, 1800463. [CrossRef]
2. Nunes-Pereira, J.; Sencadas, V.; Correia, V.; Rocha, J.G.; Lanceros-Mendez, S. Energy harvesting performance of piezoelectric electrospun polymer fibers and polymer/ceramic composites. *Sens. Actuators A Phys.* **2013**, *196*, 55–62. [CrossRef]
3. Abdul Aziz, S.A.; Mazlan, S.A.; Ubaidillah, U.; Shabdin, M.K.; Yunus, N.A.; Nordin, N.A.; Choi, S.B.; Rosnan, R.M. Enhancement of viscoelastic and electrical properties of magnetorheological elastomers with nanosized Ni-Mg cobalt-ferrites as fillers. *Materials* **2019**, *12*, 3531. [CrossRef] [PubMed]
4. Aloui, S.; Klüppel, M. Magneto-rheological response of elastomer composites with hybrid-magnetic fillers. *Smart Mater. Struct.* **2014**, *24*, 025016. [CrossRef]
5. Zaghloul, M.M.Y.; Mohamed, Y.S.; El-Gamal, H. Fatigue and tensile behaviors of fiber-reinforced thermosetting composites embedded with nanoparticles. *J. Compos. Mater.* **2019**, *53*, 709–718. [CrossRef]
6. Zaghloul, M.M.Y.; Steel, K.; Veidt, M.; Heitzmann, M.T. Wear behaviour of polymeric materials reinforced with man-made fibres: A comprehensive review about fibre volume fraction influence on wear performance. *J. Reinf. Plast. Compos.* **2022**, *41*, 215–241. [CrossRef]
7. Zaghloul, M.M.Y.M. Mechanical properties of linear low-density polyethylene fire-retarded with melamine polyphosphate. *J. Appl. Polym. Sci.* **2018**, *135*, 46770. [CrossRef]
8. Bastola, A.K.; Hossain, M. A review on magneto-mechanical characterizations of magnetorheological elastomers. *Compos. Part B Eng.* **2020**, *200*, 108348. [CrossRef]
9. Bokobza, L. Natural rubber nanocomposites: A review. *Nanomaterials* **2018**, *9*, 12. [CrossRef]

10. Zanchet, A.; Carli, L.N.; Giovanela, M.; Brandalise, R.N.; Crespo, J.S. Use of styrene butadiene rubber industrial waste devulcanized by microwave in rubber composites for automotive application. *Mater. Des.* **2012**, *39*, 437–443. [CrossRef]
11. Xu, Z.; Wu, H.; Wang, Q.; Jiang, S.; Yi, L.; Wang, J. Study on movement mechanism of magnetic particles in silicone rubber-based magnetorheological elastomers with viscosity change. *J. Magn. Magn. Mater.* **2020**, *494*, 165793. [CrossRef]
12. Shit, S.C.; Shah, P. A review on silicone rubber. *Natl. Acad. Sci. Lett.* **2013**, *36*, 355–365. [CrossRef]
13. El-Hag, A.H.; Jayaram, S.H.; Cherney, E.A. Fundamental and low frequency harmonic components of leakage current as a diagnostic tool to study aging of RTV and HTV silicone rubber in salt-fog. *IEEE Trans. Dielectr. Electr. Insul.* **2003**, *10*, 128–136. [CrossRef]
14. Kumar, V.; Lee, G.; Choi, J.; Lee, D.J. Studies on composites based on HTV and RTV silicone rubber and carbon nanotubes for sensors and actuators. *Polymer* **2020**, *190*, 122221. [CrossRef]
15. Abdul Aziz, S.A.; Mazlan, S.A.; Nik Ismail, N.I.; Choi, S.B.; Ubaidillah; Yunus, N.A.B. An enhancement of mechanical and rheological properties of magnetorheological elastomer with multiwall carbon nanotubes. *J. Intell. Mater. Syst. Struct.* **2017**, *28*, 3127–3138. [CrossRef]
16. Khimi, S.R.; Pickering, K.L. Comparison of dynamic properties of magnetorheological elastomers with existing antivibration rubbers. *Compos. Part B Eng.* **2015**, *83*, 175–183. [CrossRef]
17. Ismail, R.; Ibrahim, A.; Hamid, H.A.; Mahmood, M.R.; Adnan, A. Dynamic mechanical behavior magnetorheological nanocomposites containing CNTs: A review. In *AIP Conference Proceedings*; AIP Publishing LLC: Melville, NY, USA, 2016; Volume 1733, p. 020060.
18. Chen, D.; Yu, M.; Zhu, M.; Qi, S.; Fu, J. Carbonyl iron powder surface modification of magnetorheological elastomers for vibration absorbing application. *Smart Mater. Struct.* **2016**, *25*, 115005. [CrossRef]
19. Burgaz, E.; Goksuzoglu, M. Effects of magnetic particles and carbon black on structure and properties of magnetorheological elastomers. *Polym. Test.* **2020**, *81*, 106233. [CrossRef]
20. Manikkavel, A.; Kumar, V.; Kim, J.; Lee, D.J.; Park, S.S. Investigation of high temperature vulcanized and room temperature vulcanized silicone rubber based on flexible piezo-electric energy harvesting applications with multi-walled carbon nanotube reinforced composites. *Polym. Compos.* **2022**, *43*, 1305–1318. [CrossRef]
21. Kumar, V.; Lee, G.; Singh, K.; Choi, J.; Lee, D.J. Structure-property relationship in silicone rubber nanocomposites reinforced with carbon nanomaterials for sensors and actuators. *Sens. Actuators A Phys.* **2020**, *303*, 111712. [CrossRef]
22. Alam, M.N.; Kumar, V.; Ryu, S.R.; Choi, J.; Lee, D.J. Anisotropic magnetorheological elastomers with carbonyl iron particles in natural rubber and acrylonitrile butadiene rubber: A comparative study. *J. Intell. Mater. Syst. Struct.* **2021**, *32*, 1604–1613. [CrossRef]
23. Aziz, S.A.A.; Mazlan, S.A.; Ismail, N.I.N.; Choi, S.B.; Nordin, N.A.; Mohamad, N. A comparative assessment of different dispersing aids in enhancing magnetorheological elastomer properties. *Smart Mater. Struct.* **2018**, *27*, 117002. [CrossRef]
24. Alam, M.N.; Choi, J. Highly reinforced magneto-sensitive natural-rubber nanocomposite using iron oxide/multilayer graphene as hybrid filler. *Compos. Commun.* **2022**, *32*, 101169. [CrossRef]
25. Kumar, V.; Kumar, A.; Song, M.; Lee, D.J.; Han, S.S.; Park, S.S. Properties of silicone rubber-based composites reinforced with few-layer graphene and iron oxide or titanium dioxide. *Polymers* **2021**, *13*, 1550. [CrossRef]
26. Moucka, R.; Sedlacik, M.; Cvek, M. Dielectric properties of magnetorheological elastomers with different microstructure. *Appl. Phys. Lett.* **2018**, *112*, 122901. [CrossRef]
27. Nawaz, K.; Khan, U.; Ul-Haq, N.; May, P.; O'Neill, A.; Coleman, J.N. Observation of mechanical percolation in functionalized graphene oxide/elastomer composites. *Carbon* **2012**, *50*, 4489–4494. [CrossRef]
28. Budzien, J.; McCoy, J.D.; Adolf, D.B. Solute mobility and packing fraction: A new look at the Doolittle equation for the polymer glass transition. *J. Chem. Phys.* **2003**, *119*, 9269–9273. [CrossRef]
29. Cilento, F.; Martone, A.; Carbone, M.G.P.; Galiotis, C.; Giordano, M. Nacre-like GNP/Epoxy composites: Reinforcement efficiency vis-a-vis graphene content. *Compos. Sci. Technol.* **2021**, *211*, 108873. [CrossRef]
30. Alam, F.; Choosri, M.; Gupta, T.K.; Varadarajan, K.M.; Choi, D.; Kumar, S. Electrical, mechanical and thermal properties of graphene nanoplatelets reinforced UHMWPE nanocomposites. *Mater. Sci. Eng. B* **2019**, *241*, 82–91. [CrossRef]
31. Dias, M.M.; Mozetic, H.J.; Barboza, J.S.; Martins, R.M.; Pelegrini, L.; Schaeffer, L. Influence of resin type and content on electrical and magnetic properties of soft magnetic composites (SMCs). *Powder Technol.* **2013**, *237*, 213–220. [CrossRef]
32. Aluko, O.; Gowtham, S.; Odegard, G.M. The development of multiscale models for predicting the mechanical response of GNP reinforced composite plate. *Compos. Struct.* **2018**, *206*, 526–534. [CrossRef]
33. Yaghtin, M.; Taghvaei, A.H.; Hashemi, B.; Janghorban, K. Effect of heat treatment on magnetic properties of iron-based soft magnetic composites with Al_2O_3 insulation coating produced by sol–gel method. *J. Alloy. Compd.* **2013**, *581*, 293–297. [CrossRef]
34. Arif, M.F.; Alhashmi, H.; Varadarajan, K.M.; Koo, J.H.; Hart, A.J.; Kumar, S. Multifunctional performance of carbon nanotubes and graphene nanoplatelets reinforced PEEK composites enabled via FFF additive manufacturing. *Compos. Part B Eng.* **2020**, *184*, 107625. [CrossRef]
35. Li, J.; Zhang, X.; Geng, L. Improving graphene distribution and mechanical properties of GNP/Al composites by cold drawing. *Mater. Des.* **2018**, *144*, 159–168. [CrossRef]
36. Lahiri, D.; Hec, F.; Thiesse, M.; Durygin, A.; Zhang, C.; Agarwal, A. Nanotribological behavior of graphene nanoplatelet reinforced ultra high molecular weight polyethylene composites. *Tribol. Int.* **2014**, *70*, 165–169. [CrossRef]

37. Psarras, G.C. Hopping conductivity in polymer matrix–metal particles composites. *Compos. Part A Appl. Sci. Manuf.* **2006**, *37*, 1545–1553. [CrossRef]
38. Alam, M.N.; Kumar, V.; Ryu, S.R.; Choi, J.; Lee, D.J. Magnetic response properties of natural-rubber-based magnetorhelogical elastomers with different-structured iron fillers. *J. Magn. Magn. Mater.* **2020**, *513*, 167106. [CrossRef]
39. Alam, M.N.; Kumar, V.; Lee, D.J.; Choi, J. Magnetically active response of acrylonitrile-butadiene-rubber-based magnetorheological elastomers with different types of iron fillers and their hybrid. *Compos. Commun.* **2021**, *24*, 100657. [CrossRef]
40. Hou, X.; Liu, Y.; Wan, G.; Xu, Z.; Wen, C.; Yu, H.; Zhang, J.X.; Li, J.; Chen, Z. Magneto-sensitive bistable soft actuators: Experiments, simulations, and applications. *Appl. Phys. Lett.* **2018**, *113*, 221902. [CrossRef]
41. Wang, S.; Luo, H.; Linghu, C.; Song, J. Elastic Energy Storage Enabled Magnetically Actuated, Octopus-Inspired Smart Adhesive. *Adv. Funct. Mater.* **2021**, *31*, 2009217. [CrossRef]
42. Qi, S.; Yu, M.; Fu, J.; Zhu, M. Stress relaxation behavior of magnetorheological elastomer: Experimental and modeling study. *J. Intell. Mater. Syst. Struct.* **2018**, *29*, 205–213. [CrossRef]
43. Zhu, M.; Yu, M.; Qi, S.; Fu, J. Investigations on response time of magnetorheological elastomer under compression mode. *Smart Mater. Struct.* **2018**, *27*, 055017. [CrossRef]

MDPI
St. Alban-Anlage 66
4052 Basel
Switzerland
www.mdpi.com

Polymers Editorial Office
E-mail: polymers@mdpi.com
www.mdpi.com/journal/polymers

Disclaimer/Publisher's Note: The statements, opinions and data contained in all publications are solely those of the individual author(s) and contributor(s) and not of MDPI and/or the editor(s). MDPI and/or the editor(s) disclaim responsibility for any injury to people or property resulting from any ideas, methods, instructions or products referred to in the content.

www.ingramcontent.com/pod-product-compliance
Lightning Source LLC
LaVergne TN
LVHW070655100526
838202LV00013B/968